Fundamental Physics and Physics Education Research

Burra G. Sidharth • Jesús Carnicer Murillo •
Marisa Michelini • Carmen Perea
Editors

Fundamental Physics and Physics Education Research

Springer

Editors
Burra G. Sidharth
Birla Science Centre
Hyderabad, India

Jesús Carnicer Murillo (deceased)
Miguel Hernández University
MUDIC-VBS-CV
Orihuela (Alicante), Spain

Marisa Michelini
Physics Education Research Unit, DMIF
University of Udine
Udine, Italy

Carmen Perea
Miguel Hernández University
MUDIC-VBS-CV
Orihuela (Alicante), Spain

ISBN 978-3-030-52922-2 ISBN 978-3-030-52923-9 (eBook)
https://doi.org/10.1007/978-3-030-52923-9

This Springer imprint is published by the registered company Springer Nature Switzerland AG.
The registered company address is: Gewerbestrasse 11, 6330 Cham, Switzerland

We dedicate this book to Jesús Carnicer Murillo, who left us last October, when we are preparing together this book. He wrote us the last letter from the Hospital. He promoted and organized the Frontiers of Fundamental Physics (FFP15) in Orihuela, offering us a perfect organization and hospitality.
We remember him as a very active colleague with particular attention to science education and informal learning. Since 2012 he was Pedagogical Director of Teaching and Interactive Museum of Science Vega Baja del Segura of Valencia (MUDIC-VBS-CV) and head of the Department of Physics and Chemistry of the IES "Thader" of Orihuela (Alicante). He received the award for Ciencia en Acción. Granada. 2009 with the work Stand on contest: Pesando astronautas. Promoter and responsible for many project in Physics and Chemistry Education, he carried out many teacher education courses and initiatives to promote the MUDIC Museum.

He never spared himself in his work and has always been willing to promote social initiatives for scientific education. He was passionate about optics and astronomy. His sympathy and generosity made him a man loved and appreciated by all. We will always remember him as the one who gave us a lot on the human and scientific level.

Jesús Carnicer Murillo during the opening ceremony of the Fundamental Frontiers of Physics 15 in Orihuela (Spain)

Preface

International Frontiers of Fundamental Physics Symposium Series

(Fifteenth in the Series, Orihuela)

For over a decade the International Symposium Series *Frontiers of Fundamental Physics* has attracted some of the greatest physicists in the world as well as many other eminent physicists. The broad objective of the series has been to enable scholars working in slightly different areas to meet on a single platform and exchange ideas and status reports and even dissenting views.

The areas covered have included Astronomy and Astrophysics, Particle Physics, Theoretical Physics, Gravitation and Cosmology, Computational Physics, and related areas. The symposia have been held in India (multiple), Italy (multiple), Spain, Canada, Australia, and France (multiple).

The eminent physicists who have delivered special lectures over the years, sometimes more than once, have included Nobel Laureates, professors G.'t Hooft, S. Chu, Charles Townes, Klaus von Klitzing, Pierre de Gennes, Douglas D Osheroff, Sir Harry Kroto, Sir Tony Leggett, and also the likes of Prof. Yuval Ne'eman, Prof. J. Pati, Prof. John Ellis, Prof. Asoke Sen, and several other prominent scholars. There have also been contributed papers and posters.

The selected papers books of almost all the symposia in the series have been published by the Universities Press (Orient Longman), Kluwer Academic, Springer, and the American Institute of Physics.

Increasingly over the years, sessions involving students and physics education research have also been included. In the past physicists from India and other Asian countries including Japan and the Middle East, Europe, Russia, the USA, South America, and elsewhere have presented papers.

Contents

Part II Physics Education Research

Part I
Fundamental Physics

Chapter 1
Symmetries in the Standard Model

Jose Bernabeu

Abstract Symmetries in the Physical Laws of Nature lead to observable effects. Beyond the regularities and conserved magnitudes, the last decades in Particle Physics have seen the identification of symmetries, and their well-defined breaking, as the guiding principle for the elementary constituents of matter and their interactions. Flavour SU(3) symmetry of hadrons led to the Quark Model and the antisymmetry requirement under exchange of identical fermions led to the colour degree of freedom. Colour became the generating charge for flavour-independent strong interactions of quarks and gluons in the exact Colour SU(3) local gauge symmetry. Parity violation in weak interactions led to consider the chiral fields of fermions as the objects with definite transformation properties under the weak isospin SU(2) gauge group of the unifying electroweak SU(2) $\times$ $U(1)$ symmetry, which predicted novel weak neutral current interactions. CP violation led to three families of quarks opening the field of Flavour Physics. Time-reversal violation has recently been observed with entangled neutral mesons, compatible with CPT-invariance. The cancellation of gauge anomalies, that would invalidate the gauge symmetry of the quantum field theory, leads to quark-lepton symmetry. The experimental discovery of quarks and leptons and the mediators of their interactions, with physical observables in spectacular agreement with this standard theory, is the triumph of symmetries. The gauge symmetry is exact only when the particles are massless. One needs a subtle breaking of the symmetry, providing the origin of mass, without affecting the excellent description of the interactions. This is the Brout–Englert–Higgs mechanism which produces the Higgs boson as a remnant discovered at CERN in 2012. Open present problems are addressed with the search of New Physics Beyond-the-Standard-Model.

J. Bernabeu (✉)
Department of Theoretical Physics, University of Valencia, Valencia, Spain

IFIC, Joint Centre UV-CSIC, Burjassot, Spain
e-mail: Jose.Bernabeu@uv.es

B. G. Sidharth et al. (eds.), *Fundamental Physics and Physics Education Research*,
https://doi.org/10.1007/978-3-030-52923-9_1

1.1 Symmetry as Guiding Principle for Particles and Interactions

In ordinary life we observe symmetry of objects, like characteristic features of geometrical forms, material objects or biological bodies. The concept is related to the invariance of the object under definite transformations: One object is symmetric if, after a transformation is applied, the result remains the same, i.e. it remains "invariant". But we also observe symmetry breaking, which is particularly of interest when it is not a random effect but follows a definite pattern. In Fig. 1.1 we show the three-span arch of the FermiLab entrance, near Chicago, which appears perfectly symmetric when viewed from below, but has a calculated asymmetry from its other views. Symmetry and symmetry breaking are very important concepts in the field of elementary particle physics, however not referring to objects but to the fundamental laws of physics.

We show here how symmetry has acted as a guiding principle for both the existence of new particles and the formulation of interactions. One can claim that "Symmetry dictates Interaction", as stated by Yang. In Quantum Mechanics, the symmetry is implemented by a unitary transformation $\widehat{U}$ acting on states and observables. If the dynamics, described by the Hamiltonian $\widehat{H}$, is invariant under the transformation one has

$$\left[\widehat{H}, \widehat{U}\right] = 0 \tag{1.1}$$

Under infinitesimal transformations generated by $\widehat{G} = \widehat{G}^{\dagger}$, one obtains immediately

Fig. 1.1 Symmetry Breaking

$$\frac{\mathrm{d}}{\mathrm{d}t}\left\langle \widehat{G} \right\rangle = i\left\langle \left[\widehat{H}, \widehat{G} \right] \right\rangle = 0 \tag{1.2}$$

As $\widehat{G}$ is Hermitian, it corresponds to an observable that satisfies a conservation law if $\widehat{U}$ is symmetry of $\widehat{H}$. Well-known examples are momentum for translations, angular momentum for rotations or charge for gauge symmetry. For local gauge symmetry, the requirement of invariance leads to a covariant derivative with a mediator field responsible of interactions. This is valid for either QED with the Abelian $U(1)$ gauge group or non-Abelian gauge groups with the interaction field transforming as the adjoint representation.

In Sect. 2, we develop the ideas leading from hadrons to quarks and the symmetries of strong interactions. In Sect. 3, a parallel discussion is made for electroweak interactions starting from parity violation leading to the standard model with neutral currents and the need of charm plus the third family, including quark-lepton symmetry. Section 4 presents the Brout–Englert–Higgs mechanism for the origin of mass breaking the electroweak gauge symmetry. Some conclusions and outlook are given in Sect. 5.

1.2 Quarks and Strong Interactions

The proliferation of non-strange and strange Hadrons in the 60s of the twentieth century led to the Eightfold Way of Gell Mann and Ne'eman with the use of the Flavour SU(3) symmetry. The fundamental representations $3, \overline{3}$ are the elementary building blocks for arbitrary higher-dimensional representations. Mesons are $q - \overline{q}$ states $3 \times \overline{3} = 1 + 8$, Baryons are q-q-q states $3 \times 3 \times 3 = 1 + 8_s + 8_a + 10$, with three quark $q = u, d, s$ states. In Fig. 1.2, the octet and decuplet representations of

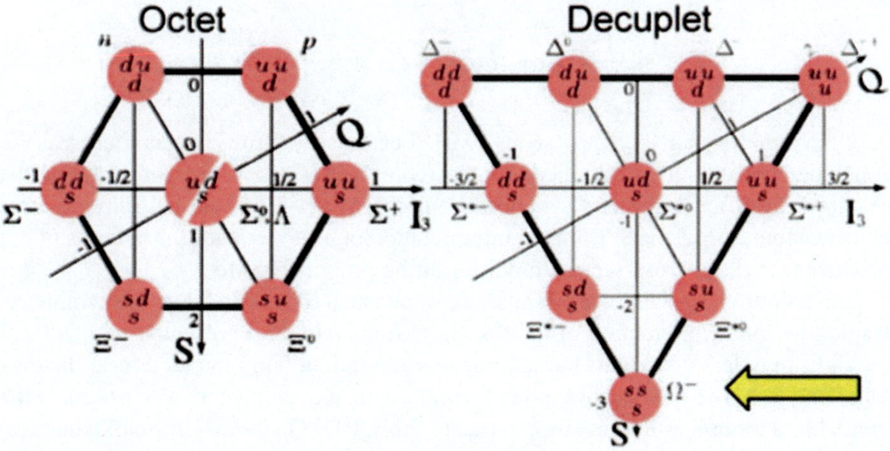

Fig. 1.2 Octet and decuplet of baryons

Baryons are given in terms of third component of Isospin I3 and hypercharge Y axes. According to the Gell Mann–Nishijima rule, the electric charge is $Q = \mathrm{I3} + Y/2$, with $Y = B + S$, B the baryonic number and S strangeness.

At the time of this formulation, the Ω^- had not been detected. Its later discovery was a great triumph of the whole scheme.

For some time, however, the quark model for hadrons (Gell Mann 1964) was considered by the scientific community as a mere theoretical construct to describe the classification of hadrons in the SU(3) symmetry. The question was "Are Quarks real?". Since 1969, deep inelastic scattering experiments (Bloom et al. 1969) at SLAC showed that the proton contained much smaller, point-like constituents and was therefore not an elementary particle. Physicists were reluctant to firmly identify these objects with quarks at the time, instead calling them "partons"—a term coined by Feynman. The partons that were observed at SLAC would later be identified as up and down quarks. Nevertheless, "parton" remains in use as a collective term for the constituents of hadrons (quarks, antiquarks and gluons). We do know at present that leptons (electrons, muons, neutrinos) find partons in the proton with high momentum transfer events.

A "jet" is a narrow cone of hadrons produced by the hadronization of a parton. Jets were observed for the first time in the $e^+ e^-$ annihilation into hadrons at the SPEAR storage ring (Hanson et al. 1975) and interpreted in terms of quarks. Quarks therefore exist, but they cannot propagate asymptotically. Quarks are then confined!

One of the reasons why the idea of real quarks was seen with scepticism was the problem of quarks with the exchange symmetry associated with the spin-statistics connection. It is easily realized with the Δ^{++} puzzle: The state $u^{\uparrow}u^{\uparrow}u^{\uparrow}$ with third component of spin $S_3 = +3/2$ is evidently symmetric under exchange of flavour (u), spin ($S_3 = +1/2$) and space ($L = 0$) degrees of freedom of the three quarks!.

If quarks are real and satisfy the exchange symmetry, a new degree of freedom is necessary for quarks, the "colour" (r, g, b) being antisymmetric for its exchange in baryons. Precisely the singlet colour wave function

$$\Psi_{\mathrm{c}}^{qqq} = \frac{1}{\sqrt{6}}(rgb - rbg + gbr - grb + brg - bgr) \tag{1.3}$$

is antisymmetric, so that qqq states exist, but these hadrons are colourless. We conclude that colour is confined, so that colourful quarks are confined. For the requirement of antisymmetry, we need a number $N_{\mathrm{c}} = 3$ of colours. Experimental evidence that $N_{\mathrm{c}} = 3$ came from the interpretation of $e^+ e^- \rightarrow$ hadrons in terms of $q\,\overline{q}$ production, with a cross-section predicted to be proportional to N_{c}.

The colour charge appears as generator of an exact $\mathrm{SU(3)_c}$ local gauge symmetry, leading to colour interaction of quarks in the fundamental representation, mediated by eight massless gluons in the adjoint representation. This interaction is flavour-blind and only the quark mass terms break flavour independence. The origin of the quark mass terms should then be external to this QCD (Quantum ChromoDynamics) theory. The field tensor is covariant ($A = 1, \ldots, 8$) leading to self-interaction of the vector gluon field A_{μ}^{A} in the Lagrangian term $-\frac{1}{4}\, F_{\mu\nu}^{\mathrm{A}}\, F^{\mathrm{A}\mu\nu}$

$$F^A_{\mu\nu} = \partial_\mu A^A_\nu - \partial_\nu A^A_\mu - g_s f_{ABC} A^B_\mu A^C_\nu \tag{1.4}$$

All coloured objects have strong interaction with gluons, so that quarks with gluons, gluons with themselves. Gluons have colour, so they are confined like quarks. Gluon jets were first observed in the annihilation $e^+e^- \rightarrow q\,\overline{q}\,g$ to three jets by the TASSO experiment (Brandelik et al. 1979) at the PETRA accelerator at the DESY laboratory.

The QCD coupling constant $\alpha_S = g_S{}^2/(4\pi)$ is dimensionless, therefore the classical field theory in the chiral (massless) limit is scale invariant. There is a conformal symmetry. However, in the perturbative treatment of the QCD quantum theory, predictions for observables are made in terms of the renormalized coupling $\alpha_S\left(\mu_R^2\right)$, which is a function of the renormalization scale. Taking it close to the momentum transfer Q^2, $\alpha_S\,(Q^2)$ indicates the effective strength of the interaction.

The coupling runs with the renormalization scale μ_R^2 and this running coupling satisfies the renormalization group equation controlled by the QCD $\beta(\alpha_S)$ function. The 1 loop β function coefficient has contributions to the gluon self-energy from gluon self-couplings and fermion couplings with opposite signs. The dominance of the first term gives to QCD, distinct to QED, the property of ASYMPTOTIC FRE EDOM (Gross and Wilczek 1973; Politzer 1973). The approximate analytic solution is

$$\alpha_s\left(\mu_R^2\right) = \left(b_0 \ln\left({}^{\mu_R^2}\!/\!_{\Lambda^2}\right)\right)^{-1}, \quad b_0 = \left(33 - 2n_f\right)/(12\,\pi) \tag{1.5}$$

with Λ a constant of integration, representing the non-perturbative scale of QCD. The running coupling has been experimentally demonstrated with $\Lambda \sim 250$ MeV. The dimensional transmutation from α_S to Λ is thus originated in the quantum conformal anomaly breaking the conformal symmetry. This Λ is responsible of the nucleon mass and, as a consequence, the baryonic mass of the Universe!

1.3 Chirality and Electroweak Interaction

Parity violation by weak interactions was postulated (Lee and Yang 1956) in the 50s of the twentieth century to solve the puzzle of the different parities of the decay products of neutral kaons. It was then observed in nuclear beta decay and later in charged pion decays.

Parity (P) $\overline{r} \rightarrow -\overline{r}$, charge conjugation (C) $q \rightarrow\ -q$ and time reversal (T) $\Delta t \rightarrow\ -\Delta t$ are discrete symmetries. In Fig. 1.3, we illustrate P and C transformations taking as reference the observed $\pi^+ \rightarrow \mu^+\,\nu_\mu$ decay.

Whereas the P-transformed and C-transformed processes do not exist in nature, the

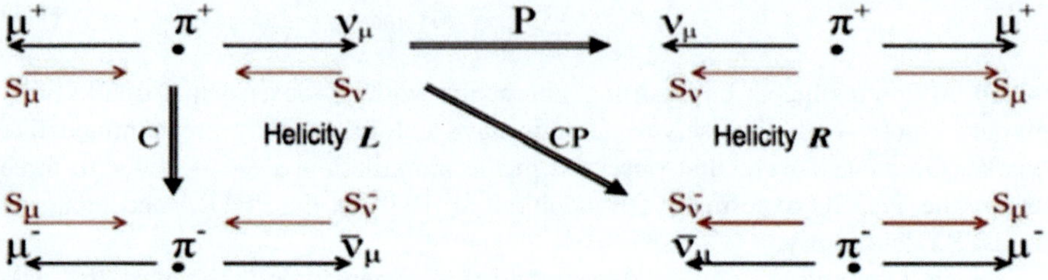

Fig. 1.3 The P, C and CP transformations in pion decays

CP-transformed decay $\pi^- \to \mu^- \ \overline{\nu}_\mu$ is observed with the same decay rate. We conclude that parity, as well as charge conjugation, is maximally violated, whereas CP is a good symmetry for pion decays.

We call a chiral phenomenon to one which is not identical to its mirror image. The spin component of a particle along its momentum may be used to define a handedness, or helicity. For massless fermions, the helicity is invariant and this intrinsic property is the "chirality". The invariance under parity for a Dirac fermion state ψ is called "chiral symmetry" and the transformation in Dirac space is implemented by the γ_5 Dirac matrix. Using projectors, left and right chiral fermions, with definite chirality −1 and +1, are given by $\frac{1}{2}(1-\gamma_5)\psi$, $\frac{1}{2}(1+\gamma_5)\psi$. There are observables, like the vector and axial vector charges that conserve chirality of the fermions, whereas other observables, like the mass or dipole moments, connect the two chiralities.

In the unified electroweak theory (Glashow 1961; Weinberg 1967; Salam 1968) based on the $SU(2)_L \times U(1)_Y$ gauge group, the fermion building blocks are not the Dirac fields ψ, but the chiral fields and the gauge group transformation distinguishes them: whereas the left fields transform as doublets under $SU(2)_L$, the right fields transform as singlets under $SU(2)_L$. We say that this unified field theory is a CHIR AL GAUGE THEORY.

The electroweak gauge group $SU(2)_L \times U(1)_Y$ symmetry demands three gauge bosons W_1, W_2, W_3 of weak isospin from $SU(2)_L$ and the B boson of weak hypercharge Y from $U(1)_Y$. The gauge symmetry is here broken by the mass terms and the physical fields with definite mass and charge are $W^\pm$, γ, Z given by

$$\begin{pmatrix} \gamma \\ Z \end{pmatrix} = \begin{pmatrix} \cos\theta_w & \sin\theta_w \\ -\sin\theta_w & \cos\theta_w \end{pmatrix} \begin{pmatrix} B \\ W_3 \end{pmatrix}, \quad M_z = \frac{M_W}{\cos\theta_w} \tag{1.6}$$

with θ_w the weak mixing angle. The theory predicts the existence of weak neutral currents mediated by the Z boson and they were discovered (Hasert et al. 1973) by the Gargamelle Bubble Chamber Collaboration at CERN in 1973 with muon neutrino interactions without muons in the final state. Ten years later, in 1983, the UA1 and UA2 experiments in the $S\overline{p}pS$ Collider at CERN discovered the massive W, Z bosons as real particles reconstructed from their $W^+ \to l^+ \nu_l$, $Z \to l^+ l^-$ (Arnison et al. 1983; Bagnaia et al. 1983) decays. These CERN discoveries established the triumph of the standard model of electroweak interactions.

1.3.1 GIM Mechanism: Need of Charm

With *u*, *d*, *s* quarks only, the Cabibbo *d-s* mixing in the charged weak current leads, by the $SU(2)_L$ symmetry of the standard model, to strangeness-changing-neutral current at tree level implying, for example, fast $K_L \rightarrow \mu^+ \mu^-$ decay, against experiment. In 1970, Glashow–Iliopoulos–Maiani (Glashow et al. 1970) solved this problem with an additional fourth quark flavour c completing two families of quark doublets

$$\begin{Bmatrix} u \\ d \end{Bmatrix} \begin{Bmatrix} c \\ s \end{Bmatrix}, \quad U = \begin{pmatrix} \cos\theta & -\sin\theta \\ \sin\theta & \cos\theta \end{pmatrix} \tag{1.7}$$

and interpreting the Cabibbo mixing as a unitary mixing matrix U in *d-s* space, exhibiting the mismatch between weak eigenstates and mass eigenstates, with charged currents relative to both *u* and *c* quarks. $SU(2)_L$ then dictates that neutral currents are governed by $U + U = I$, so they are diagonal and universal. Neutral currents are flavour-conserving at tree level! At higher orders, flavour-changing-neutral currents can be induced from c-u mass difference. The $K_L \rightarrow \mu^+ \mu^-$ is suppressed—not only by higher orders—by the GIM additional factor $(m_c^2 - m_u^2)/M_W^2$.

The discovery (Aubert et al. 1974; Agustin et al. 1974) of the $c\,\bar{c}$ J/ψ meson in 1974 at BNL and SLAC is coined as the November Revolution of particle physics. Charmed $c\,\bar{d}, c\,\bar{s}, c\,u\,d$ ... hadrons were discovered later.

1.3.2 CP Violation

CP symmetry would imply that the Laws of Physics should be invariant in form when a particle is interchanged with its antiparticle (C) while its spatial coordinates are inverted (P). For the neutral kaon system with mixing $\Delta S = 2\ K^0 - \overline{K}^0$ by weak interactions, the physical states of definite mass and lifetime K_L, K_S should be CP eigenstates leading to conservation laws: the decay $K_L \rightarrow \pi\,\pi$ should be forbidden. Its unexpected observation (Christenson et al. 1964) in 1964 opened the entire new field of CP violation in Flavour Physics.

Can CP violation be described in the standard model? In 1973, Kobayashi and Maskawa discovered (Kobayashi and Maskawa 1973) such a possibility by breaking the CP symmetry in the standard model Lagrangian by means of enlarging the particle content of the theory. By going to, at least, three families of fermions the

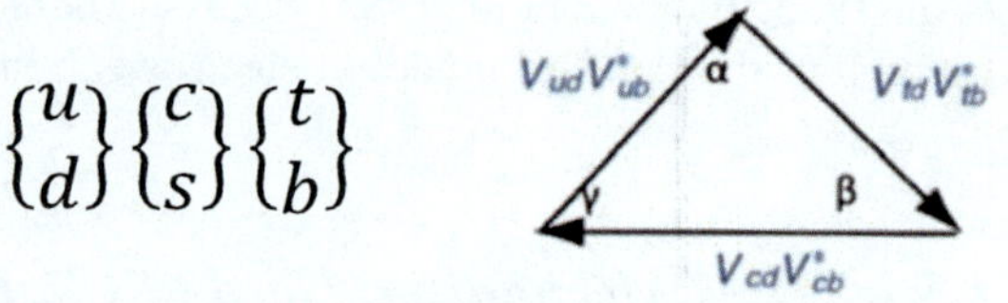

Fig. 1.4 Three quark families and unitary triangle for B_d physics

most general mismatch mixing matrix U between weak and mass eigenstates for *d-s-b* quarks contains a physical relative phase such that for antiquarks becomes its complex conjugate U^*. One would need three families of fermions at least!

The $b\,\bar{b}\,\Upsilon$ meson was discovered in 1977 at FermiLab and B-mesons later. Since then all known laboratory experimental results on CP violation for K, B and D physics are in agreement with the unitary mixing matrix paradigm U (CKM) with three active families of quarks. In Fig. 1.4, the three families are written and the corresponding "unitarity triangle" relation for B_d physics represented

One should notice: (a) the three upper u, c, t quarks have to be involved; (b) the three angles α, β, γ are CP violating observable phases, the first two involving the virtual $B^0 - \overline{B}^0$ mixing through the heavier t quark, whereas γ is a signal of direct CP violation in the decays to i and u quarks.

However, this standard model description of CP violation is not enough to explain the matter-antimatter asymmetry in the Universe!

1.3.3 Top Quark physics

The top quark is the most massive of all observed elementary particles. With a mass of 172.44 GeV/c^2, it weighs like an atom of tungsten!. It decays by weak interaction $t \rightarrow bW$ with a lifetime of 5×10^{-25} s. Such a short life is 1/20 of the timescale for quark hadronization, allowing "bare" quark studies with its entire spin density matrix in the production as well as in the decay.

The top quark was first indirectly "seen" with non-decoupling virtual quantum effects in $B^0 - \overline{B}^0$ mixing (Albajar et al. 1987a, b; Albrecht et al. 1987) measured by UA1 and ARGUS in 1987, in the universal Z boson self-energy (Veltman 1980) and in the specific $Z\ b\,\bar{b}$ vertex (Bernabeu et al. 1988; Bernabeu et al. 1991), the last two observed in the LEP experiments. The direct detection of top quarks was then made in 1995 at the $p\,\bar{p}$ Tevatron (Abe et al. 1995; Abachi et al. 1995). The $p\,p$ collider LHC facility is at present a top quark factory by means of its strong $g\,g \rightarrow t\,\bar{t}$ and weak $u\,\bar{d} \rightarrow t\,\bar{b}$ production mechanisms.

1.3.4 Time Reversal

A symmetry transformation T that changes the dynamics of a physical system into another with an inverted sense of time evolution is called time reversal (reversal-in-time). It is implemented in the space of states by an antiunitary operator implying that its study has to be made with asymmetries built under the exchange of in, out states.

The decay is an irreversible process indicating that a true TRV observable, needing a definite preparation and filtering of the appropriate initial and final particle state, looks impossible for transitions in the case of decaying particles. A bypass to this NO-GO argument was found (Banuls and Bernabéu 1999, 2000) using entangled systems of unstable particles with the ingredients: (a) The decay as a filtering measurement; (b) Entanglement implying the information transfer from the decayed particle to its living partner. For the entangled $B^0 - \overline{B}^0$ system produced by $e^+ e^-$ collisions at the $\Upsilon(4S)$ peak, one may study the time dependence in meson transitions associated to the definite flavour-CP eigenstate decay products. There are 2(Flavour) × 2(CP)—2(time ordering) = 8 transitions of this kind which can be connected by different separate genuine T, CP, CPT symmetry transformations.

In Fig. 1.5 the experimental steps to measure the time-dependent TRV asymmetry for the $\overline{B}^0 \to B_-$ and $B_- \to \overline{B}^0$ meson transitions between flavour and CP eigenstates are given.

Using these concepts, the BABAR collaboration observed (Lees et al. 2012) in 2012 a true TRV effect with 14 σ significance.

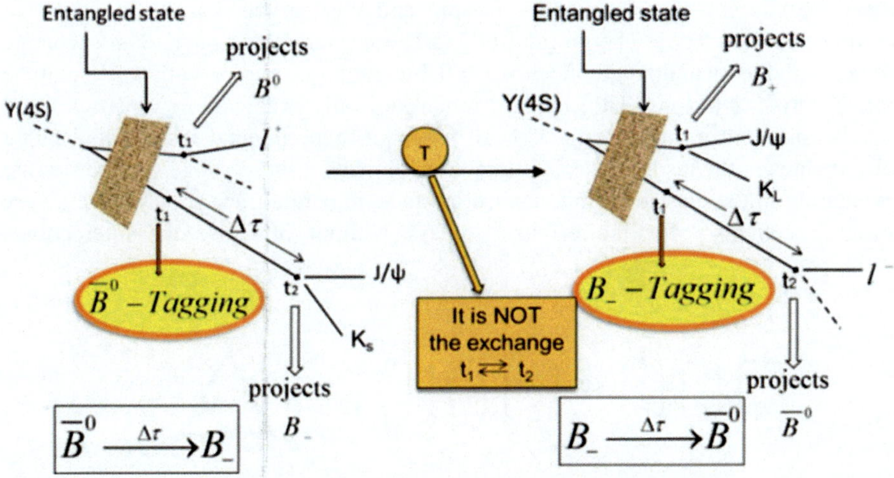

Fig. 1.5 Experimental steps to observe TRV in the entangled B_d-system

1.3.5 Gauge Anomalies: Quark-Lepton Symmetry

A gauge anomaly is a feature of quantum physics, a one-loop diagram, invalidating the gauge symmetry of a quantum field theory. All gauge anomalies must cancel out. Anomalies in gauge symmetries would destroy the required cancellation of unphysical degrees of freedom, such as a photon polarization in time direction.

Are gauge anomalies cancelled in the standard model? Anomalies appear in even D spacetime dimensions with CHIRAL fermions running in the loop with $n = 1 + D/2$ vertices. For $D = 4$, $n = 3$, it corresponds to vector-vector-axial couplings! The condition for cancellation involves the particle content and the relations among their couplings (Fujikawa and Suzuki 2004): the symmetrized trace over the generators of the gauge group vanishes

$$\mathrm{tr}\left(\left\{\tau_i, \tau_j\right\}\tau_k\right) = 0 \tag{1.8}$$

Such a cancellation operates within each family of quarks and leptons establishing an intriguing connection between the two sectors announcing a grand unification. For the three families required to incorporate CP violation, we then write the symmetry between quarks and leptons (Fig. 1.6).

1.4 The Brout–Englert–Higgs Mechanism

The standard model of particle physics contains as elementary constituents the three families of fermions with the quark-lepton symmetry. Their interactions appear as a requirement of the local gauge symmetries $\mathrm{SU}(3)_c \times \mathrm{SU}(2)_L \times U(1)_Y$ generated by the three charges of colour, weak isospin and weak hypercharge. The last two combine to the electric charge for $U(1)_{em}$. These interactions operate as exchange forces with the mediators gluon with $m = 0$, but confined, photon with $m = 0$ and the massive $W^{\pm}$, Z bosons. The standard model not only predicted new particles and interactions, but its agreement with all precision experimental results of detailed observables in the last decades is impressive. However, these gauge symmetries are exact only in the massless limit, against the facts in nature. One should have a very subtle mechanism for the origin of mass without affecting the interactions,

$$\begin{Bmatrix} u \\ d \end{Bmatrix} \begin{Bmatrix} c \\ s \end{Bmatrix} \begin{Bmatrix} t \\ b \end{Bmatrix} \leftrightarrow \begin{Bmatrix} \nu_e \\ e \end{Bmatrix} \begin{Bmatrix} \nu_\mu \\ \mu \end{Bmatrix} \begin{Bmatrix} \nu_\tau \\ \tau \end{Bmatrix}$$

Fig. 1.6 Quark-Lepton Symmetry requested for the cancellation of Eq. (1.8)

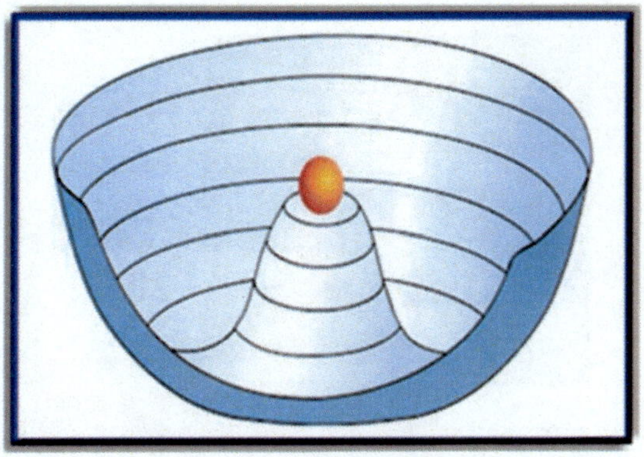

Fig. 1.7 Interaction of the complex scalar field

responsible of the $SU(2)_L \times U(1)_Y$ gauge symmetry breaking into $U(1)_{em}$. This is the Brout–Englert–Higgs mechanism (Englert and Brout 1964; Higgs 1964).

The Spontaneous ElectroWeak Symmetry Breaking (SEWSB) is based on the possibility that a symmetric Law of Physics can lead to asymmetric solutions. One should be aware that a quantum field theory needs for its precise definition not only the Lagrangian (the physical law) but also the quantum vacuum, the lowest energy state from which particles are created and annihilated. SEWSB means that the physical law is symmetric and the vacuum is asymmetric. How?

The spacetime is filled with a "medium", a complex scalar field with the interaction being like a "mexican hat" (Fig. 1.7).

This behaviour is obtained from a negative "mass square" quadratic term plus a positive quartic term. We observe that, instead of a unique symmetric lowest energy state, there are many possible vacua and one choice breaks the symmetry. This "spontaneous symmetry breaking" could be called a hidden symmetry because the results are independent of the chosen vacuum.

The physical particle created from the new vacuum is the Higgs boson, a remnant of the Brout–Englert–Higgs Mechanism, hence its importance. There is a crystal clear signature of the Higgs particle: its coupling to all particles, including to itself, is proportional to their mass, a property that breaks the gauge symmetry. The origin of mass comes from the asymmetry of the new vacuum.

On 4 July 2012, the ATLAS and CMS experiments at CERN's Large Hadron Collider announced (Aad et al. 2012; Chatrchyan et al. 2012) they had each observed a new particle in the mass region around 125 GeV. In Fig. 1.8, we show these original data together with the comparison of the measured partial decay rates to different channels to the expected theoretical predictions in the standard model.

As seen, the couplings are consistent with hose expected for a Higgs particle.

On 8 October 2013, the Nobel prize in physics was awarded jointly to François Englert and Peter Higgs "for the theoretical discovery of a mechanism that contributes to our understanding of the origin of mass of subatomic particles, and which recently was confirmed through the discovery of the predicted fundamental particle, by the ATLAS and CMS experiments at CERN's Large Hadron Collider".

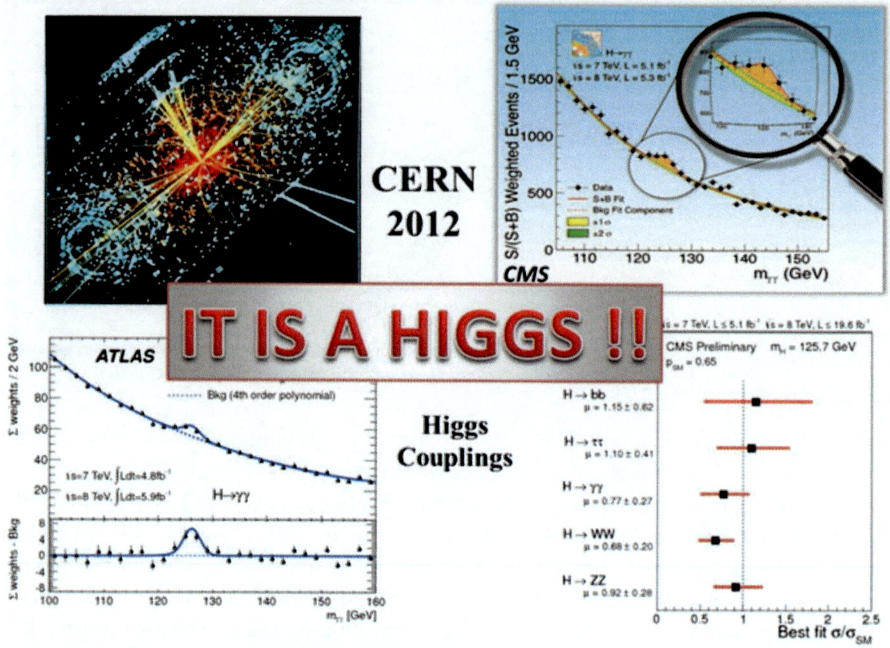

Fig. 1.8 Production and Decay of the Higgs boson observed in the ATLAS and CMS experiments

1.5 Conclusions and Outlook

The three sectors of the standard model—strong, electroweak and Higgs—represent a tribute to the concept that symmetry, and symmetry breaking, is the guiding principle for particles and interactions.

We have emphasized the role of different definite patterns for the breaking of symmetries, like

- Mass terms are incompatible with both gauge and chirality symmetries.
- Quantum loop anomalies break conformal symmetry for vector theories and gauge symmetry for chiral field theories.
- The particle content of the theory controls the breaking of discrete symmetries CP and T.
- A gauge asymmetric vacuum leads to spontaneous symmetry breaking with hidden gauge symmetry and explaining the origin of mass for elementary particles.

What next? There are theoretical and observational reasons for searching Beyond-Standard-Model Physics at LHC experiments and in other facilities. I list some of them:

- Why the quantization of electric charge, magnetic monopoles?

- The principle of "Threeality" in fundamental physics.
- The hierarchy problem for scalars, supersymmetry?
- Grand unification, p-decay?
- Neutrino Mass, mixing, CPV, majorana?
- Charged lepton flavour violation?
- Baryon asymmetry of the Universe
- Dark matter
- Dark energy

Most ideas tackling these points are linked to the Minkowski spacetime paradigm that symmetries will continue to be the guiding principle for fundamental physics. Among the discrete symmetries, CPT is protected by the "CPT-Theorem" in quantum field theory formulated in with interactions satisfying Lorentz invariance, locality and unitarity. But there is nothing at the level of quantum mechanics which forbids to have CPT-violation and there are sound quantum gravity arguments in favour of this ultimate symmetry breaking.

Acknowledgement I would like to thank the organizers of Frontiers of Fundamental Physics 2017 for the opportunity of presenting these ideas and the warm atmosphere of the Conference. This research has been supported by MINECO Project FPA 2014-54459-P, Generalitat Valenciana Project GV PROMETEO/2017/033 and Severo Ochoa Excellence Centre Project SEV 2014-0398.

References

Aad G et al (2012) Phys Lett B716:1
Abachi S et al (1995) Phys Rev Lett 74:2622
Abe F et al (1995) Phys Rev Lett 74:2626
Agustin JE et al (1974) Phys Rev Lett 33:1406
Albajar C et al (1987a) Phys Lett B186:247
Albajar C et al (1987b) Erratum Phys Lett B197:565
Albrecht H et al (1987) Phys Lett B192:245
Arnison G et al (1983) Phys Lett B126:398
Aubert JJ et al (1974) Phys Rev Lett 33:1404
Bagnaia P et al (1983) Phys Lett B129:130
Banuls MC, Bernabéu J (1999) Phys Lett B464:117
Banuls MC, Bernabéu J (2000) Nucl Phys B590:19
Bernabeu J, Pich A, Santamaría A (1988) Phys Lett B200:569
Bernabeu J, Pich A, Santamaría A (1991) Nucl Phys B363:326
Bloom ED et al (1969) Phys Rev Lett 23:930
Brandelik R et al (1979) Phys Lett B86:243
Chatrchyan S et al (2012) Phys Lett B716:30
Christenson JH, Cronin JW, Fitch VL, Turlay R (1964) Phys Rev Lett 13:138
Englert F, Brout R (1964) Phys Rev Lett 13:321
Fujikawa K, Suzuki H (2004) Path integrals and quantum anomalies. Clarendon Press, Oxford., ISBN 0-19-852913-9
Gell Mann M (1964) Phys Lett 8:214
Glashow SL (1961) Nucl Phys 22:579
Glashow SL, Iliopoulos J, Maiani L (1970) Phys Rev D2:1285

Gross DJ, Wilczek F (1973) Phys Rev Lett 30:1343
Hanson G et al (1975) Phys Rev Lett 35:1609
Hasert FJ et al (1973) Phys Lett B46:138
Higgs PW (1964) Phys Rev Lett 13:508
Kobayashi M, Maskawa T (1973) Prog Theor Phys 49:652
Lee TD, Yang CN (1956) Phys Rev Lett 104:254
Lees JP et al (2012) Phys Rev Lett 109:211801
Politzer HD (1973) Phys Rev Lett 30:1346
Salam A (1968) Conf Proc C680519:367
Veltman MJG (1980) Phys Lett B91:95
Weinberg S (1967) Phys Rev Lett 19:1264

Chapter 2
Going Beyond the Standard Model

B. G. Sidharth

Abstract In this communication we had argued that we could account for the shortcomings of the standard model by including noncommutative geometry which could lead to a non-zero (electron) neutrino mass.

At that point in time it was widely accepted that the standard model of particle physics is the most complete theory and yet there have been frantic efforts to go beyond the standard model to overcome its shortcomings. Some of these are:

1. In the theory prevalent at that time, it was stated that it fails to deliver the mass to the neutrino which thus remains a massless particle.
2. This apart, it did not include gravity, which is otherwise one of the four fundamental interactions.
3. We had to keep in mind the hierarchy problem viz., the wide range of masses for the elementary particles or even for the quarks.
4. It appears that other as of yet undiscovered particles exist which could change the picture, for example, in supersymmetry in which the particles have their supersymmetric counterparts.
5. The standard model has no place for dark matter, which on the other hand has not yet been definitely found. Nor is there place for dark energy.
6. Finally, the 18 odd arbitrary constants which creep into the theory need to be explained.

There are however obvious shortcomings which could be addressed in a relatively simple manner which could enable us to go beyond the standard model. Let us start with the standard model Lagrangian

B. G. Sidharth (✉)
Birla Science Centre, Hyderabad, India

B. G. Sidharth et al. (eds.), *Fundamental Physics and Physics Education Research*,
https://doi.org/10.1007/978-3-030-52923-9_2

$$
\begin{aligned}
L_{\mathrm{GWS}} = & \sum_f (\overline{\Psi}_f (i\gamma^\mu \partial_\mu - m_f)\psi_f - eQ_f\psi_f\gamma^\mu\psi_f A_\mu) + \\
& + \frac{g}{\sqrt{2}} \sum_i \left(a_L^{-1} \gamma^\mu b_L^i W_\mu^+ + \overline{b}_L^i \gamma^\mu a_L^i W_\mu^- \right) + \frac{g}{2C_\omega} \sum_f \overline{\Psi}_f \gamma^\mu \left(I_f^3 - 2S_\omega^2 Q_f - I_f^3 \gamma_5 \right) \Psi_f Z_\mu + \\
& - \frac{1}{4} \left| \partial_\mu A_\nu - \partial_\nu A_\mu - \mathrm{ie} \left(W_\mu^- W_\nu^+ - W_\mu^+ W_\nu^- \right) \right|^2 - \frac{1}{2} \mid \partial_\mu W_\nu^+ - \partial_\nu W_\mu^+ + \\
& - \mathrm{i.e.} \left(W_\mu^+ + A_\nu - W_\nu^+ + A_\mu \right) + ig' c_\omega \left(W_\mu^+ Z_\nu - W_\mu^+ W_\nu^- \right) \Big|^2 + \\
& - \frac{1}{4} \mid \partial_\mu Z_\nu - \partial_\nu Z_\mu + ig' c_\omega \left(W_\mu^- W_\nu^+ - W_\mu^+ W_\nu^- \right) \Big|^2 + \\
& - \frac{1}{2} M_\eta^2 \eta^2 - \frac{g M_\eta^2}{8 M_W} \eta^3 - \frac{g'^2 M_\eta^2}{32 M_W} \eta^4 + \left| M_W W_\mu^+ + \frac{g}{2} \eta \, W_\mu^+ \right|^2 + \\
& + \frac{1}{2} \left| \partial_\mu \eta + i M_Z Z_\mu + \frac{ig}{2C_\omega} \eta \, Z_\mu \right|^2 - \sum_f \frac{g \, m_F}{2 \, M_W} \overline{\Psi}_f \Psi_f \eta
\end{aligned}
\tag{2.1}
$$

which includes the Dirac Lagrangian amongst other things.

We pointed out that all these have been on the basis of the usual point spacetime which is what may be called commutative. But in recent years several authors including in particular the present author has worked with a noncommutative spacetime which originates back to Snyder in the late forties itself. (This was an attempt to overcome the divergences.)

We first observed that it was Dirac (1958) who pointed out two intriguing features of his equation: (1) The Compton wavelength and (2) Zitterbewegung.

For the former, his intuition was that we can never make measurements at space or time points. We need to observe over an interval to get a meaningful definition of momentum for example. This interval was the Compton region (Sidharth and Das 2017). Next, his solution was rapidly oscillatory, what is called Zitterbewegung. This oscillatory behaviour disappears on averaging over spacetime intervals over the Compton region. Once this is done while meaningful physics appears, we are left with not points but minimum intervals.

This leads to a noncommutative geometry. One model for this is that of Snyder (1947). Applied at the Compton wavelength this leads to the so-called Snyder–Sidharth dispersion relation, the geometry being given by Sidharth (2008)

$$
\left[x_i, x_j\right] = \beta_{ij} . l^2 \tag{2.2}
$$

As described in detail in Sidharth (2010), this leads to a modification in the Dirac and also the Klein–Gordon equation. This is because Eq. (2.2) in particular leads to the following energy momentum relation (cf. Sidharth 2008)

$$E^2 - p^2 - m^2 + \alpha l^2 p^4 = 0 \tag{2.3}$$

where α is a scalar constant, $|\alpha| \approx 10^{-3}$ (Sidharth et al. 2015, 2016). Though the value of α is of no consequence for the present work, it may be mentioned that α gives the Schwinger term. If we work with this energy momentum relation (2.3) and follow the usual process, we get as in the usual Dirac theory

$$\{\gamma^\mu p_\mu - m\}\psi \equiv \{\gamma^o p^o + \Gamma\}\psi = 0 \tag{2.4}$$

We now include the extra term in the energy momentum relation (2.3). It can be easily shown that this leads to

$$p_o^2 - \left(\Gamma\Gamma + \{\Gamma\beta + \beta\Gamma\} + \beta^2 \alpha l^2 p^4\}\right) \psi = 0 \tag{2.5}$$

Whence the modified Dirac equation

$$\left\{\gamma^o p^o + \Gamma + \gamma^5 \alpha^2\right\} \psi = 0 \tag{2.6}$$

The modified Dirac equation contains an extra term. The extra term gives a slight mass for the neutrino which is roughly of the correct order viz., $10^{-8} m_e$, m_e being the mass of the electron. The behaviour too is that of the neutrino (Sidharth 2010, 2017).

To sum up the introduction of the noncommutative geometry described in Eq. (2.2) leads to a Dirac like Eq. (2.6) and a Lagrangian that leads to the electron neutrino mass.

It must be pointed out that the modified Lagrangian differs from the usual Lagrangian in that the γ^o matrix is now replaced by a new matrix

$$\gamma^{o\prime} = \gamma^o + \gamma^o . \gamma^5 l p^2$$

that includes the term giving rise to the neutrino mass. We could verify that the modified Lagrangian gives back the modified Dirac equation (2.6). Further as has been discussed in detail, the extra term arising out of the noncommutative geometry is the direct result of the dark energy which thus also features in the modified standard model Lagrangian. This apart, this argument has been shown to lead to a mass spectrum for elementary particles that includes all the elementary particles, giving the masses with about 5% or less error (Sidharth 2008).

References

Dirac PAM (1958) The principles of quantum mechanics. Clarenden Press, Oxford
Sidharth BG (2008) The thermodynamic universe. World Scientific, Singapore

Sidharth BG (2010) The mass of the neutrinos. Int J Mod Phys E19(11):1–8
Sidharth BG (2017) Report: "twenty years" of dark energy. New Adv Phys 11(2):85–97
Sidharth BG, Das A (2017) Int J Mod Phys A32:1750173ff
Sidharth BG, Das A, Arka R (2015) Electron J Theor Phys 12(34):139–152
Sidharth BG, Das A, Arka R (2016) Int J Theor Phys 55(2):801–808
Snyder HS (1947) Phys Rev 72(1):68–71

Chapter 3
Using "Enhanced Quantization" to Bound the Cosmological Constant, (for a Bound-on Graviton Mass), by Comparing Two Action Integrals (One Being from General Relativity) at the Start of Inflation

Andrew Walcott Beckwith

Abstract **The first result from 2018** is looking at two action integrals and also a Lagrangian multiplier as a constraint equation (on cosmological expansion). In doing so, with Padmanabhan's version version of an inflaton, we then have a bound upon the cosmological constant. For the record, this is in fidelity with the author's publication, in **JHEPGC, entitled "Using 'Enhanced Quantization' to Bound the Cosmological Constant, and Computing Quantum Number *n* for Production of 100 Relic Mini Black Holes in a Spherical Region of Emergent Space-Time"** which was in 2018. And was the genesis of the two integral comparison idea. The **second result from 2018** is to use the inflaton results and conflate them with John Klauder's action principle for a way to have the idea of a potential well, generalized by Klauder, with a wall of space-time in the pre-Planckian regime to ask what bounds the cosmological constant prior to inflation, and get an upper bound on the mass of a graviton. **The third result from 2018** and the first cited reference is a redo of a multiverse version of the Penrose cyclic conformal cosmology to show how this mass of a heavy graviton inconsistent from cycle to cycle. **The fourth result from 2020** is to ask if we can, using an idea from a publication by Diosi, in the Dice 2018 physics conference use a high energy comparison of Planck Length, and a De Broglie wavelength, to find out if we can extract from our estimate of Planck mass a statement as to entropy of the early universe, and the **fifth result from 2020** is to comment upon a comparison of the power of the entropy result so obtained with the number of e foldings arising in inflation. This last question we view as essential for answering if there is a foundation of inflation which is linked to quantum gravity. **We wish to avoid the anthropic principle in setting initial conditions for the massive graviton, which is why we referenced the modification of the Penrose CCC theory in part of our manuscript.**

A. W. Beckwith (✉)
Physics Department, College of Physics, Chongqing University, Chongqing, People's Republic of China
e-mail: abeckwith@uh.edu

B. G. Sidharth et al. (eds.), *Fundamental Physics and Physics Education Research*,
https://doi.org/10.1007/978-3-030-52923-9_3

3.1 Basic Idea, Can Two First Integrals Give Equivalent Information? This Is Due to the First Reference

We use a construction for a mass of a graviton from Beckwith (2018). Instead of Hamber's (Beckwith 2018; Hamber 2009) first integral, we use what John Klauder wrote in Klauder (2015) to form a first integral in Beckwith (2018) so as to make a 1 to 1 equivalence with the first integral of general relativity (Dalarsson and Dalarsson 2005; Weinberg 1972). As in Beckwith (2018) we have a 1 to 1 relationship between two first action integrals. We avoid starting with a cosmic singularity, and use the cosmic bounce, as given in Rovelli and Vidotto (2015). In doing so, from Beckwith (2017, 2018) we use a barrier between interior and exterior space-time, to set up a cosmological constant. Our goal is to set up arguments which have no connection to the anthropic principle (Barrow and Tipler 1988) which the author regards as crackpot science of the worst sort. Having a clean break from the anthropic principle is a main goal of our enterprise.

3.1.1 This Is Our Argument for the GR First Integral: From Beckwith (2018)

The first integral used in Beckwith (2018) is also in a first integral (Padmanabhan 2005) of the form given by Eq. (3.1), having a Ricci scalar (Padmanabhan 2006). Also, the curvature $\aleph$ is small. Therefore

$$\begin{aligned} S_1 &= \frac{1}{2\kappa}\int \sqrt{-g}\cdot \mathrm{d}^4x \cdot (\mathfrak{R} - 2\Lambda) \\ \& -g &= -\det g_{uv} \\ \& \mathfrak{R} &= 6\cdot \left(\frac{\ddot{a}}{a} + \left(\frac{\dot{a}}{a}\right)^2 + \frac{\aleph}{a^2}\right) \end{aligned} \tag{3.1}$$

Also, $\delta g_{tt} \approx a_{\min}^2 \phi$ (Beckwith 2016; Giovannini 2008) uses inflaton, ϕ (Padmanabhan 2006). Which if we abide by Beckwith (2018) and Padmanabhan (2006)) we have

$$a(t) = a_{\min}\cdot t^{\gamma} \tag{3.2}$$

Padmanabhan, in (2006), states Eq. (3.2) above leads to the Beckwith (2018) and Padmanabhan (2006)) inflaton.

$$\phi \approx \sqrt{\frac{\gamma}{4\pi G}} \cdot \ln\left\{\sqrt{\frac{8\pi G V_0}{\gamma \cdot (3\gamma - 1)}} \cdot t\right\} \tag{3.3}$$

Here $a_{\min}$ is a minimum value of the scale factor (Beckwith 2018; Camara et al. 2004) which may have an electromagnetic contribution as given in Camara et al. (2004).

3.2 Next from Klauder Details Which Are Used to Give More Q.M. Structure

Note (Beckwith 2018) we are using a restricted quantum action principle (Klauder 2015) S_2, Eq. (3.4), when we assume for simplicity that Λ is a constant.

$$\begin{aligned} S_2 &= \int_0^T \mathrm{d}t \cdot [p(t)\dot{q}(t) - H_N(p(t), q(t))] \\ &\approx S_1 = \frac{1}{2\kappa} \int \sqrt{-g} \cdot \mathrm{d}^4 x \cdot (\mathfrak{R} - 2\Lambda) \end{aligned} \tag{3.4}$$

Hence

$$\begin{aligned} \Lambda \approx & \frac{-[p(\tilde{t})\dot{q}(\tilde{t}) - H_N(p(\tilde{t}), q(\tilde{t}))]}{\frac{1}{\kappa}\int \sqrt{-g} \cdot \mathrm{d}^3 x} \\ & + \frac{\frac{1}{2\kappa}\int \sqrt{-g} \cdot \mathrm{d}^3 x \cdot \left(\mathfrak{R} = 6 \cdot \left(\frac{\ddot{a}}{a} + \left(\frac{\dot{a}}{a}\right)^2 + \frac{\aleph}{a^2}\right)\right)\Big|_{t=\tilde{t}}}{\frac{1}{2\kappa}\int \sqrt{-g} \cdot \mathrm{d}^3 x} \end{aligned} \tag{3.5}$$

3.3 Filling in the Details of the Above Using Details from Klauder (2015) with Explanations

First

(a) That $S_2 \approx S_1$ from pre-Planckian to Planckian space-time when we avoid the cosmic singularity (Beckwith 2018). We ask that our model be consistent with the cosmology associated with a cosmic bounce, instead of a traditional singularity (Beckwith 2017)

(b) Also curvature $\aleph$ will be a small part of Ricci scalar $\mathfrak{R}$ (Novello n.d.)

$$\mathfrak{R} = 6 \cdot \left(\frac{\ddot{a}}{a} + \left(\frac{\dot{a}}{a} \right)^2 + \frac{\aleph}{a^2} \right) \sim 6 \cdot \left(\frac{\ddot{a}}{a} + \left(\frac{\dot{a}}{a} \right)^2 \right) \tag{3.6}$$

Furthermore, assume that there is a barrier between the pre-Planckian and Planckian physics regimes, so that we have a quantum mechanical potential well, using Beckwith (2017) which has Klauder's (2015) notation that N represents the strength of the wall.

$$\begin{aligned} \frac{p_0^2}{2} &= \frac{p_0^2(N)}{2} + N; \quad \text{for } 0 < N \leq \infty \\ q &= q_0 \pm p_0 t \\ V_N(x) &= 0; \quad \text{for } 0 < x < 1 \\ V_N(x) &= N; \quad \text{otherwise} \\ H_N(p(t), q(t)) &= \frac{p_0^2}{2} + \frac{(\hbar - \pi)^2}{2} + N; \quad \text{for } 0 < N \leq \infty \end{aligned} \tag{3.7}$$

We set $q = q_0 \pm p_0 t \sim \phi$ and assume small time steps, and the scale factor is given by Camara et al. (2004)

3.4 Why This Is Linked to Gravity/Massive Gravitons

Klauder's program is to isolate a regime of space-time for a canonical quantization of a classical system. That is, what we did is to utilize the ideas of Klauder (2015) to make the identification of Eq. (3.7) which when combined with enhanced quantization of the, as given in Eq. (3.3). That is, assume $\sqrt{-g}$ is a constant. And this is for extremely small-time intervals (in the boundary between pre-Planckian to Planckian physical boundary regime). As given by Giovannini (2008) in his comprehensive review of cosmological production of gravitons and early universe entropy. This approximation is why $g_{tt}\tilde{\delta} g_{tt} \approx a_{\min}^2 \phi$.

If so, the mass of a graviton is referred to, as given by Novello (n.d.). We can then write a bound, based upon the early universe conditions so set forth, as a way to ascertain a bound to the effective heavy graviton (Beckwith 2014, 2018; Novello n.d.)

$$m_g^2 = \left(\frac{\hbar \cdot \sqrt{\Lambda}}{c}\right)^2$$
$$\approx \frac{\hbar^2}{c^2}\left[\frac{-\left[\frac{V_0}{3\gamma - 1} + 2N + \frac{\gamma\cdot(3\gamma-1)}{8\pi G\cdot\tilde{t}^{\,2}}\right]}{\frac{1}{\kappa}\int\sqrt{-g}\cdot \mathrm{d}^3x} + \left(6\cdot\left(\frac{\ddot{a}}{a} + \left(\frac{\dot{a}}{a}\right)^2\right)\right)\bigg|_{t=\tilde{t}}\right]_{\tilde{t}=t(\text{Planck})} \tag{3.8}$$

Our next step is to review the input of parameters which may affect Eq. (3.10). To do so we consider a multi universe generalization of the CCC as given below. This is done specifically to kill off references to the anthropic principle as far as Eq. (3.8) (Barrow and Tipler 1988).

3.5 Reviewing Multiverse Generalization of the CCC of Penrose, and How This Relates to Beckwith's (2018) Conclusions

We are extending Penrose's suggestion of cyclic universes, black hole evaporation, and the embedding structure our universe is contained within. The following is largely taken from Beckwith (2014, 2018) and Penrose (2011) and has relevance to the final part of the conclusion. That there are no fewer than *N*(Penrose style ccc) universes undergoing Penrose "infinite expansion" (Penrose) (Beckwith 2014, 2018; Penrose 2011) contained in a mega universe structure. Furthermore, each of the *N* (Penrose style ccc) universes has black hole evaporation, with the Hawking radiation from decaying black holes. If each of the *N*(Penrose style ccc) universes is defined by a partition function, called $\{\Xi_i\}_{i\equiv N}^{i\equiv 1}$, then there exist an information ensemble of mixed minimum information correlated as about $10^7 - 10^8$ bits of information per partition function in the set $\{\Xi_i\}_{i\equiv N}^{i\equiv 1}\big|_{\text{before}}$, so minimum information is conserved between a set of partition functions per universe

$$\{\Xi_i\}_{i\equiv N}^{i\equiv 1}\big|_{\text{before}} \equiv \{\Xi_i\}_{i\equiv N}^{i\equiv 1}\big|_{\text{after}} \tag{3.9}$$

However, there is non-uniqueness of information put into each partition function $\{\Xi_i\}_{i\equiv N}^{i\equiv 1}$. Furthermore, Hawking radiation from the black holes is collated via a strange attractor collection in the mega universe structure to form a new big bang for each of the *N* (Penrose CCC style) universes represented by $\{\Xi_i\}_{i\equiv N}^{i\equiv 1}$. The n_f value will be using Ng (2008) $S_{\text{entropy}} \sim n_f$ (Beckwith 2014, 2018; Ng 2008). This assumes an energy expression as given by Beckwith (2014, 2018) and Poplawski (2011) and as by Beckwith (2018) will be an energy conservation equation before and right after the big bang, for the structure of our local universe, so formed in our modified CCC argument. Then the following holds (Beckwith 2014, 2018; Penrose 2011),

Claim 3.1

$$\frac{1}{N} \cdot \sum_{j=1}^{N} \Xi_j \Big|_{j\text{-before-nucleation-regime}} \xrightarrow[\text{vacuum-nucleation-tranfer}]{} \Xi_i \Big|_{i\text{-fixed-after-nucleation-regime}} \tag{3.10}$$

And

$$\Xi_j \Big|_{j\text{-before-nucleation-regime}} \approx \sum_{k=1}^{\text{Max}} \widetilde{\Xi}_k \Big|_{\text{black-holes-}j\text{th-universe}}$$

For N (Penrose style ccc) number of universes, with each $\Xi_j|_{j\text{ - before - nucleation - regime}}$ for $j = 1$ to N(Penrose style ccc) (Beckwith 2014, 2018; Penrose 2011) being the partition function of each universe just before the blend into the RHS of Eq. (3.11) above for our present universe. Also, each of the independent universes given by $\Xi_j|_{j\text{ - before - nucleation - regime}}$ is constructed by the absorption of black holes taking in energy. **I.e., (Penrose)** (Beckwith 2014, 2018; Penrose 2011). Furthermore, Eq. (3.11) uses the idea of Dye (1965) in terms of general ergodic mixing.

This is after we make the following identification, i.e., look at the formation of a nontrivial gravitational measure as a new big bang for each of the N universes as by $n(E_i)\cdot$ the density of states at a given energy E_i for a partition function (Beckwith 2014, 2018; Poplawski 2011).

Claim 3.2

$$\{\Xi_i\}_{i\equiv 1}^{i\equiv N} \propto \left\{ \int_0^\infty \mathrm{d}E_i \cdot n(E_i) \cdot \mathrm{e}^{-E_i} \right\}_{i\equiv 1}^{i\equiv N} \tag{3.11}$$

What is done in Claim 3.1 **and** Claim 3.2 is to come up with a protocol as to how a multi-dimensional representation of black hole physics enables continual mixing of space-time (Hawkings n.d.) largely as a way to avoid the anthropic principle (Barrow and Tipler 1988), as to a preferred set of initial conditions.

Claim 3.1 **which uses** Claim 3.2 is important. The idea here is to use what is known as CCC cosmology (Beckwith 2014, 2018; Penrose 2011), which can be thought of as the following. First. Have a big bang (initial expansion) for the universe. After redshift $z = 10$, a billion years ago, SMBH formation starts. Matter-energy is vacuumed up by the SMBHs, which at a much later date than today (present era) gather up all the matter-energy of the universe and recycle it in a cyclic conformal translation. As given, by the transformations alluded to in Eqs. (3.12) and (3.13) below. So we can thereby understand the change in CCC space-time geometry.

Note that the main methodology in the Penrose proposal (Beckwith 2014, 2018; Penrose 2011) has been evaluating a change in the metric g_{ab} by a conformal mapping $\widehat{\Omega}$ to

$$\widehat{g}_{ab} = \widehat{\Omega}^2 g_{ab} \tag{3.12}$$

Penrose's suggestion has been to utilize the following (Beckwith 2014, 2018; Penrose 2011)

$$\widehat{\Omega} \xrightarrow[ccc]{} \widehat{\Omega}^{-1} \tag{3.13}$$

In fall into cosmic black hopes has been the main mechanism which will be useful for the recycling apparent in Eq. (3.14) above with ħ kept constant from cycle to cycle as represented by Beckwith (2014, 2018)

$$\hbar_{\text{old-cosmology-cycle}} = \hbar_{\text{present-cosmology-cycle}} \tag{3.14}$$

We claim that Eq. (3.14) with Eq. (3.8) and Eq. (3.10) above gives a uniform mass to a graviton, per cycle, if Equation (3.14) holds we have the ability to make gravitons keep a constant non zero mass from cycle to cycle of the CCC paradigm. The idea is to keep consistency in physical law during each cycle of CCC dynamics. If the laws of laws of physics remain invariant , then graviton mass will not be altered. This also involves the physics of Ng (2008), Poplawski (2011), and Hawkings (n.d.). In doing so, from Beckwith (2018) consider the dynamics of the scale factor *a*(*t*), which is nearly zero, in the pre-Planckian regime of space-time. i.e., as to how *a*(*t*) changes and evolves, we look at the treatment given in Roos (2003) as well as Dye (1965). In addition by Beckwith (2018) and Hamber (2009).

$$\begin{gathered}\int \mathrm{d}t\sqrt{g_{tt}}V_3(t) = V_4(t) \sim 8\pi^2 r^4/3 \\ \& V_3(t) = 2\pi^2 a(t)^3/3 \\ \& k_2 = 9\left(2\pi^2\right)^{2/3}\end{gathered} \tag{3.15}$$

These are volume elements of the Hamber (2009) first integral. i.e., see Ambjorn et al. (2010), Karabulut (2006), Spiegel (1980). A Lagrangian multiplier is a constraint of how a "minimal surface" is obtained by constraining a physical process so as to use Ambjorn et al. (2010), Karabulut (2006), and Spiegel (1980). In the case of Karabulut (2006), the minimization process is if *a*(*t*) a scale factor as defined by Dye (1965) and g_{tt} a time component of a metric tensor. Here, the subscripts 3 and 4 in the volume refer to 3- and 4-dimensional spatial dimensions, and this will lead to, via Hamber (2009), a first integral as defined by Beckwith (2018) and Hamber (2009), in the form, if *G* is the gravitational constant,

$$S_1 = \frac{1}{24\pi G} \cdot \left(\int dt \sqrt{g_{tt}} \left(\frac{g_{tt} \dot{V}_3^2}{V_3(t)} + k_2 V_3^{1/3}(t) - \lambda V_3(t) \right) \right) \tag{3.16}$$

This should be compared against the Padmanabhan first integral (Barrow and Tipler 1988; Padmanabhan 2005), with the third entry of Eq. (3.3) having a Ricci scalar defined via Weinberg (1972) and usually the curvature ℵ (Weinberg 1972) set as extremely small, with the general relativity version of from Beckwith (2018) of Eq. (3.1). Note that our write-up actually uses all this and aligns it with the ideas of the Klauder enhanced quantization (Klauder 2015) for what we think is a better extension of the same idea. In order to obtain maximum results, we will be stating that the following will be assumed to be equivalent, i.e., in the spirit of Beckwith (2018)

$$\sqrt{g_{tt}}(\lambda V_3(t)) \sim \frac{1}{2\kappa} \int \sqrt{-g} \cdot d^3x \cdot (2\Lambda) \tag{3.17}$$

So, from Beckwith (2018), a relationship of the Lagrangian multiplier:

$$\lambda \sim \frac{1}{\kappa} \sqrt{\frac{-g}{\left(\delta g_{tt} \approx a_{\min}^2 \phi\right)}} \cdot \Lambda \tag{3.18}$$

We are obtaining the exact same physics, as in Beckwith (2018) for when we appeal to Eq. (3.8) as a bound to the enhanced quantization, hence we have extended our basic idea via use of Beckwith (2018) and Klauder (2015). To conclude with this mini section, this is included in as a way to set up a statistical averaging procedure as to avoid the anthropic principle. After having said this, if our Eq. (3.11) has successfully set up a program to avoid the anthropic principle (Barrow and Tipler 1988), we can relate this to massive gravitons, next and early universe entropy.

3.6 Why This Is Linked to Gravity/Massive Gravitons, and Possibly Early Universe Entropy

Klauder's program (Klauder 2015) is to embed via Eq. (3.8) as a quantum mechanical well for a pre-Planckian system for inflaton physics as given in Beckwith (2018) where we go to the idea as given in Klauder's treatment of the action integral as of page 87 of Klauder (2015) where Klauder talks of the weak correspondence principle, where an enhanced classical Hamiltonian is given 1–1 correspondence with quantum effects, in a non-vanishing fashion. If so, by Novello (Camara et al. 2004) and Eq. (3.8) we have then for early universe conditions, that we will be leading up to using an algorithm for massive gravitons, as we were working with in Beckwith (2018) with the result that we write, for a Plank time value is to go back to our Eq. (3.8) which we subsequently turn into Eq. (3.11) where we use the convention for scale factors as given in Camara et.al. (Giovannini 2008) so that we have a tentative value of the cosmological constant and then by extension the graviton mass via Novello (Camara et al. 2004) of Eq. (3.11). The long and short of it is to tie this value of the cosmological constant, and the production of gravitons due to early universe conditions, to a relationship between De Broglie wavelength, Planck length, and if the velocity v gets to a partial value close to the speed of light, that, we have, say by using Landau and Lifshitz (2005) as given in DICE 2018 and also part of the JHEPGC publication for quantum systems, if we have instead of a velocity much smaller than the speed of light, a situation where the particle moves very quickly (a fraction of the speed of light) that instead of the slow massive particle postulated in Landau and Lifshitz (2005)

$$\begin{aligned}
&\lambda_{\text{De-Broglie}} \approx \frac{2\pi\hbar}{m_g v}\cdot\sqrt{1-\frac{v^2}{c^2}} \cong \ell_{\text{Planck}} \approx \sqrt{\frac{\hbar G}{c^3}} \\
&\Rightarrow \text{if } v(\text{particle}) \to c - \xi^{+}; \text{then} \\
&\varepsilon(\text{energy-particle}) \approx E_{\text{Planck}}(\text{Planck-energy})
\end{aligned} \tag{3.19}$$

Let us start with a specified value of mass of a graviton, say of the order of 10–62 g and also the application of the Ng "infinite quantum statistic" counting algorithm with S (entropy) being equivalent to the number of generated gravitons, which we call n. We will then use the construction of cyclic conformal cosmology (ccc) given by Penrose (2011) so that

$$\begin{aligned}
&\text{If } c \equiv 1, m_g \approx 10^{-62}\,\text{g} \\
&E_{\text{Planck}}(\text{Planck-energy}) \cong 2.18 \times 10^{-5}\,\text{g} \\
&\cong \left(m_g \approx 10^{-62}\,\text{g}\right) \times N(\text{entropy-number}) \\
&\Rightarrow N(\text{entropy-number}) \cong 10^{58} \equiv 10^{\mathbb{N}}, \text{and} \therefore \mathbb{N} \cong 58
\end{aligned} \tag{3.20}$$

3.7 Can This Tie in with Early Universe e Folds? That Is, from Chongchitnan (n.d.) e Folds Are Between 55 and 60

E folds in cosmology are a way of delineating if we have enough expansion of the universe in line with inflation. As seen in Chongchitnan (n.d.), we can have

$$\mathbb{N}(\text{e-fold}, \cos\text{mol}) \approx -\int \mathrm{d}t \cdot H(\cos\text{mol}) \tag{3.21}$$

where H(cosmol) is a value of the Friedman equation, and if we use the idea that the potential energy, V, of initial inflation is initially over shadowed by the contributions of the Friedman equation, H, at the onset of inflation. Then

$$\mathbb{N}(\text{e-fold, cosmol}) \approx 55 - 60 \tag{3.22}$$

What we wish to explore is if Eq. (3.22) is consistent with Eq. (3.23) and what the consequences will be of this identification

$$N(\text{entropy-number}) \cong 10^{58} \equiv 10^{\mathbb{N}}, \text{and} \therefore \mathbb{N} \cong 58 \tag{3.23}$$

Doing so may involve use of the Corda article, as given in Corda (2018).

3.8 Conclusion, Does Our Bound as to the Graviton Mass, and Its Input Variables Due to Klauder Enhanced Quantization Argue in Favor of a Quantum Gravity Linkage to e Folds and Inflation? This Needs to Be Determined Next

We argue that we may have inputs into the building of a bound to the mass of a graviton if a multiverse may contribute to the construction of the graviton mass. That is the input side of the phenomena used for getting a bound to the massive graviton, and the use of the Novello (Camara et al. 2004) supposition of a linkage between massive gravity and an allowed cosmological constant. An output version of this phenomena after we create a necessary condition for massive gravity is in the issue raised by Eq. (3.8). And also the intriguing possibility of more overlap between Eqs. (3.22) and (3.23). If there is an overlap of these two Eqs. (3.22) and (3.23), which raises the intriguing question of if a mass of a graviton, i.e., a quantum gravity lodestone is in fidelity with the e folds of cosmology, then we have the distinct possibility of quantum gravity having at least a partial linkage to e fold inflationary cosmology. If this is not true, then tensor-scalar version of gravity and other models need a very hard look over. And while we are on the subject, Appendix A, as given below is yet another datum which needs experimental vetting. All these together would be needed to be confirmed via experimental gravitational data sets. A side note which is to consider is, if this happens, does it in any way have linkage to the idea of forming symmetries in space-time which could lead to SU(n) type group thinking? See Dyson (1966) as the gold standard. We can only go there though if we understand the physical input phenomena as to the creation of a burst of inflationary energies later, and if we have quantifiable data sets to come up with readily understood models. With falsifiable input parameters. This is our hope and our aspiration as of the twenty-first century as far as gravitational experimental science.

Finally what if there are many micro black holes created almost at the start of the Universe? What would this do in terms of GW generation? See Appendix B, as a future works project which may add more context as far as the varying mechanisms as far as GW, and our project is so delineated. This is in terms of entropy, itself and is a much smaller contribution of entropy than what is already delineated. But if we develop GW physics as an experimental science, this too may be something not to ignore and will involve the physics of Chongchitnan (n.d.), Corda (2018), and Dyson (1966) fully.

Appendix A: Infinite Quantum Statistics as Given by Jack Ng

Jack Ng changes conventional statistics: he outlines how to get $S \approx N$, which with additional arguments we refine to be $S \approx \; <n>$ (where $<n>$ is graviton density). Begin with a partition function (Beckwith 2014, 2018; Ng 2008)

$$Z_N \sim \left(\frac{1}{N!}\right) \cdot \left(\frac{V}{\lambda^3}\right)^N \tag{3.24}$$

This, according to Ng, leads to entropy of the limiting value of, if $S = (\log[Z_N])$

$$\begin{aligned} S &\approx N \cdot \left(\log\left[V/N\lambda^3\right] + 5/2\right) \xrightarrow[\text{Ng-inf inite-Quantum-Statistics}]{} N \\ &\quad \cdot \left(\log\left[V/\lambda^3\right] + 5/2\right) \\ &\approx N \end{aligned} \tag{3.25}$$

Appendix B: Micro Black Hole, at the Start of the Universe and Their Contribution to Early Universe GW Generation Via an Entropy Count

This is in partial fidelity to Beckwith (2018) and may be added as a factor in future entropy contributions as a secondary effect which may affect future analysis of this phenomenon. We begin first with what would transpire for micro black holes, at the start of the space-time regime. That is the first level of analysis we have done in the main document purports to create graviton type disturbances at or before the electroweak regime of space-time. The addition of entropy so given here is meant to be included, pending an evaluation as to how many primordial black holes may be considered to be created. If the number is large (many micro black holes), then what we have below will significantly add to entropy. Or it may not be a decisive factor. The analysis is included for this

document as a secondary effect for entropy generation. The idea would be that we would have, for a quantum level, n, specified for a black hole, due to what Corda developed the following temperature distribution which would ALSO add into more entropy. We include it in as a future works project. The first term below comes from Dyson (1966) and is linkable to a way to also, in addition to the mechanism so brought up a way to also add more early universe entropy, which is linkable to gravitons.

$$\begin{aligned}
&T_{\mathrm{H}} = m_D \cdot \left(\frac{m_D \cdot (n+2)}{m_{bh} \cdot 8 \cdot \Gamma\left(\frac{n+3}{2}\right)}\right)^{(1/n+1)} \cdot \left(\frac{n+1}{4\sqrt{\pi}}\right) \\
&m_D \doteq 1\,\mathrm{TeV} \equiv 10^{12}\ \mathrm{eV} = 1.783 \times 10^{-30}\,\mathrm{g} \\
&m_{bh} \doteq 1.22 \times 10^{21}\,\mathrm{TeV} = 2.175 \times 10^{-9}\ \mathrm{g} \\
&T_{\mathrm{H}} \propto 1\,\mathrm{TeV} \cdot \left(\frac{(n+2)}{1.22 \times 10^{21} \cdot 8 \cdot \Gamma\left(\frac{n+3}{2}\right)}\right)^{(1/n+1)} \cdot \left(\frac{n+1}{4\sqrt{\pi}}\right)
\end{aligned} \tag{3.26}$$

We look at the initial state of created gravitons, and use the physics given in Hawking (1974) with the specified Hawking temperature, as the main physics phenomenon of interest to our analysis. We then can, if we have this Hawkings temperature, as given in Eq. (3.7) consider the question of first, if the black holes have classical or quantum behavior as well as Γ being a gamma function, i.e., look at what is given in Beckwith (2018), in its conclusion which we will cite here. That is, the idea is based upon the formation of a finite number of black holes, which decay. Quoting Beckwith (2018) we have that we will be looking at the following:

Quote, Beckwith (2018)

"Our physics is simplified if we change Planck length to be scaled as 1 and we look at a 'unit' evaluated space-time volume." Then we can set n of Eq. (3.26) equal to zero initially and obtain the following from Corda (2018) and Beckwith (2018)

$$\begin{aligned}
&16\pi\left(\frac{\widetilde{n}_{\mathrm{qm}}}{4 \times 10^4}\right) \cdot \left|\sqrt{1-\left(\frac{\widetilde{n}_{\mathrm{qm}}-1}{\widetilde{n}_{\mathrm{qm}}}\right)}\right| \approx 10^6 \\
&m_{bh} \approx 10^2 \times m_{\mathrm{planck}}(4-\mathrm{dim}) \\
&\Delta V_{\mathrm{total}} \simeq 10^2 \times \Delta V_{\widetilde{n}_{\mathrm{qm}}-1 \to \widetilde{n}_{\mathrm{qm}}} \approx 10^2 \times 16\pi\left(\frac{\widetilde{n}_{\mathrm{qm}}}{4 \times 10^4}\right) \cdot \left|\sqrt{1-\left(\frac{\widetilde{n}_{\mathrm{qm}}-1}{\widetilde{n}_{\mathrm{qm}}}\right)}\right| \approx 10^8
\end{aligned} \tag{3.27}$$

This puts a serious restriction on the number of allowed quantization levels $\widetilde{n}_{\mathrm{qm}}$, but it also means that within this horizon space we may be seeing mini black holes created which could release gravitons. We will then discuss what may be pertinent to characterizing if the black holes are behaving classically or quantum mechanically. Note, if n in Eqs. (3.26) and (3.31) is set equal to zero, we have that if we literally interpreted Eq. (3.26), with n equal zero (4 dimensions) with a 10^2 plank mass (in four dimensions) black hole, that we would have

$$T_{\mathrm{H}} \propto 1\,\mathrm{TeV} \cdot \left(\frac{1}{1.22 \times 10^{21} \cdot 2 \cdot \sqrt{\pi}}\right) \cdot \left(\frac{1}{4\sqrt{\pi}}\right) \tag{3.28}$$

Noticeably we have a set of reality problems to attend to. Hawking radiation and the Ng (2008) paradigm alter our process so that we obtain the following results as given in Eq. (3.29) for Hawking blackbody style results.

Diosi (2019) and Calmet (n.d.) will add further refinement as to the physics of Eq. (3.29) and hopefully alter them to reflect more of the known observational physics, since Eq. (3.29) is a greatly over simplified version of the gravitational physics input into observational gravitational astronomy. We hope in doing so we obtain data sets as to confirm or falsify our hypothesis as given in this document.

$$\begin{aligned}
T_{\mathrm{H}}(\mathrm{normal}-4\mathrm{dim}) &= \frac{\hbar c^3}{8\pi G m_{bh} k_{\mathrm{B}}} = 6.17 \times 10^{-8}\,\mathrm{K} - \mathrm{if}\, m_{bh} = 1 - \text{solar-mass} \\
T_{\mathrm{H}}(\mathrm{normal}-4\mathrm{dim}) &= \frac{\hbar c^3}{8\pi G m_{bh} k_{\mathrm{B}}} \simeq 6.17 \times 10^{28}\,\mathrm{K} - \mathrm{if}\, m_{bh} = 10^2 \times m_{\mathrm{planck}}(4-\mathrm{dim}) \\
T_{\mathrm{H}}(\mathrm{normal}-4\mathrm{dim}) &= \frac{6.17 \times 10^{28}}{11{,}604}\,\mathrm{eV} - \mathrm{if}\, m_{bh} = 10^2 \times m_{\mathrm{planck}}(4-\mathrm{dim}) \\
T_{\mathrm{H}}(\mathrm{normal}-4\mathrm{dim}) &= 5.317 \times 10^{12}\,\mathrm{TeV} - \mathrm{if}\, m_{bh} = 10^2 \times m_{\mathrm{planck}}(4-\mathrm{dim})
\end{aligned} \tag{3.29}$$

The sun is 9.8 times 10^{37} Planck masses, so this means that the Hawkings temperature of a 10^2 times four-dimensional Planck mass black hole, as postulated here would be then about 10^{36} times higher, 1 eV in Kelvin is 11,604 K, hence if we have only $n = 0$ we really would prefer not to use Eq. (3.28). Hence, using Eq. (3.29). we can then go to Hawking (1974) for entropy, i.e., we have then that

$$\begin{aligned}
S_{bh} &= \frac{1+n}{2+n} \cdot \frac{m_{bh}}{T_{bh}} = \frac{1+n}{2+n} \cdot \frac{10^2 \times m_{\mathrm{planck}}(4-\mathrm{dim})}{5.317 \times 10^{12}\,\mathrm{TeV}} \approx \frac{1+n}{2+n} \cdot \frac{10^2 \times 1.22 \times 10^{16}\,\mathrm{TeV}}{5.317 \times 10^{12}\,\mathrm{TeV}} \\
S_{bh} &\approx \frac{1+n}{4+2n} \cdot 10^4 \underset{n \to 0}{\longrightarrow} 10^3
\end{aligned} \tag{3.30}$$

That is, if we had 10^2 black holes, of mass about 10^2 times a four-dimensional black hole, we probably would be looking at an initial entropy of about 10^5. Then using Kolb and Turner (1991), we would see

$$s(\text{entropy-density}) = \frac{2\pi^2}{45} g_* \cdot (T_{\text{universe}}/T_{\text{Planck}})^2 \tag{3.31}$$

This postulates that relic, initially formed black holes would be formed, say in a one-meter radius ball about 10^{-27} s, after the onset of inflation and that we would see the rapid decay of these micro-sized black holes in less than 10^{-27} s.

End of quote from Beckwith (2018)

This construction above, with suitable work later on, will be useful in removing the anthropic principle from serious consideration in cosmology (Barrow and Tipler 1988) Once again, the relevance to the analysis given in the main text is heavily flavored as to precisely how many primordial black holes are created. If it is just 10^2, the number of primordial black holes, then this is a very minor addition to the entropy. If the number of primordial black holes is significantly higher, then the contribution is different and then alters the calculations in potentially significant ways. See Ruutu et al. (1996) as to another model of what may transpire as yet another effect if there are many primordial black holes, creating early universe turbulence which may contribute (the turbulence) to entropy generation, i.e., in a cosmological scale replicating (Ruutu et al. 1996) and cosmic string representations of black holes (t'Hooft n.d.) in the early universe.

References

Ambjorn J, Jurkiewicz J, Loll R (2010) Quantum gravity as sum over space-times. In: Boob-Bavnbek B, Esposito G, Lesch M (eds) New paths towards quantum gravity. Lecture notes in physics 807. Springer-Verlag, Berlin, pp 59–124

Barrow JD, Tipler FJ (1988) The anthropic cosmological principle. Oxford University Press, Oxford. ISBN 978-0-19-282147-8. LCCN 87028148

Beckwith A (2014) Analyzing black hole super-radiance emission of particles/energy from a black hole as a Gedanken experiment to get bounds on the mass of a graviton. Adv High Energy Phys 2014:230713. arXiv:1404.7167 [physics.gen-ph]

Beckwith A (2016) Gedanken experiment for refining the unruh metric tensor uncertainty principle via Schwarzschild geometry and Planckian space-time with initial nonzero entropy and applying the Riemannian-Penrose inequality and initial kinetic energy for a lower bound to graviton mass (massive gravity). J High Energy Phys Gravit Cosmol 2:106–124. https://doi.org/10.4236/jhepgc.2016.21012

Beckwith A (2017) How to determine a jump in energy prior to a causal barrier, with an attendant current, for an effective initial magnetic field. In the pre Planckian to Planckian space-time. http://vixra.org/abs/1707.0250

Beckwith AW (2018) Using "enhanced quantization" to bound the cosmological constant, and computing quantum number n for production of 100 relic mini black holes in a spherical region of emergent space-time. J High Energy Phys Gravit Cosmol 4:549–566. https://doi.org/10.4236/jhepgc.2018.4303

Calmet X (n.d.) Quantum black holes. https://www.mpifr-bonn.mpg.de/1311186/Calmet.pdf

Camara CS, de Garcia Maia MR, Carvalho JC, Lima JAS (2004) Nonsingular FRW cosmology and non linear dynamics. Arxiv astro-ph/0402311 Version 1, Feb 12, 2004

Chongchitnan S (n.d.) Inflationary e-folding and the implications for gravitational-wave detection. https://arxiv.org/abs/1705.02712

Corda C (2018) Space-time can be neither discrete nor continuous. Mod Phys Lett A 33:1850069

Dalarsson M, Dalarsson N (2005) Tensors, relativity and cosmology. Elsevier/Academic Press, London

Diosi L (2019) Planck length challenges of non relativistic quantum mechanics of large masses. J Phys Conf Ser 1275:011001. https://arxiv.org/abs/1903.04852

Dye HA (1965) On the ergodic mixing theorem. Trans Am Math Soc 118:123–130

Dyson F (1966) Lecture 3, the SU(6) theory. In: Dyson F (ed) Symmetry groups in nuclear and particle physics. W.A. Benjamin, Inc., New York, pp 17–22. (USA) with the 1966 volume labeled as W.A. Benjamin, Inc, New York, Amsterdam, 1966

Giovannini M (2008) A primer on the physics of the cosmic microwave background. World Press Scientific, Hackensack. https://doi.org/10.1142/6730

Hamber H (2009) Quantum gravitation, the Feynman path integral approach. Springer, Heidelberg

Hawking SW (1974) Black hole explosions?. Nature 248(5443):30–31. Bibcode: 1974Natur.248...30H. https://doi.org/10.1038/248030a0. → Hawking's first article on the topic

Hawkings SW (n.d.) Informational preservation and weather forecasting for black holes. http://arxiv.org/abs/1401.5761

Karabulut H (2006) The physical meaning of Lagrange multipliers. Eur J Phys 27:709–718. https://arxiv.org/abs/0705.0609

Klauder J (2015) Enhanced quantization, particles, fields & gravity. World Press Scientific, Singapore

Kolb E, Turner (1991) The early universe. Addison and Westley Advanced book program, Menlo Park, California, USA. http://power.itp.ac.cn/~guozk/books/The_Early_Universe(Kolb_Turner_1988).pdf

Landau LD, Lifshitz EM (2005) Mechanics, 3rd edition. In: Course in theoretical physics, Volume 1. Elsevier Books, Boston. (printing dates)

Ng YJ (2008) Spacetime foam: from entropy and holography to infinite statistics and nonlocality. Entropy 10(4):441–461

Novello M (n.d.) The mass of the graviton and the cosmological constant puzzle. https://arxiv.org/abs/astro-ph/0504505

Padmanabhan T (2005) Understanding our universe; current status, and open issues. In: 100 years of relativity, space-time, structure: Einstein and beyond. World Scientific, P.T.E. LTD, Singapore, pp 175–204. http://arxiv.org/abs/gr-qc/0503107

Padmanabhan T (2006) An invitation to astrophysics, world scientific series in astronomy and astrophysics: volume 8. World Press Scientific, Singapore

Penrose R (2011) Cycles of time—an extraordinary new view of the universe. Alfred A. Knopf, New York

Poplawski N (2011) Cosmological constant from QCD vacuum and torsion. Annalen der Physik 523:291–295

Roos M (2003) Introduction to cosmology, 3rd edn. Wiley Scientific, Hoboken

Rovelli C, Vidotto F (2015) Covariant loop quantum gravity. Cambridge University Press, Cambridge

Ruutu V, Eltsov V, Gill A, Kibble T, Krusius M, Makhlin YG, Placais B, Volvik G, Wen Z (1996) Vortex formation in neutron – irradiated ^{3}He as an analog of cosmological defect formation. Nature 382:334–336

Spiegel M (1980) Theory and problem of theoretical mechanics: Schaum's outline series. McGraw Hill, San Francisco

t'Hooft (n.d.) The black hole representation of string theory. http://www.staff.science.uu.nl/~hooft101/gthpub/BH_interpretation_stringtheory.pdf

Weinberg S (1972) Gravitation and cosmology: principles and applications of the general theory of relativity. Wiley, Cambridge

Chapter 4
"Equat Causa Effectum"

Peter Enders

Abstract In contrast to the Newtonian force of gravity of a spherically symmetric body ("cause"), the Kepler orbits ("effects") are *not* spherically symmetric. Does this contradict and thus disprove the metaphysical rule "equat causa effectum" (the effect equals its cause)? I will show this contradiction to be resolved through taking into account the initial conditions. Spherically symmetric sets of Kepler orbits are constructed by means of the conserved quantities energy, angular momentum, and Laplace–Runge–Lenz vector. The quantum analogs (Bohr orbitals) are considered as well. The fact, that the Kepler hodographs and orbits are plane, is shown by means of qualitative and intuitive arguing. The old rule "equat causa effectum" thus leads to new insights due to another point of view. This should be exploited in teaching physics, too.

4.1 Introduction

There are external frontiers:

- Space ("the final frontier")
- Novel materials like graphene
- Quantum computing
- Novel experiments
- ...

    and internal frontiers:

- Better understanding of what we know
- Improving existing experimental methods
- ...

This talk is rather about the latter ones.

The relationship between cause and effect is an old and lasting issue. "Aequat causa effectum" (the cause corresponds to its effect) is originally a scholastic rule

P. Enders (✉)
Taraz State Pedagogical University, Taraz, Kazakhstan

B. G. Sidharth et al. (eds.), *Fundamental Physics and Physics Education Research*,
https://doi.org/10.1007/978-3-030-52923-9_4

about the relationship between reason and consequence.[1] It has been traced back to Nicholas of Autrecourt (1298–1369) (Wöhler 2006, p. 163). Leibniz (1677, 1695, 1989, pp. 201, 441, 443) changed it to a metaphysical law for motion. Mayer (1842) exploited it to advance the energy conservation law, although in too much an absolute sense (Helmholtz 1884, p. 70). Typical modern statements are these:

> From given Equal, Equal follows. (Newton, n.d., see also 1999, Book III, 2nd Rule)

> La symétrie caractéristique d'un phénomène est la symétrie maxima compatible avec l'existence du phénomène. Un phénomène peut exister dans un milieu qui possède sa symétrie caractéristique ou celle d'un des intergroupes de sa symétrie caractéristique. ...C'est la dissymétrie qui crée le phénomène. (Curie 1894, p. 400)[2]

> If an ensemble of causes is invariant with respect to any transformation, the ensemble of their effects is invariant with respect to the same transformation. (Renaud 1935, as quoted in Rosen 1982, p. 26)

> The symmetry group of the cause is a subgroup of the symmetry group of the effect. Or less precisely: The effect is at least as symmetric as the cause. Equivalent states of a cause are mapped to (*i.e.*, are correlated with) equivalent states of its effect... Also less precisely: Equivalent causes are associated with equivalent effects. (Rosen 2005, pp. 308f.)

In agreement with Helmholtz (1884, p. 69), I will use the rule "aequat causa effectum" rather heuristically and concentrate myself on the benefit of Newton's (1999, Axioms) and Euler's (1750, Ch. 7) notions of (stationary) state, cf (Enders 2008).

To be specifically, let us consider the Newtonian model potential $\sim 1/r$ as "cause" of the "effect" Kepler orbits. Now, while the "cause" is spherically symmetric, the "effect" is obviously *not* spherically symmetric, see Fig. 4.1. Does this difference contradict and thus disprove that rule? In this contribution, this contradiction will be shown to be seemingly only, because the initial conditions have not been taken into account.

Generally speaking, this issue is not new. Zee (1999, pp. 13f.) explains,

> It is crucial to distinguish between the symmetry of physical laws and the symmetry imposed by a specific situation... This distinction ... was one of Newton's great intellectual achievements, and it enabled physics as we know it to take shape.

In his Nobel lecture, Wigner (1963, p. 7) defined "law" and "situation" as follows.

> The regularities in the phenomena which physical science endeavors to uncover are called the laws of nature... The elements of the behavior which are not specified by the laws of

[1] *Cf* https://de.wikipedia.org/wiki/Aequat_causa_effectum.—For the similar maxims "The same causes will always produce the same effects" and "like causes produce like effects," see Maxwell (1877, para. 19).

[2] For broad reviews, see e.g., Conway et al. (2008), Rosen (2008).

Fig. 4.1 Plane of ecliptic

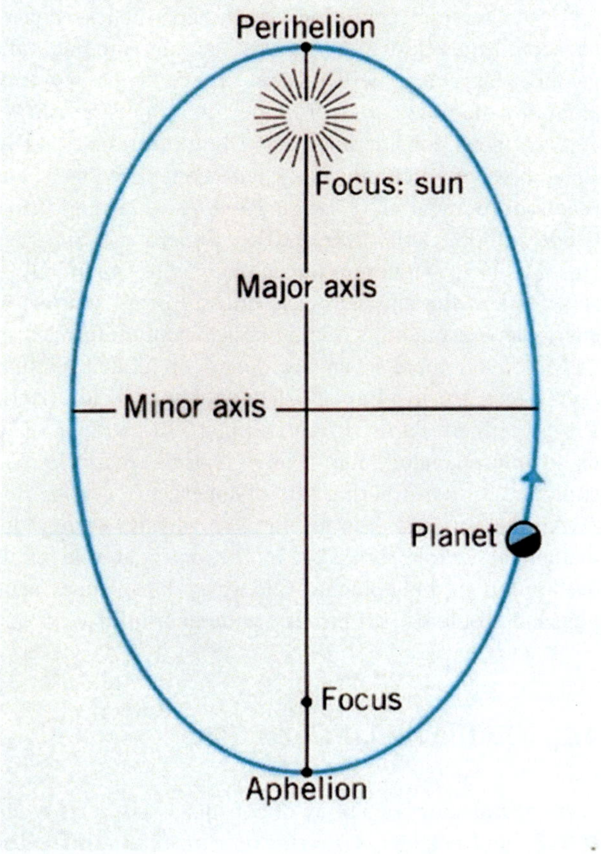

nature are called initial conditions. These, then, together with the laws of nature, specify the behavior as far as it can be specified at all...[3]

New in this contribution are the point of view and the finding of new commonalities between classical and quantum systems.[4]

The trajectory of a single body is also determined by the initial conditions, and these are *not* spherically symmetric. As a consequence, the trajectory of a single body moving in a spherically symmetric force field exhibits a *lower* symmetry than that of the force field.

---

[3]In view of the Titius–Bode rule (Titius 1766; Bode 1772), the initial conditions of the planets appear not to be independent, *cf* (Weizsäcker 1943; Chandrasekhar 1946). Einstein (1923, p. 359) also noticed that the initial conditions "are not free, but also have to obey certain laws." This makes the notion "initial condition" relative. However, this issue is beyond the scope of this contribution.

[4]There is a very short notice in Enders (2012).

The old rule "equat causa effectum" will be restored through constructing sets of orbits with initial conditions on spheres. These constructions are guided by the conserved quantities total energy, angular momentum, and Laplace–Runge–Lenz vector. Moreover, I will present criteria to be posed on the initial conditions for obtaining the trajectories of highest possible symmetry.

Analogous considerations will be performed for Bohr orbitals, ψ_{nlm}. Here, the squared wave functions, $|\psi_{nlm}|^2$, are considered to be analogs to the stationary-state functions momentum (Newton 1999, Axioms) and velocity (Euler 1750, Ch. 7), see (Enders 2006, 2008, 2009, 2013; Enders and Suisky 2005; Suisky 2009). Then, Unsöld's (1927) theorem paves the way to spherically symmetric sets of stationary states. As in the classical case, the set of all stationary-state functions with given energy and/or modulus of angular momentum forms a spherically symmetric figure.

Moreover, there is an accidental or hidden symmetry of the $1/r$-potential as represented by the Laplace–Runge–Lenz vector (Hermann 1710, 1732; Laplace 1799, Premiere Partie, Livre II, pp. 165ff.; Runge 1921; Lenz 1924, Eq. (2)[5]) and its quantum analog, the "Pauli vector" (Pauli 1926, Eq. (50)), which may be exploited for constructing sets of spherically symmetric orbits, too.

All that suggests the teaching of physics to pay due attention not only to the mathematical role, but also to the physical role of the initial conditions in the description of phenomena. Old rules like "equat causa effectum" do not solve physical problems, but provide another fruitful view on them.

4.2 On the Initial Conditions

A single trajectory of a body of constant mass, m, is specified by the initial position, $\mathbf{r}(t = 0)$, and velocity, $\mathbf{v}(t = 0)$, or momentum, $\mathbf{p}(t = 0)$, of the body. The developments of canonical, statistical and quantum mechanics have led to treating position and velocity/momentum on more or less equal footing. For our purpose, it is more appropriate to consider their different relationships to the external force, in order to understand the different symmetry properties of the curves $\mathbf{r}(t)$ on the one hand and $\mathbf{v}(t)$ respectively $\mathbf{p}(t)$ on the other hand.

In contrast to the curves $\mathbf{v}(t)$ and $\mathbf{p}(t)$, the curve $\mathbf{r}(t)$ (the trajectory) is not immediately connected with the force. According to Newton's Law 2, the immediate "effect" of the force, $\mathbf{F}$,—the "cause"—is the "momentum trajectory," $\mathbf{p}(t)$, hence—at constant mass—the hodograph,[6] $\mathbf{v}(t)$ (Hamilton 1847). Following Euler (1750, para. 56, 70; 1752, para. 20), we have $\mathbf{r} = \mathbf{v}\mathrm{d}t$ and $\mathrm{d}\mathbf{v} = \mathbf{F}\mathrm{d}t/m$ (the mass, m, being

[5]For more sources and newer developments, see https://en.wikipedia.org/wiki/Laplace–Runge–Lenz_vector, and links therein.

[6]The literal meaning of the word "hodograph" is *path describer*, where "path" means not a trajectory, but the trace of the top of a parameter-dependent vector with fixed origin when the parameter (here, the time, t) is changing. For developments till Today, see Cariñena et al. (2016).

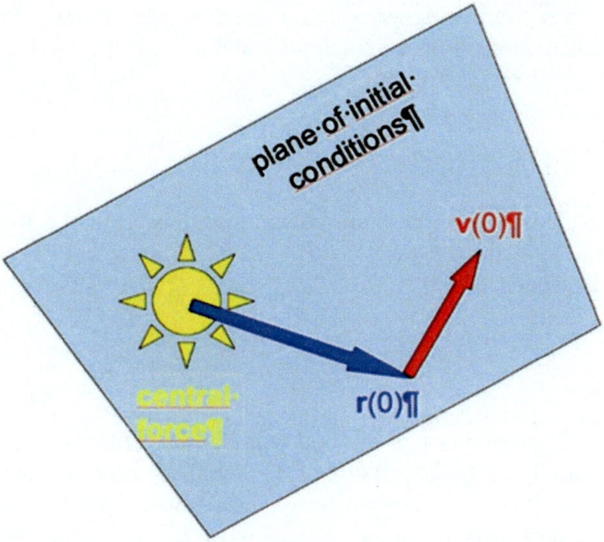

Fig. 4.2 Plane defined by the central force (origin), initial position, **r**(0), and initial velocity, **v**(0)

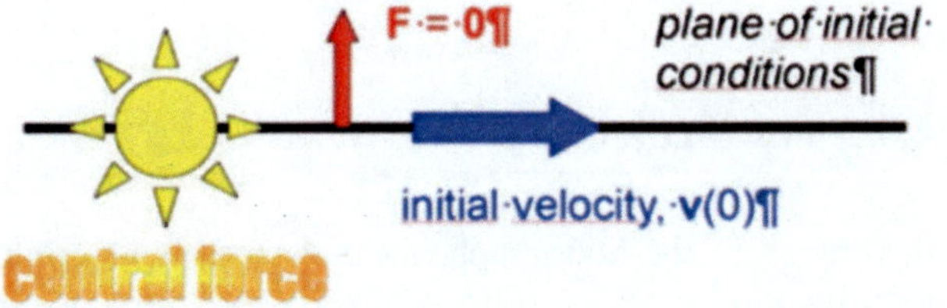

Fig. 4.3 Vanishing net force perpendicular to the plane of initial conditions

constant): The force affects the path, **r**(t), only indirectly, mediated via the velocity, **v** (t) (see also (Suisky 2009, Sect. 4.1)). For this, let us examine the hodograph first.

4.3 The Hodograph

The fact, that the hodograph is plane (despite of motions through the origin), can most easily be derived from the following consideration.

For each initial velocity, **v**(0), there is a plane containing this vector and the origin being the position of the force center, see Fig. 4.2. By the symmetry of the force field, there is nowhere on this plane a net force perpendicular to it, see Fig. 4.3. Hence, the velocity vector, **v**(t), will stay in this plane for all times.

The most general central force reads

$$\mathbf{F} = f(r)\frac{\mathbf{r}}{r}$$

where $f(r)$ is a rather arbitrary function. Generalizing Sommerfeld's (1994, p. 154) calculations, I obtain the hodograph as

$$(\dot{x}(t) - \dot{x}_0)^2 + (\dot{y}(t) - \dot{y}_0)^2 = \left(\frac{f(r(t))r(t)^2}{Cm}\right)^2$$

This is a "circle with changing radius." For the Kepler orbits, it is a common circle with constant radius, GM/C (Maxwell 1877, para. 133; Goodstein and Goodstein 1996).

$$(\dot{x}(t) - \dot{x}_0)^2 + (\dot{y}(t) - \dot{y}_0)^2 = \left(\frac{GM}{C}\right)^2$$

The symmetry of this hodograph is the conjunction of the symmetries of the force field (here: spherical) and of the initial velocity (a straight directed line). The lower symmetry of the latter one makes the hodograph to exhibit *non*-spherical symmetry. In turn, sets of initial velocities of higher symmetry lead to figures of higher symmetry.

4.4 The Trajectory

In contrast to the hodograph, the trajectory is determined by the initial values of velocity *and* position. The symmetry of the trajectory is determined by their interplay.[7]

Nevertheless, the fact, that the trajectory lies in the plane of the hodograph, can most easily be derived as follows, see Fig. 4.4. If the velocity vector, $\mathbf{v}(t)$, stays within a plane for all times, then, there is never a change of position off this plane.

Kinematically, the direction of the velocity vector represents a distinguished vector in space. For this, initial position vectors being parallel or anti-parallel to the initial velocity vector are distinguished. The resulting orbits describe straight lines towards or away from the force center and exhibit a correspondingly high symmetry.

Dynamically, Newton's Law 2 postulates equilibrium between the field force (centripetal force) and the inertial force of the planet, $-m\mathrm{d}\mathbf{v}/\mathrm{d}t$. The latter can be decomposed in the centrifugal force, which is related to the change of the direction of motion, and the tangential component, which is related to the change of the modulus of the velocity.

[7]It is known from Bloch electrons in non-cubic crystals, that the complexity of their stationary states (band structure) depends on the angle between the direction of motion (quasi-momentum, $\mathbf{k}$) and the symmetry axes of the crystal lattice (Enders et al. 1995).

Fig. 4.4 The trajectory is plane

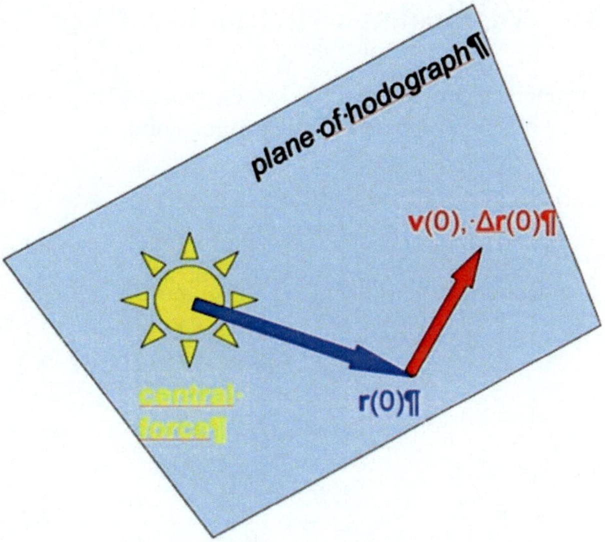

Fig. 4.5 Motion away from force center

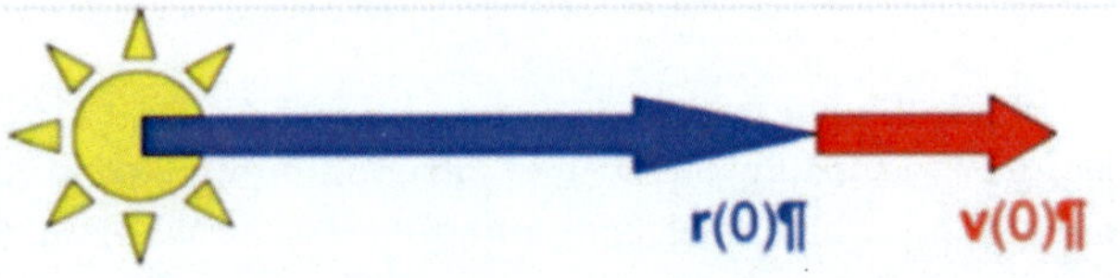

There are two distinguished cases.

1. The centrifugal force vanishes. This is the case of motion towards or away from the center discussed above, see Fig. 4.5.
2. The centrifugal force exactly counterbalances the central force, so that the radius is constant; accordingly, the tangential component of the inertial force vanishes, so that the tangential velocity is constant. For $1/r$-potentials, this is the case of circular orbits. Their symmetry is the highest among all Kepler orbits.

The set of *all* possible Kepler orbits forms a spherically symmetric figure as it fills the whole space but the origin. Here, no single initial condition plays a role. The same holds true for the sets of all Kepler orbits, which are obtained through rotations of a given orbit about all angles around its center. This is the case, if the values of the following quantities are given:

- (total) energy or modulus of Laplace–Runge–Lenz vector, i.e., major semi-axis, or
- modulus of angular momentum, or
- energy *and* modulus of angular momentum, i.e., minor semi-axis.

4.5 The Quantum Mechanical Analog

In spherical coordinates (r, θ, φ), the potential depends only on r, and the stationary wave functions separate as (discarding spin)

$$\psi_{nlm}(\mathbf{r}) = R_{nl}(r)\Psi_{lm}(\theta, \varphi); \quad n = 1, 2, 3, \ldots; \quad l = 0, 1, \ldots, n-1; \quad m = -l, 1-l, \ldots, l-1, l$$

According to Unsöld's (1927) theorem,

$$\sum_{m=-l}^{l} |\Psi_{lm}(\theta, \varphi)|^2 \equiv \frac{2l+1}{4\pi}$$

the expression

$$F_{nl}(\mathbf{r}) \equiv \frac{1}{2l+1} \sum_{m=-l}^{l} |\psi_{nlm}(\mathbf{r})|^2 = \frac{1}{4\pi} R_{nl}^2(r)$$

is spherically symmetric. Following Schrödinger's (1926, para. 7) interpretation of $|\psi_{nlm}(\mathbf{r})|^2$ as being the weight of configuration $\mathbf{r}$ in state nlm, F_{nl} describes the average weight of the configurations $\{\mathbf{r}| \, |\mathbf{r}| = r\}$ in the $2l + 1$ stationary states $\{nlm| - l \leq m \leq l\}$.

The additional or accidental symmetry of the quantum Kepler problem shows up in the existence of the quantum Laplace–Runge–Lenz vector, which I propose to term *Pauli vector* (Pauli 1926, Eq. (50)).

Thus, the set of all possible values of $|\psi_{nlm}(\mathbf{r})|^2$ ($n = 1,2,\ldots$; $l = 0, 1, \ldots, n$; $m = -l, -l + 1, \ldots, 0, \ldots, l$) forms a spherically symmetric figure. Here, no single initial state ("initial condition") plays any role. The same holds true for the set of all possible values of $|\psi_{nlm}(\mathbf{r})|^2$ under the condition of given (expectation) value(s) of

- main quantum number, n, or energy, E_n, iff $V(r) \sim 1/r$, or
- angular quantum number, l, or modulus of angular momentum, $\sqrt{l(l+1)}\hbar$, or
- main and angular quantum numbers, n and l, or energy, E_{nl}, and/or modulus of angular momentum, $\sqrt{l(l+1)}\hbar$, in case of general spherically symmetric potentials, $V(r)$, or
- Pauli vector.

The latter condition has been posed to the auditorium for examination in their classes.

4.6 Summary and Conclusions

The metaphysical postulate "equat causa effectum" does not solve physical problems, but it helps to explore them more profoundly and hence should be exploited in teaching physics. Here, it leads to the examination of the role of the initial conditions as another "cause" of a trajectory, additionally to the force. If the initial conditions "equal" the cause, the single effect does so as well.

Because the force acts directly upon the momentum and velocity, respectively, but only indirectly upon the position, the hodograph is more symmetric than the trajectory. For a central force field, both are plane. This has been shown by means of qualitative and intuitive arguments, which are much more easily accessible than angular momentum conservation.

In both the classical and quantum cases, two different initial conditions lead to equivalent trajectories/wave functions, if the one can be obtained from the other one through a symmetry transformation of the force field/potential. The set of all such trajectories respectively wave functions squared forms a figure that exhibits the symmetry of the force field/potential.

For Kepler orbits, there are no boundary conditions; for Bohr orbitals, the symmetry of the boundary conditions equals that of the force field. For this, they play no role for the symmetry of the motion. If their symmetry is lower than that of the force field, their influence on the symmetry of the motion can be treated analogously to that of the initial conditions.

Acknowledgment I would like to thank Michael Erdmann, Joseph Rosen, Dieter Suisky, and a former colleague not wishing to be named for numerous enlightening discussions. I have benefited from numerous excitations in science as well in arts during my stay at Orihuela/Oriola and thus feel highly indebted to the organizers of FFP15.

References

Bode JE (1772) Deutliche Anleitung zur Kentniss des Gestirnten Himmels. Harmsen, Hamburg

Cariñena JF, Rañada MF, Santander M (2016) A macroscopic test of the Aharonov-Bohm effect. arXiv:0708.2428v1

Chandrasekhar S (1946) On a new theory of Weizsäcker on the origin of the solar system. Rev Mod Phys 18:94–102

Conway JH, Burgiel H, Goodman-Strauss C (2008) The symmetries of things. A K Peters/CRC Press, New York

Curie P (1894) Sur la symétrie dans les phénomènes physiques, symétrie d'un champ électrique et d'un champ magnétique. J Phys (Paris) 3(3):393–415. https://hal.archives-ouvertes.fr/jpa-00239814. Engl.: On symmetry in physical phenomena, symmetry of an electric field and of a magnetic field, In: Rosen (1982, pp 17–25)

Einstein A (1923) Bietet die Feldtheorie Möglichkeiten für die Lösung des Quantenproblems? Sitzungsber Preuss Akad Wiss phys-math Kl 13:XXXIII

Enders P (2006) Von der klassischen Physik zur Quantenphysik. Eine historisch-kritische deduktive Ableitung mit Anwendungsbeispielen aus der Festkörperphysik. Springer, Berlin

Enders P (2008) Equality and identity and (in)distinguishability in classical and quantum mechanics from the point of view of Newton's notion of state. ICFAI Univ J Phys I:71–78

Enders P (2009) Gibbs' paradox in the light of Newton's notion of state. Entropy 11:454–456. http://www.mdpi.com/1099-4300/11/3/454

Enders P (2012) "Equat causa effectum" and the Kepler orbits. Lat Am J Phys Educ 6:661. http://www.lajpe.org/dec2012/24_LAJPE_714_Peter_Enders_preprint_corr_f.pdf

Enders P (2013) Quantization as selection rather than eigenvalue problem. In: Bracken P (ed) Advances in quantum mechanics (Ch. 23). InTech, Rijeka, pp 543–564. http://cdn.intechopen.com/pdfs/41542/InTech-Quantization_as_selection_rather_than_eigenvalue_problem.pdf

Enders P, Suisky D (2005) Quantization as selection problem. Int J Theor Phys 44:161–194

Enders P, Bärwolff A, Wörner M, Suisky D (1995) Kp theory of energy bands, wave functions, and optical selection rules in strained tetrahedral semiconductors. Phys Rev B 51:16695–16704

Euler L (1750) Anleitung zur Naturlehre. In: Opera Omnia III, vol 1, pp 17–178; Opera posthuma 2 (1862), pp 449–560 (E842); Engl.: An introduction to natural science. http://17centurymaths.com/contents/contentse842.html

Euler L (1752) Decouverte d'un nouveau principe de Mecanique. Mém. acad. Berlin 6:185–217. In: Opera Omnia II, 5:81–108. http://eulerarchive.maa.org//tour/tour_08.html (E177)

Goodstein DL, Goodstein JR (1996) Feynman's lost lecture. The motion of planets around the sun. Norton, New York

Hamilton WR (1847) The hodograph or a new method of expressing in symbolic language the Newtonian law of attraction. Proc Roy Irish Acad 3:344–353. http://www.emis.de/classics/Hamilton/Hodo.pdf

Helmholtz HV (1884) Über die Wechselwirkung der Naturkräfte und die darauf bezüglichen neuesten Ermittelungen der Physik (talk deliverd 1854) (Appendix 1) Robert Mayer's Priorität (added 1883). In: Reden und Vorträge, vol 1. Vieweg, Braunschweig. https://ia802606.us.archive.org/7/items/vortrgeundreden02helmgoog/vortrgeundreden02helmgoog.pdf

Hermann J (1710) d'investigare l'Orbite de' Pianeti, nell'ipotesi che le forze centrali o pure le gravit'a delgi stessi Pianeti sono in ragione reciproca de' quadrati delle distanze, che i medesimi tengono dal Centro, a cui si dirigono le forze stesse. Giorn Lett D'Italia 2:447–467

Hermann J (1732) Extrait d'une lettre de M. Herman à M. Bernoulli datée de Padoüe le 12. Juillet 1710. Hist l'acad roy sci (Paris), 519–521

Laplace PS (1799) Traité de mécanique celeste. Crapelet. Tome I, Paris. https://catalog.hathitrust.org/Record/100242426

Leibniz GW (1677) On a method of arriving at a true analysis of bodies and the causes of natural things. Engl. in Leibniz (1989, no. 15)

Leibniz GW (1695) Specimen Dynamicum; Engl. in Leibniz (1989, no. 46)

Leibniz GW (1989) Philosophical papers and letters (selected, translated, edited and introduced by L. E. Loemker). Kluwer, Dordrecht (2nd print of 2nd ed)

Lenz W (1924) Über den Bewegungsverlauf und die Quantenzustände der gestörten Keplerbewegung. Zs Physik 24:197–207

Maxwell JC (1877) Matter and motion. Extended reprint (J. Larmor, Ed.). Society for the Promotion of Christian Knowledge 1920, London (reprint: Dover, New York, 1952, 1991)

Mayer JR (1842) Bemerkungen über die Kräfte der unbelebten Natur. Ann Chemie Pharmacie 43:233–240; Engl.: Remarks on the forces of inorganic nature (transl. by G. C. Foster) (1862), Philos Mag 24(4):371; reprint in: Magie WF (ed) (1935) A source book in physics. McGraw-Hill, New York. http://web.lemoyne.edu/~giunta/mayer.html

Newton I (1999) The principia. mathematical principles of natural philosophy (A new translation by I. Bernhard Cohen and Anne Whitman assisted by Julia Buden, Preceded by A guide to Newton's principia by I. Bernhard Cohen (Among others, this 370 (!) p. Guide contains not only historical notes, but also explanations of and completions to Newton's text)). University of California Press, Berkeley etc.

Newton I (n.d.) De Gravitatione et aequipondio fluidorum. Cambridge Univ. Lib., Ms. Add 4003. http://www.newtonproject.ox.ac.uk/view/texts/diplomatic/THEM00093. Engl.: Bennett J (transl.) (2017) Descartes, space and body. http://earlymoderntexts.com/assets/pdfs/newton1666.pdf. Stein H (transl. and comments) (2013) De gravitatione et æquipondio fluidorum. http://strangebeautiful.com/other-texts/newton-de-grav-stein-trans.pdf. Latin and German: Böhme G (transl. and introd.) (1988) Über die Gravitation ...: Texte zu den philosophischen Grundlagen der klassischen Mechanik. Klostermann, Frankfur (unpublished)

Pauli W (1926) Über das Wasserstoffspektrum vom Standpunkt der neuen Quantenmechanik. Zs Physik 36:336–363. http://www.chemie.unibas.ch/~steinhauser/documents/Pauli_1926_36_336-363.pdf. Engl. in van der Waerden (1967, no. 16)

Renaud P (1935) Sur une généralisation du principe de symétrie de Curie, Comptes Rendus 200, 531–534. Engl.: On a generalization of Curie's symmetry principle. In Rosen (1982)

Rosen J (ed) (1982) Symmetry in physics: selected reprints. AAPT, New York

Rosen J (2005) The symmetry principle. Entropy 7:308–313. http://www.mdpi.com/1099-4300/7/4/308

Rosen J (2008) Symmetry rules: how science and nature are founded on symmetry. Springer, Berlin

Runge CDT (1921) Praxis der Gleichungen. de Gruyter, Berlin. Engl.: Vector analysis (1923). Methuen, London

Schrödinger E (1926) Quantisierung als Eigenwertproblem. Vierte Mitteilung. Ann Phys 81 (4):109–139. http://home.tiscali.nl/physis/HistoricPaper/HistoricPapers.html

Sommerfeld A (1994) Vorlesungen über theoretische Physik, Bd. I Mechanik. Thun. Deutsch, Frankfurt am Main

Suisky D (2009) Euler as Physicist. Springer, Berlin

Titius JD (1766) Betrachtungen über die Natur. Johann Friedrich Junius, Leipzig

Unsöld A (1927) Beiträge zur Quantenmechanik des Atoms. Thesis, LMU Munich; Ann Phys 82 (4):355–393 (see also R. Binder, Unsöld's theorem). http://www.demonstrations.wolfram.com/UnsoeldsTheorem/

van der Waerden BL (1967) Sources of quantum mechanics. North-Holland, Amsterdam. Reprint: Dover, New York, 1968. https://archive.org/details/SourcesOfQuantumMechanics

Weizsäcker CFv (1943) Über die Entstehung des Planetensystems. Z Astrophys 22:319–355. http://adsabs.harvard.edu/full/1943ZA.....22..319W

Wigner EP (1963) Events, laws of nature, and invariance principles. Nobel Lecture. https://www.nobelprize.org/nobel_prizes/physics/laureates/1963/wigner-lecture.pdf

Wöhler H-U (2006) Dialektik in der mittelalterlichen Philosophie. Akademie-Verlag, Berlin

Zee A (1999) Fearful symmetry. Princeton University Press, Princeton

Chapter 5
(Non-)Uniqueness of Einstein–Palatini Gravity

Bert Janssen, Alejandro Jiménez-Cano, José Alberto Orejuela, and Pablo Sánchez-Moreno

Abstract We analyse the most general connection allowed by Einstein–Hilbert theory in Palatini formalism. We also consider a matter lagrangian independent of the affine connection. We show that any solution of the equation of the connection is essentially Levi-Civita up to a term that contains an undetermined 1-form. Finally, it is proved that these connections and Levi-Civita describe a completely equivalent physics.

Talk given by A. J. C. and based on the publication (Bernal et al., Phys Lett B 768:280–287, 2017).

5.1 Introduction and Mathematical Notions

Since the publication of the Einstein's theory of General Relativity in 1915, we understand gravitation as a geometrical effect. Many extensions of this theory have been formulated in order to solve various problems in theoretical physics, such as dark matter or the first corrections to General Relativity that could come from the quantum gravity regime.

In the geometrical framework introduced by Einstein, the *spacetime* is defined as a differentiable manifold $\mathcal{M}$. Omitting some mathematical details, a *D-dimensional manifold* is essentially a topological space that looks, locally, as the Euclidean space $\mathbb{R}^D$. For example, spheres, planes and hyperboloids are 2-dimensional manifolds.

B. Janssen · A. Jiménez-Cano (✉) · J. A. Orejuela
Departamento de Física Teórica y del Cosmos, Universidad de Granada, Granada, Spain

Centro Andaluz de Física de Partículas Elementales, Universidad de Granada, Granada, Spain
e-mail: bjanssen@ugr.es; alejandrojc@ugr.es; josealberto@ugr.es

P. Sánchez-Moreno
Departamento de Matemática Aplicada, Universidad de Granada, Granada, Spain

Instituto Carlos I de Física Teórica y Computacional, Universidad de Granada, Granada, Spain
e-mail: pablos@ugr.es

B. G. Sidharth et al. (eds.), *Fundamental Physics and Physics Education Research*,
https://doi.org/10.1007/978-3-030-52923-9_5

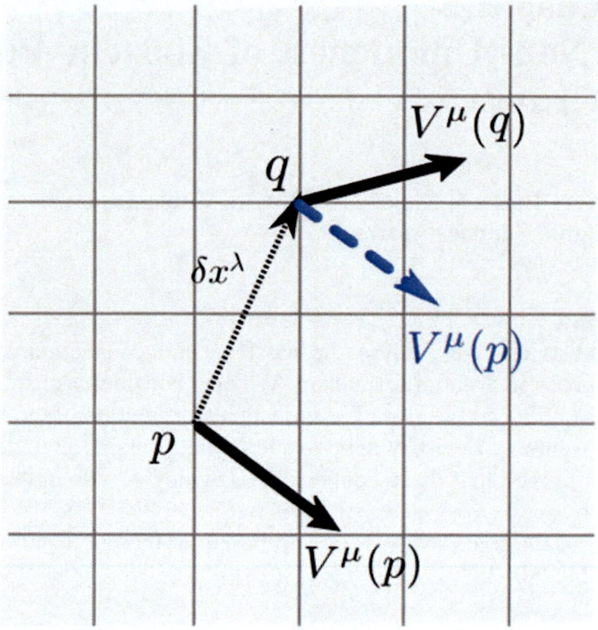

Fig. 5.1 Comparison of vectors in Euclidean space (natural notion of parallelism)

Additionally, we include a *lorentzian metric tensor*, $g_{\mu\nu}$, which allows to measure lengths, volumes and so on. Hence it is possible to talk about the module of a vector that is not necessarily non-negative, due to the lorentzian signature. Those vectors that are not trivial but have zero norm determine the lightlike paths and, then, light cones that define the casual structure of the spacetime.

Another fundamental notion that can be defined, even in the absence of metric, is *parallelism*. The motivation for this additional concept is the following. Consider the Euclidean space $\mathbb{R}^D$ and a couple of vectors in different points, p and q (Fig. 5.1). If we want to compare them, we simply take, for example, the one in p and move it to q keeping the vector parallel to itself and without changing the module. And, finally, we subtract both vectors to see the difference.

However, if the manifold is general, the initial vectors live in different spaces (the tangent spaces at p and q, respectively, $T_p\mathcal{M}$ and $T_q\mathcal{M}$) and there is no natural way to relate them (Fig. 5.2). In the Euclidean space, both tangent spaces can be identified making the comparison trivial. In the general case, we need to introduce an additional structure that carries the information about parallelism, the *affine structure*, whose fundamental object is the *(affine) connection* $\Gamma^\sigma_{\mu\nu}$. Once we have a connection, given a curve between two points, we have a rule to relate vectors in them: the *parallel transport* associated to the connection.

The connection permits the definition of a *covariant derivative* ("covariant" means that once applied to a tensor, the result is also a tensor), and other intrinsic geometrical properties of the spacetime, such as the *curvature* and the *torsion*, respectively:

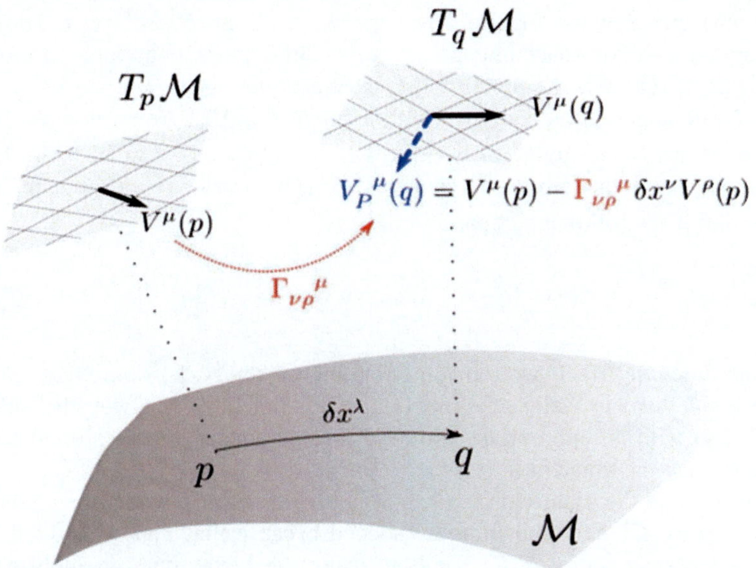

Fig. 5.2 An affine connection represents a notion of parallel in the manifold and allows to parallely transport vectors along curves. The dashed line is, by definition, the parallel transport of $V^\mu(p)$ from p to q along the x^λ direction. This vector and $V^\mu(q)$ can be compared since both live in $T_q\mathcal{M}$

$$R_{\mu\nu\rho}{}^\lambda \equiv \partial_\mu \Gamma^\lambda_{\nu\rho} - \partial_\nu \Gamma^\lambda_{\mu\rho} + \Gamma^\lambda_{\mu\sigma}\Gamma^\sigma_{\nu\rho} - \Gamma^\lambda_{\nu\sigma}\Gamma^\sigma_{\mu\rho},$$

$$T_{\mu\nu}{}^\lambda \equiv \Gamma^\lambda_{\mu\nu} - \Gamma^\lambda_{\nu\mu}.$$

Consider a manifold with a metric structure $g_{\mu\nu}$, then it can be proved that there is only one connection, ${\Gamma^{(g)}}^\lambda_{\mu\nu}$, compatible with the metric and torsionless, namely

$$\nabla_\lambda g_{\mu\nu} = 0, \quad T_{\mu\nu}{}^\lambda = 0.$$

This connection is called the *Levi-Civita connection* of $g_{\mu\nu}$, and it is completely determined by the metric:

$${\Gamma^{(g)}}^\lambda_{\mu\nu} = \frac{1}{2} g^{\lambda\sigma}\left(\partial_\mu g_{\sigma\nu} + \partial_\nu g_{\mu\sigma} - \partial_\sigma g_{\mu\nu}\right).$$

In fact, given a metric, the Levi-Civita affine structure is the simplest connection we can deal with. The metric compatibility and the nullity of the torsion simplify many geometrical identities. Moreover, we are not introducing extra degrees of freedom in the theory, just the ones that come from the metric.

Before starting with the physics, let us introduce a few useful definitions. A *curve* $\gamma(\alpha)$ is a differentiable function $\gamma : I \longrightarrow \mathcal{M}$, where I is an interval of the real line. The image of a curve in the manifold is what we will call *trajectory* or *path*. So a trajectory is a set of spacetime points joined in a continuous and differentiable way, while the curve is the function that generates this set.

Let $\mathcal{M}$ be a spacetime equipped with a connection $\Gamma^{\rho}_{\mu\nu}$. An *autoparallel* of this affine structure is the image of a curve whose velocity is parallel to itself (with respect to $\Gamma^{\rho}_{\mu\nu}$). In other words, given a curve $\gamma(\alpha)$ with velocity $v^{\mu}(\alpha)$, its image is an autoparallel if the following equation holds:

$$v^{\mu}\nabla_{\mu}v^{\rho} \equiv \frac{dv^{\rho}}{d\alpha} + \Gamma^{\rho}_{\mu\nu}v^{\mu}v^{\nu} = f(\alpha)v^{\rho}.$$

for some function $f(\alpha)$. If we reparametrize the trajectory, by doing $\alpha \longrightarrow \beta(\alpha)$, we change the velocity as well as the function f. This function can always be set to zero (identically) with an appropriate choice of the parameter. Those are called *affine parameters* for the trajectory.

Consider that the manifold also has a metric structure. The autoparallels of the associated Levi-Civita connection are special because they can be derived from a completely metric approach, i.e. without using the Levi-Civita parallelism. If the velocity is timelike or spacelike, they correspond to trajectories that are critical points of the length functional:

$$s[\gamma](\alpha) = \int_0^{\alpha} \sqrt{\left|g_{\mu\nu}v^{\mu}v^{\nu}\right|}d\alpha'.$$

The lightlike case should be treated separately, because the length functional cannot be varied smoothly. However, they can be seen as critical points of other functionals that, again, only involve the metric structure. For these reasons, in general, we will call the autoparallels of Levi-Civita *critical trajectories*. The associated affine parameters have a special meaning, because their changes are proportional to the length between the considered points. In the timelike case, it is essentially the *proper time*, so these parameters represent the rhythm of a proper (i.e. freely falling) clock along them.

5.2 Einstein's Equations and Variational Principles

General relativity is a geometric theory of the spacetime whose dynamics is described by the *Einstein's equation*:

$$R_{\mu\nu}(g) - \frac{1}{2}g_{\mu\nu}R(g) = -\kappa\mathcal{T}_{\mu\nu}, \tag{5.1}$$

where $R_{\mu\nu}(g) \equiv R_{\mu\lambda\nu}{}^{\lambda}(g)$ is the Levi-Civita Ricci tensor, $R(g) \equiv g^{\mu\nu}R_{\mu\nu}(g)$ is the Levi-Civita Ricci scalar, $\kappa \equiv 8\pi G$ (G is the Newton's constant) and $\mathcal{T}_{\mu\nu}$ is the (Hilbert) *energy-momentum tensor* that contains the information about the matter and energy content.

This equation can be obtained from a more fundamental object through a variational principle, the *Einstein–Hilbert action*:

$$S[g,\psi] = \frac{1}{2\kappa}\int R(g)\sqrt{|g|}d^Dx + S_{\text{matter}}[g,\psi].$$

The energy-momentum tensor is then defined by

$$\mathcal{T}_{\mu\nu} \equiv \frac{2}{\sqrt{|g|}}\frac{\delta S_{\text{matter}}}{\delta g^{\mu\nu}}.$$

Notice that we are assuming (from the start) a particular affine structure, the one fixed by the metric. We are selecting the Levi-Civita connection by hand and this can be considered natural because it is the simplest one. When we obtain from a gravitational action the equations of motion admitting that the affine structure is Levi-Civita, we are using the so-called *metric formalism*, because the metric determines everything related to the gravitational field.

Another approach, which is called *Palatini* or *metric-affine formalism*, consists in considering the metric and the connection as independent fields. Now the connection is general and the corresponding equations of motion should determine whether it is Levi-Civita or not. The action in this formalism is

$$S[g,\Gamma,\psi] = \frac{1}{2\kappa}\int g^{\mu\nu}R_{\mu\nu}(\Gamma)\sqrt{|g|}d^Dx + S_{\text{matter}}[g,\psi].$$

It is worth remarking that we are assuming that the matter part of the action does not depend on the affine connection Γ.

This formalism is interesting because we expect Levi-Civita connection to be fixed by the corresponding equations of motion, in contrast with the metric formalism in which it is selected artificially.

5.3 Palatini Solutions of the Einstein–Hilbert Action

If we vary the Einstein–Hilbert action in the metric-affine formalism, we obtain the following equations of motion for the metric and the connection, respectively:

$$0 = \frac{1}{2}\left(R_{\mu\nu}(\Gamma) + R_{\nu\mu}(\Gamma)\right) - \frac{1}{2} g_{\mu\nu} R(\Gamma) + \kappa \mathcal{T}_{\mu\nu}, \tag{5.2}$$

$$0 = \nabla_\lambda g^{\mu\nu} - \nabla_\sigma g^{\sigma\nu}\, \delta^\mu_\lambda + \frac{1}{2} g^{\mu\nu} g^{\rho\tau} \nabla_\lambda g_{\rho\tau} - \frac{1}{2} g^{\rho\tau} \nabla^\nu g_{\rho\tau} \delta^\mu_\lambda + T_{\lambda\sigma}{}^\sigma g^{\mu\nu} - T_{\rho\sigma}{}^\sigma g^{\rho\nu} \delta^\mu_\lambda + T_{\sigma\lambda}{}^\mu g^{\sigma\nu}. \tag{5.3}$$

Equation (5.3) can be simplified if the dimension of the spacetime is $D > 2$.[1] We then obtain:

$$0 = \nabla_\lambda g_{\mu\nu} - T_{\nu\lambda}{}^\sigma g_{\mu\sigma} + \frac{1}{D-1} T_{\lambda\sigma}{}^\sigma g_{\nu\mu} + \frac{1}{D-1} T_{\nu\sigma}{}^\sigma g_{\lambda\mu}. \tag{5.4}$$

Clearly, Levi-Civita is a solution, because in that case each term of the right hand side vanishes. However, let us try for other solutions. Consider only those that are torsionless, then, necessarily, $\nabla_\lambda g_{\mu\nu}$ should be zero, so Levi-Civita is the only possibility. The same happens for metric-compatible solutions. In fact, when Palatini formalism is presented (in textbooks for example), one of these two conditions is assumed. Consequently, we lose the information about the general solution and it reduces to Levi-Civita.

The most general solution of Eq. (5.4) has the form:

$$\Gamma^\sigma_{\mu\nu} = {\Gamma^{(g)}}^\sigma_{\mu\nu} + \mathcal{A}_\mu \delta^\sigma_\nu,$$

where $\mathcal{A}_\mu$ is an arbitrary 1-form (Bernal et al. 2017). We will call it *Palatini connection* from now on.

The associated covariant derivative of the metric (also called *non-metricity* tensor) and torsion are

$$\nabla_\lambda g_{\mu\nu} = -2\mathcal{A}_\lambda g_{\mu\nu},$$
$$T_{\mu\nu}{}^\sigma = \mathcal{A}_\mu \delta^\sigma_\nu - \mathcal{A}_\nu \delta^\sigma_\mu.$$

Here, we clearly notice what we stated before: switching off one of them implies $\mathcal{A}_\lambda = 0$ and, then, Levi-Civita as the only possibility.

The Palatini Riemann tensor, Ricci tensor and Ricci scalar are given by

$$R_{\mu\nu\rho}{}^\lambda(\Gamma) = R_{\mu\nu\rho}{}^\lambda(g) + \mathcal{F}_{\mu\nu}\delta^\lambda_\rho,$$
$$R_{\mu\nu}(\Gamma) = R_{\mu\nu}(g) + \mathcal{F}_{\mu\nu},$$

[1] For the particular case $D = 2$ see (Deser 1996).

$$R(\Gamma) = R_{\mu\nu}(g),$$

respectively, where $\mathcal{F}_{\mu\nu} = \partial_\mu \mathcal{A}_\nu - \partial_\nu \mathcal{A}_\mu$. As a consequence of these expressions, the equation of motion of the metric (5.2) becomes the Einstein's Eq. (5.1), see (Dadhich and Pons 2012). We now present some properties of these solutions.

5.3.1 Projective Relation Between Solutions

Any two Palatini connections, for example

$$\Gamma^\rho_{\mu\nu} = \Gamma^{(g)\sigma}{}_{\mu\nu} + \mathcal{A}_\mu \delta^\sigma_\nu, \quad \Gamma'^\rho_{\mu\nu} = \Gamma^{(g)\sigma}{}_{\mu\nu} + \mathcal{A}'_\mu \delta^\sigma_\nu,$$

are related by a transformation:

$$\Gamma^\rho_{\mu\nu} \to \Gamma'^\rho_{\mu\nu} = \Gamma^\rho_{\mu\nu} + k_\mu \delta^\sigma_\nu,$$

for certain covector k_μ. This transformation is a *projective transformation* which means that preserves autoparallels. This can be proved easily. First, consider an autoparallel trajectory for the connection $\Gamma'^\rho_{\mu\nu}$,

$$\frac{dv^\rho}{d\beta} + \Gamma'^\rho_{\mu\nu} v^\mu v^\nu = 0.$$

Then, imposing the projective relation between both connections and defining $-k_\mu v^\mu \equiv f(\beta)$ we get to the expression:

$$\frac{dv^\rho}{d\beta} + \Gamma^\rho_{\mu\nu} v^\mu v^\nu = f(\beta) v^\rho.$$

And this is the equation of an autoparallel for the connection $\Gamma^\rho_{\mu\nu}$ with a non-affine parametrization. If we parametrize the path affinely ($\beta \to \alpha$ and $v^\rho \to u^\rho$) we obtain:

$$\frac{du^\rho}{d\alpha} + \Gamma^\rho_{\mu\nu} u^\mu u^\nu = 0.$$

Q.E.D.

Consequently, *the whole set of Palatini connections shares the same autoparallels*, up to reparametrizations, which have no physical meaning. As a matter of fact, since Levi-Civita is a particular Palatini connection (the case with $\mathcal{A}_\mu = 0$) we conclude: *the autoparallels of any Palatini connection are critical trajectories of the metric*.

5.3.2 *Homothety Property*

Let $\gamma(\tau)$ be a general curve in the spacetime and v^μ its velocity, and a vector W^μ. If we compare the change of the vector along $\gamma(\tau)$ under Palatini and Levi-Civita parallel transport, we see that the difference between both transports is proportional to W^μ:

$$\left(v^\mu \nabla_\mu - v^\mu \nabla^{(g)}_\mu\right) W^\rho = -\mathcal{A}_\mu v^\mu W^\rho \equiv \lambda(\tau) W^\rho.$$

The module is not conserved but the direction does. Due to this, we say the Palatini parallel transport is *homothetic with respect to the Levi-Civita transport*. It can be proved that the only connections with this property are the Palatini connections (Bernal et al. 2017). Any other connection would generate a perturbation in the direction of W^μ.

5.4 Observability and Physical Implications

We introduced the Palatini formalism in order to see if the dynamics could fix Levi-Civita as the fundamental connection of the theory, in contrast with the metric formalism in which it is selected by hand. However, we have obtained a family of connections that differ in a vector field, with Levi-Civita as a particular case. In this section, we analyse the physical implications of this field. Indeed, we will see that it is undetectable or, equivalently, that metric and Palatini formalism describe the same physics.

The main point is that the gravitational dynamics is the same in both formalisms. The equation of the matter is clearly the same, because the corresponding action does not depend on the connection, so the difference between formalisms does not affect this part of the total action. And, as we previously showed, Palatini connections imply the reduction of the equation of the metric to the Einstein's equation. The resulting dynamics for the metric and the matter content is given by

$$R_{\mu\nu}(g) - \frac{1}{2} g_{\mu\nu} R(g) = -\kappa \mathcal{T}_{\mu\nu}, \qquad \frac{\delta S_{\text{matter}}}{\delta \psi} = 0,$$

in both formalisms.

Furthermore, in Einstein–Hilbert gravity the distinction between critical and autoparallel trajectories disappears due to the projective symmetry. In fact, defining the trajectory of a test particle is often presented as a basic problem of metric-affine theories. Those of critical length and those with covariantly constant velocity are candidates because both of them infinitesimally reduce to straight lines. The critical paths are the simplest approach, but there are authors who defend the description with autoparallels (Kleinert and Pelster 1999) and others who state that only the conserved currents determine the test paths (Hehl and Obukhov 2007).

In Einstein–Hilbert gravity, the conservation of the energy-momentum tensor selects the critical paths. However, fortunately, the autoparallels set by the Palatini dynamics coincide, as a consequence of the projective relation between Palatini connections (a family that includes Levi-Civita). Indeed, we have seen that the field $\mathcal{A}_\mu$ can be eliminated by the freely falling observer with an appropriate choice of the parameter.

All of these ideas point in the same direction: the field $\mathcal{A}_\mu$ has no physical effects (Bernal et al. 2017).

5.5 Equivalence in Other Theories

Finally, we add a few remarks about Palatini connections in other theories. One example is *Lovelock theory* in Palatini formalism:

$$S[g,\Gamma,\psi] = S_{\text{Lov}}[g,\Gamma] + S_{\text{matter}}[g,\psi], \quad S_{\text{Lov}}[g,\Gamma] \equiv \int \sum_{n=1}^{K} a_n \mathcal{L}_n[g,\Gamma]\sqrt{|g|}d^D x,$$

where a_n are certain dimensionful parameters, $\mathbb{N} \ni K \leq \text{ceiling}(D/2-1)$ and the nth-order Lovelock lagrangian is defined by

$$\mathcal{L}_n[g,\Gamma] = \delta^{[\mu_1}_{\nu_1} \dots \delta^{\mu_{2n}]}_{\nu_{2n}} g^{\rho_1\nu_1} \dots g^{\rho_n\nu_{2n-1}} R_{\mu_1\mu_2\rho_1}{}^{\nu_2}(\Gamma) \dots R_{\mu_{2n-1}\mu_{2n}\rho_n}{}^{\nu_{2n}}(\Gamma).$$

It was shown in Borunda et al. (2008) that Levi-Civita is a solution for the equation of the connection in any of these theories. As far as we know, the general solution remains unknown, but we have found that the action presents the projective symmetry, $\Gamma^\sigma_{\mu\nu} \to \Gamma^\sigma_{\mu\nu} + \mathcal{A}_\mu \delta^\sigma_\nu$. The proof is the following. Under the projective transformation, the Riemann tensor is modified:

$$R_{\mu\nu\rho}{}^{\lambda} \to R_{\mu\nu\rho}{}^{\lambda} + \mathcal{F}_{\mu\nu}\delta^\lambda_\rho.$$

Then, the lagrangian transforms:

$$\mathcal{L}_n \to \delta^{\mu_1}_{[\nu_1} \dots \delta^{\mu_{2n}}_{\nu_{2n}]} (R_{\mu_1\mu_2}{}^{\nu_1\nu_2} + \mathcal{F}_{\mu_1\mu_2} g^{\nu_1\nu_2}) \dots (R_{\mu_{2n-1}\mu_{2n}}{}^{\nu_{2n-1}\nu_{2n}} + \mathcal{F}_{\mu_{2n-1}\mu_{2n}} g^{\nu_{2n-1}\nu_{2n}})$$

and the ν's antisymmetrization cancels all the terms proportional to the metric, so

$$\delta_{\text{proj}}\mathcal{L}_n = 0, \quad \forall n \;\Rightarrow\; \delta_{\text{proj}}\{S_{\text{Lov}}[g,\Gamma] + S_{\text{matter}}[g,\psi]\} = 0.$$

Q.E.D.

Consequently, since Levi-Civita (${\Gamma^{(g)}}^{\sigma}_{\mu\nu}$) is a solution, then ${\Gamma^{(g)}}^{\sigma}_{\mu\nu} + \mathcal{A}_\mu \delta^\sigma_\nu$ is also a solution. Indeed, the whole set of solutions can be separated into equivalence classes of projectively related connections. So if a new solution ${\Gamma^{\text{sol}}}^{\sigma}_{\mu\nu}$ of a Lovelock theory that has not the form ${\Gamma^{(g)}}^{\sigma}_{\mu\nu} + \mathcal{A}_\mu \delta^\sigma_\nu$ (for some $\mathcal{A}_\mu$) is found, then we could build a family of new solutions just adding a term $\mathcal{B}_\mu \delta^\sigma_\nu$ where $\mathcal{B}_\mu$ is arbitrary. This property also holds for any other lagrangians with projective invariance, such as $f(R)$ gravity.

Other theories we have tested are those with quadratic torsion corrections to the Einstein–Hilbert lagrangian. The torsion corrections we consider are only those with even parity:

$$S[g,\Gamma,\psi] = \frac{1}{2\kappa}\int g^{\mu\nu}R_{\mu\nu}(\Gamma)\sqrt{|g|}d^Dx + S_{matter}[g,\psi] + \frac{1}{2\kappa}$$
$$\times \int \left[b_1 T^{(1)}_{\mu\nu\rho}T^{(1)\mu\nu\rho} + b_2 T^{(2)}_{\mu\nu\rho}T^{(2)\mu\nu\rho} + b_3 T^{(3)}_{\mu\nu\rho}T^{(3)\mu\nu\rho}\right]\sqrt{|g|}d^Dx,$$

where b_i are arbitrary dimensionless real constants and $T^{(i)}_{\mu\nu\rho}$ are the irreducible components of the torsion (see (McCrea 1992)). For these extensions, the equivalence between metric and Palatini formalism holds.

5.6 Conclusions

To conclude we summarize our results. We have seen that Einstein–Hilbert gravity in the Palatini formalism has some interesting features. If we couple this theory with a matter action through the metric (and not the connection), the result is physically indistinguishable from the dynamics obtained assuming Levi-Civita as the fundamental affine structure from the beginning (metric formalism).

The general solution of the equation of the connection is Levi-Civita plus the term $\mathcal{A}_\mu \delta^\sigma_\nu$ where $\mathcal{A}_\mu$ is an undetermined field. However, the equations of motion are the same as in metric formalism. Therefore, we get to different mathematical descriptions (related through the projective symmetry) that describe the same physics. In other words, it is not necessary to set the connection to be Levi-Civita by hand. The dynamics fixes the affine structure.

Another additional property of the Palatini connections is that they are the only affine structures whose parallel transport is homothetic with respect to the Levi-Civita transport. So the directions obtained in both cases are coincident.

We have also proved that an autoparallel of a given Palatini connection is a trajectory with critical length (autoparallel of Levi-Civita). The undetermined field $\mathcal{A}_\mu$ for a free falling observer can be absorbed in a reparametrization of its worldline, so it has no physical meaning since a particular choice of the parameter is meaningless.

We have found solutions of this kind in more general theories, for example, in Lovelock gravity. In addition, if we admit matter that does not feel the connection, the equivalence between formalisms can be extended from Einstein–Hilbert to any theory with additional quadratic torsion terms in the action. Current work involves the treatment with more general matter and with additional terms that introduce, for example, dynamics for the torsion field.

Acknowledgements We are grateful to A. Bernal and M. Sánchez for previous collaborations. We also wish to thank F. W. Hehl and J. Pons for useful comments and discussions. This work has been funded by the Junta de Andalucía (FQM101), the Universidad de Granada (PP2015-03), the MEC (FIS2016-78198-P) and the Unidad de Excelencia UCE-PP2016-02. In addition, J. A. O. was also supported by a PhD contract of the Plan Propio of the Universidad de Granada, and A. J. C. by a PhD contract of the FPU15 programme of the MEC.

References

Bernal A et al (2017) On the (non-) uniqueness of the Levi-Civita solution in the Einstein-Hilbert-Palatini formalism. Phys Lett B 768:280–287

Borunda M, Janssen B, Bastero-Gil M (2008) Palatini versus metric formulation in higher curvature gravity. J Cosmol Astropart Phys 0811:008

Dadhich N, Pons JM (2012) On the equivalence of the Einstein-Hilbert and the Einstein-Palatini formulations of general relativity for an arbitrary connection. Gen Relat Gravit 44 (9):2337–2352

Deser S (1996) Inequivalence of first and second-order formulations in d=2 gravity models. Found Phys 26(5):617–621

Hehl FW, Obukhov YN (2007) Elie Cartan's torsion in geometry and field theory, an essay. Ann Fond Broglie 32:157–194

Kleinert H, Pelster A (1999) Letter: autoparallels from a new action principle. Gen Relat Gravit 31 (9):1439

McCrea JD (1992) Irreducible decompositions of nonmetricity, torsion, curvature and Bianchi identities in metric-affine spacetimes. Class Quant Gravit 9(2):553–568

Chapter 6
Ballistic Transport in Nanowires

B. G. Sidharth and Abhishek Das

Abstract In this chapter, we endeavour to investigate the phenomenon of electrical conduction through nanowires. To be precise we derive a novel expression for the conductance resorting to the uncertainty principle.

6.1 Introduction

It is known that ballistic conduction or ballistic transport is a term associated with the transport of electrons through a medium where the electrical resistivity can be neglected. Sharvin (1965) led the foundation of the phenomenon of ballistic transport while visualizing an experiment. Since then, it has been a subject of widespread interest. There is a plethora of literature on ballistic conduction (Choi et al. 1985; Thornton et al. 1986; Wees et al. 1988; Wharam et al. 1988; Houten and Beenakker 1996; Frank et al. 1998).

However, even before Sharvin a significant nexus between conductance and transmission probability was put forward by Landauer (1957) which was essentially in contradistinction to the Drude model (Drude 1900a, b) or the quantum mechanically modified Drude-Sommerfeld model. In this chapter, we use the Heisenberg uncertainty principle to derive an expression for the conductance and then coalesce it with the Landauer–Buttiker formalism (Landauer 1957; Buttiker 1988) to find out if the transmission probability can be unity altogether, thereby allowing ballistic transport. Interestingly, very recently it has been experimentally found that electrons exhibit superballistic flow in the ballistic regime (Kumar et al. 2017).

On another track Sidharth had argued in the mid-1990s (Sidharth 1999) that in one dimensions and two dimensions, the transport would be luminal mimicking superconductivity. It must be remembered that by this time nanotubes had not yet been found and only 10 years later graphene was discovered, exhibiting the same properties. Sidharth's arguments were quantum mechanical, using the Dirac equations in one and two dimensions.

B. G. Sidharth (✉) · A. Das
B.M. Birla Science Centre, Adarsh Nagar, Hyderabad, India

B. G. Sidharth et al. (eds.), *Fundamental Physics and Physics Education Research*,
https://doi.org/10.1007/978-3-030-52923-9_6

6.2 Conductance and Transmission

Now, let us consider the case of a carbon nanotube. Here, electrons can travel freely through the axis of the nanowire without collision for a long time, i.e., the mean free path ($l_f = v_f\tau$, with Fermi velocity v_f) for the electron is very large. This type of transport is also known as ballistic transport (Datta 1997). Considering 1D nanowires, the conductance is generally given by

$$G = I/V = 1/R \tag{6.1}$$

where the symbols have their usual meanings. Since the electrons travel along the axis without any collision, there is apparently no loss during the transmission. This is a key feature of ballistic transport. Now, if any signal is transformed into an electric signal and thereby the electrons act as carriers and are transmitted through the nanowire with transmission probability nearly equal to unity, then we have a new means for telecommunications.

Now, we know that the diameter of a nanotube can be written as

$$D = \frac{a}{\pi}\sqrt{p^2 + pq + q^2}$$

where $a = 0.246$ (in nm), and 'n', 'm' are the integers that denote the unit vectors along the two directions in the honeycomb lattice. Adjusting the indices 'p' and 'q' such that

$$p = q$$

one can fabricate an *armchair nanotube* with a feasible diameter (D) which makes it metallic along the tubular axis. We would like to consider the metallic case specifically. Now, when the mean free path (l_f) of an electron is greater than the diameter (D) of the constriction or the quantum point contact, it necessitates us to consider the conductance of the medium. Now, suppose an electric pulse has to be transmitted through the nanotube medium. In that scenario, an electric field develops from which a magnetic field accumulates. This will be useful in the explanations given later.

Now, let us find an alternative expression of the conductance (G) other than that given in Eq. (6.1). We commence with the uncertainty principle, in terms of energy and time. This is known as

$$\Delta E.\Delta t \approx h$$

where 'h' is the Planck's constant. Now, the current 'I' in a quantum channel can be written as e/t, where 't' is transit time and the 'e' is electron charge. Applying a voltage V we get the energy as

$$E = eV$$

Again, if the uncertainty in energy is given by ΔE and the uncertainty in time is given by Δt then we can write

$$\Delta E.\Delta t = e^2 V/I \approx h$$

Now, exploiting relation (6.1) we have

$$\Delta E.\Delta t = e^2 G \approx h \tag{6.2}$$

Now, we know that the chemical potential of an electron is given as

$$\mu = \delta E/\delta N$$

where 'N' is the number of electrons. Suppose, that ΔE represents the uncertainty in the energy and ΔN does the uncertainty in the electron concentration, then we may write

$$\mu = \Delta E/\Delta N$$

Again, consider that the Fermi velocity of an electron can be expressed as

$$v_f = \Delta x/\Delta t$$

where we assume that Δx is the uncertainty in the position of the electron. Therefore, using the last two equations in relation (6.2) we derive

$$\Delta x.\Delta N \frac{\mu}{v_f} = e^2/G \tag{6.3}$$

Now, let the uncertainty in momentum be expressed as

$$\Delta p_x = \Delta m v_f$$

where the uncertainty in the mass arises due to the magnetic field that arises due to the electric field. Now, it has been found recently (Liu et al. 2016) that a magnetic field can impart mass to an electron. We are merely exploiting that phenomenon. Now, using the above relation in Eq. (6.3) one gets

$$(\Delta x.\Delta p_x).\frac{\mu \Delta N}{\Delta m v_f^2} = e^2 G$$

Manipulating the uncertainty principle ($\Delta x.\ \Delta p_x \approx \sim h$) again, we have finally for the conductance

$$G = \frac{(ev_f)^2}{\mu\alpha} \tag{6.4}$$

where $\alpha = \Delta N/\Delta m$. This is a novel expression to get a measure of the conductance in case of ballistic transport in nanowires. As one can see, in the following relation

$$\alpha = \Delta N/\Delta m$$

one can consider the ΔN to be the uncertainty in the concentration of transmitted electrons and Δm as defined before. Interestingly, this parameter, α can be looked upon as the effective value of

$$N/m$$

which appears in the Drude formula for conductivity of the classical Drude model. It is to be noted that α can be very significant in the derivation of the conductance of a particular nanowire sample.

Now, we know that according to the Landauer–Buttiker formalism, the current through the medium is given by

$$I = \int_{-\infty}^{+\infty} f'(E)M(E)T(E)dE \tag{6.5}$$

where $M(E)$ represents the total number of modes, $f'(E)$ is the deviation from the original electron distribution function and $T(E)$ represents the transmission probability. This leads us to an alternative definition of the conductance, also known as the Landauer formula (Landauer 1957) which is given by

$$G = G_0 - MT \tag{6.6}$$

where $G_0 = 2e^2/h$ is the quantum of conductance (Taylor and Mohr 2014). Now, taking into account the relations (6.4) and (6.6) we have

$$G_0 - MT = \frac{(ev_f)^2}{\mu\alpha}.$$

Using the expression for G_0 we have from here

$$T = \frac{hv_f^2}{2\mu\alpha M} \tag{6.7}$$

Now, from relation (6.7) we can see that if the parameters μ, α, M are adjusted in proper way then one can have

$$T \approx 1 \tag{6.8}$$

which is the intrinsic property for ballistic transport. Thus, relation (6.7) is novel and resorting to it one can achieve ballistic transport in nanowires, so that an electric pulse can be transmitted through the medium without significant losses.

6.3 Discussion

Currently, optical fibres are considered as the best medium for telecommunication and computer networking. Nevertheless, owing to attenuation loss which is lesser than in case of electric cables, communication systems experience many difficulties. In general, a light signal carries the information through the fibre which serves as an optical waveguide and the wave propagates by virtue of total internal reflection. Evidently, there are losses due to the many reflections and collisions.

But, according to this chapter, if an electrical signal carries information in lieu of a light signal through the medium of a 1D nanowire, then evidently there would be almost no loss and the communication technology can advance further taking a steady and huge leap. This could indeed be beneficial for the advancement in both the field of nanotechnology and telecommunications.

References

Buttiker M (1988) Absence of backscattering in the quantum Hall effect in multiprobe conductors. Phys Rev B 38:9375

Choi KK, Tsui DC, Palmateer SC (1985) Size effects on electron-electron interactions in GaAs-AlxGa1−xAs heterostructures. Phys Rev B 32:5540

Datta S (1997) Electronic transport in mesoscopic systems. Cambridge University Press, New York, pp 57–111. ISBN 0-521-59943-1

Drude P (1900a) Zur Elektronentheorie der Metalle. Ann Phys 306:566–613

Drude P (1900b) Zur Elektronentheorie der Metalle; II. Teil. Galvanomagnetische und thermomagnetische Effecte. Ann Phys 308:369–402

Frank S et al (1998) Carbon nanotube quantum resistors. Science 280:1744–1746

Houten HV, Beenakker C (1996) Quantum point contacts. Phys Today 49(7):22–27

Kumar RK et al (2017) Superballistic flow of viscous electron fluid through graphene constrictions. Nat Phys 13:1182. https://doi.org/10.1038/NPHYS4240

Landauer R (1957) Spatial variation of currents and fields due to localized scatterers in metallic conduction. IBM J Res Dev 1:223–231. https://doi.org/10.1147/rd.13.0223

Liu Y et al (2016) Zeeman splitting and dynamical mass generation in Dirac semimetal ZrTe5. Nat Commun 7:12516
Sharvin YV (1965) A possible method for studying Fermi surfaces. Sov Phys JETP 21:655
Sidharth BG (1999) Anomalous fermions. J Stat Phys 95(3–4):775–784
Taylor BN, Mohr PJ (2014) CODATA value: conductance quantum. National Institute of Standards and Technology
Thornton TJ et al (1986) One-dimensional conduction in the 2D electron gas of a GaAs-AlGaAs heterojunction. Phys Rev Lett 56:1198
Wees BJV et al (1988) Quantized conductance in point contacts in a two dimensional electron gas. Phys Rev Lett 60(9):848–850
Wharam DA et al (1988) One-dimensional transport and the quantization of the ballistic resistance. J Phys C 21(8):L209

Chapter 7
Gravitation by Condensation

Jonathan J. Dickau

Abstract For gravity to exert an influence, it must have a surface to act upon, or something for its pull to push things against. With everyday objects made of ordinary matter, the electromagnetic force acting through the electrons in atoms and molecules does all the pushing back. But in the degenerate cases, such as neutron stars and objects even more dense; other forces come into play or predominate as massive objects become more and more compact, until the gravitational limit sets the absolute minimum size at the Schwarzschild radius. Dvali and Gomez suggest that the event horizon of a Schwarzschild object is well modeled by the quantum critical point of a Bose–Einstein condensate, which makes gravity a process of condensation. This fits well with thermodynamic or entropic theories of gravitation, such as those proposed by Jacobson, Padmanabhan, and Verlinde, because condensation is precipitated by cooling. This connection is also clearly represented in features at the Misiurewicz point in the Mandelbrot Set near $(-1.543689, 0i)$, which the author has extensively studied, and these analogies offer many avenues for further research.

7.1 Introduction

Gravity is perhaps least understood of the natural forces, despite how sophisticated our understanding has become. We know that gravity draws other massive objects toward any center of mass, with a pull proportional to their combined masses and inversely to the square of their distance from its center. With the refinement of relativity, we learn this is an approximation because there is actually a radius of gravitation defined by any mass—rather than a simple center of mass—as was emphasized by Eddington (1920). With familiar and astrophysical objects, the gravitational radius is a fraction of a massive body's total volume. But when we consider the case of a black hole devoid of charge or spin, the radius of gravitation is at the surface, determining the object's size, and is called the Schwarzschild radius.

J. J. Dickau (✉)
Independent Researcher/ISGRG Member, Poughkeepsie, NY, USA
e-mail: jonathan@jonathandickau.com

B. G. Sidharth et al. (eds.), *Fundamental Physics and Physics Education Research*,
https://doi.org/10.1007/978-3-030-52923-9_7

Although Schwarzschild black holes are the simplest textbook example (Hawking and Ellis 1973), a simplified or toy model rather than a physically realistic one, the Schwarzschild radius is profoundly important nonetheless. This feature may be the key to understanding the quantum nature of gravity, because the surface area at that radius is quantized, and this sets its information-carrying capacity. If we treat gravity as a process of graviton condensation, information storage and processing at the event horizon can be explained as behavior at the quantum critical point of a BEC—according to recent work by Dvali and Gomez (2014)—and this is worthy to explore further. It opens up the possibility that laboratory studies of various types of quantum condensation can be used to probe the nature of BH event horizons (Steinhauer 2016) (Herdman et al. 2017), and thereby give us a better understanding of gravity itself.

Sakharov (1967) first proposed analogies, which ascribe to the action of gravity the same nature as BEC formation—but at that time Bose–Einstein condensation was seen as only a curious theoretical possibility. In the current era, we know that BECs will reliably form if we cool an appropriate sample far enough (Anderson et al. 1995). This is now within the reach of almost any University Physics department, as well, because the process has been perfected and miniaturized (Hänsel et al. 2001). But the analogy of gravity with Bose–Einstein condensation offers a window on how gravity's quantum mechanical action gives rise to well-known classical properties. The same analogy is seen in features of the Mandelbrot Set (M) at the Misiurewicz point near $(-1.543689, 0i)$, and in associated and related figures of M, which the author began studying more than 30 years ago (Dickau 2016a). This point is a solution to $((c^2 + c)^2 + c)^2 + c = (c^2 + c)^2 + c$, and it is one of the few places in M where an exact analytical solution (its precise location) is possible to obtain. When points are highlighted, where the function's iterand diminishes monotonically over 3 iterations (i.e., the Mandelbrot Butterfly), the special significance of this spot is easy to discern. But examined in the standard rendering; it is an unassuming spot where a collection of telephone poles diminishes in size to extinction, then grows on the other side in opposite phase—a feature that is easily missed. Nonetheless, since this location in M is an analogy for both black hole event horizons and Bose–Einstein condensation, it is relevant to unifying the quantum and relativistic views of Physics. The remainder of this paper discusses what we can learn from these overlapping analogies, where multiple descriptions afford a single congruent result.

7.2 Entropic Gravity, Condensation, and Dimensional Reduction

When people ask "what is gravity?" the standard answers do not mention entropy or thermodynamics, and describe it as an attractive force between massive objects, state that it is due to the curvature of space, or say it is one of the four fundamental forces; but there are other answers. It has been suggested by Jacobson (1995), Verlinde (2011), Padmanabhan (2010a, b), and others, that instead of a fundamental force of nature, gravity is a residual or consequence of the other forces. In effect, the energy left over from the rest of the universe radiating and expanding, and from quantum fluctuations in the vacuum, supplies the force driving any two objects together. This insight combines with the relation that as the universe expands it cools and the observation that mixed gases undergo sequential liquefaction or fractionation when cooled in this way. Different species condense at different rates or times. This is true for the cosmos too. Condensed matter gives gravity something to act upon—surfaces to push against and objects to push together—once the universe cools enough. We see cohesive forces of fluids that make droplets bind together and coalesce—and the quantum condensation of a BEC—are similar to gravity's action upon matter. Dvali and Gomez liken the quantum critical point of BEC formation to a Schwarzschild event horizon, which is purely gravitational (having no charge or spin). Cosmology shows that, regardless of the model we use, the universe employs this mechanism—in its evolution to the present—where objects and forces congeal from the energy soup of the early universe by phase transitions to become loci of action and attraction.

A similar pattern occurs in pure Mathematics if we consider the geometrical objects and spaces. Alain Connes, noted expert in non-commutative geometry, states (2000) that if we examine the categories of forms and spaces, they form a hierarchy:

$$\text{Sm} > \text{Top} > \text{Meas}$$

here we see transitions as funnels constraining possibilities where smooth is the largest category, topological forms and spaces a subset of those, and measurable objects and spaces (which are most familiar) the most constrained. This is similar (with some caveats) to the process described above; so we can write out in simile:

$$\text{Gases} > \text{Liquids} > \text{Solids}$$

Ideal gases are well-behaved or smooth. Liquids have a surface, so they are topological, and solids have a constant shape and size, so they are measureable. Similarly, we can look to the normed division algebras, which provide or enumerate the natural number types:

$$\mathbb{O} \supset \mathbb{H} \supset \mathbb{C} \supset \mathbb{R}$$

Here the octonions are the most general, the quaternions a limited subset (by fixing four of seven axes of rotation), the complex numbers a subset of those (by fixing two more axes), and the reals (the most commonly used) are the most restricted—with only a measure of fixity. Progressing from the octonions—with seven degrees of freedom (imaginaries) and one scalar quantity—to the reals, we see the degrees of freedom diminish until there are none, where ordinary real numbers represent specific fixed values.

Real numbers are in a sense solidified values of something initially more variable than stable, having imaginary dimensions, which then settled on a unique quantity.

But it is then correct to say that familiar real numbers are a projection of higher-dimensional entities, or their finalized and settled identity. Applying this principle to Physics means a precursor state with a higher dimension condensed into a lower-dimensional state cosmologically. This is precisely what has been proposed in DGP gravity (a braneworld formulation by Dvali et al. (2000)), and explained cosmologically in work by Deffayet (2001), in the 5-d black hole $\rightarrow$ 4-d spacetime white hole scenario of Pourhasan et al. (2014), in a similar construction by Poplawski (2010) based on Einstein-Cartan, and in modified and extended versions of DGP including cascading gravity (de Rham et al. 2008a, b), that begin in still higher dimensions. If the cosmos originates in a higher-dimensional space, this could mean cosmological transitions literally or effectively turned the universe inside-out, causing outward-pressing forces to turn inward, and this is likely what happened to gravity—which explains its weakness compared to the other fundamental forces. This is precisely what we should expect, moreover, if the familiar properties of associativity and commutativity emerged as a result of cosmological events and transitions which allow things that would be free-floating in higher dimensions to congeal in a dimensionally reduced space. We note that 3-d space allows forms to be created that are more stable than is possible in higher dimensions, and this underlies both the process of condensation and its power.

Condensation is therefore observed to be a process where variability becomes fixity as degrees of freedom possessed by a system or entity become fixed through phase transitions of various kinds. This means that variability was needed and higher dimensions might have thus facilitated the emergence of the cosmos as we know it. Cosmologies featuring a higher-dimensional precursor or bulk (in which our spacetime is embedded), including those above, are well-respected at this point. Spaces of higher dimension are dimensionally reduced, while objects and systems with freedom to vary multiple ways become constrained in various ways instead—attaining constancy, or stability and predictability. A relevant example is massive gravitons, which figure into some of the work cited above, because massive spin-2 particles have 5 polarization states $(0, \pm 1, \pm 2)$ while a massless graviton has only two (± 2). Continuation of condensation from higher dimensions beyond 3-d involves the settling of form and energy onto 2-d surfaces. In another paper by Dvali and Gomez (2013), it is suggested that quantum mechanical properties of a black hole can be completely characterized by the portrait of N gravitons condensing at the horizon. But condensation is also a feature or natural consequence of various flavors of entropic gravity theory. We see a common element of these theories is a holographic screen or local Rindler horizon; the action of gravity is calculated relative to a 2-d surface, where microcausal degrees of freedom go to a large N.

7.3 The Part Played by the Mandelbrot Set

As it turns out, the scenario I spelled out above—where a higher-dimensional volume is everted to create the 4-d spacetime bubble we now inhabit, and where the action of gravity at a Schwarzschild horizon is like Bose–Einstein condensation—is encoded in the shape of the Mandelbrot Set—and in its family of related and associated figures. This is something the author has struggled to understand for more than 30 years (Dickau 1987), wondering if it would turn out to be of use to Physics. A sketch of these ideas was presented on a poster at GR21 in 2016 (Dickau 2016b), but my talk at FFP15 is the first attempt to spell out to the Physics community how one can connect the dots to join Mandelbrot gravity with a larger body of theoretical Physics relating to gravity and cosmology. The truly remarkable thing is that M displays or reproduces features of 5-d $\rightarrow$ 4-d DGP gravity and the correspondence of Schwarzschild event horizons with Bose–Einstein condensation, with no adjusting factors or constants put in by hand. This is mainly due to the fact that M is maximally asymmetric, but contains many symmetries, and the ubiquitous role played by symmetry-breaking and asymmetry shaping the laws of Physics (Dickau 2018).

The Mandelbrot Set is symmetrical across the real axis, but grossly asymmetrical along it, having a positive maximum of $(0.25, 0i)$ and a much greater extent in the negative, with a negative extremum at $(-2, 0i)$. We know that M displays the progression to chaos, such that iterating the core equation $z \rightarrow z^2 + c$ in the reals yields a bifurcation diagram exactly mimicking the logistic map—where at each split point the boundary of M folds back on itself. The first bifurcation point is at $(-0.75, 0i)$ where the maximum of the cardioid opens into the circular disc centered at $(-1, 0i)$. The feature at $(-0.75, 0i)$ resembles the folding of space in a braneworld model like DGP gravity, or the 5-d black hole $\rightarrow$ 4-d spacetime bubble cited above. That space can fold back on itself is the crucial generalization here, but dimensionality can be higher before and lower after such a transition. Though the Mandelbrot Set is seen as a two-dimensional object, living in the complex numbers, it is defined up to the octonions. The familiar representation of M is a projection of the set onto any one imaginary dimension, but the hypercomplex numbers ($\mathbb{H}$ and $\mathbb{O}$), with 3 and 7 imaginaries, respectively, afford M with more range for variations. What makes the transition at $(-0.75, 0i)$ more than a curious visual metaphor, however, is that M encodes Cartan's rolling ball analogy for G_2 symmetries (Fig. 7.1), where G_2 is the smallest exceptional Lie group and the exceptional Lie groups all come from the octonions.

In regard to what M tells us about gravity though, the bifurcation map in the Reals gives us the essential clue. While a bifurcation diagram mainly shows how the trajectories diverge, we observe there is a spot where all of the divergent trajectories appear to converge, which is a Misiurewicz point. The corresponding location in M $\sim (-1.543689012692, 0i)$ is relatively unassuming, but it has a special dual significance. It represents both the quantum critical point of Bose–Einstein condensation and the event horizon of a Schwarzschild black hole. This is where gravitation by condensation is depicted in M. We note that this value satisfies the equation

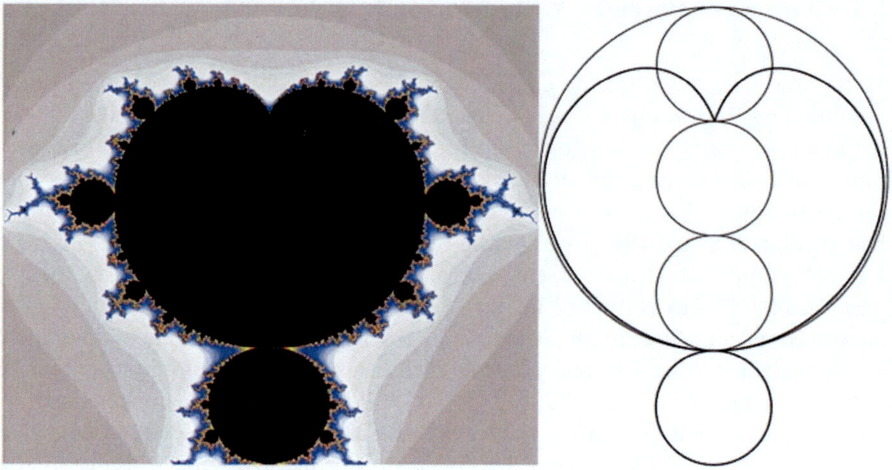

Fig. 7.1 The Mandelbrot Set illustrates Cartan's G_2 rolling ball analogy in its major geometry

$((c^2 + c)^2 + c)^2 + c = (c^2 + c)^2 + c$, representing 3 and 2 iterations of the Mandelbrot formula on the left and right hand side, respectively. But this location may have more to teach us than most, because it is an archetypal example of principles that apply in a more complex way to a broad class of Misiurewicz points in M, and these points appear to be the most relevant places in M to Physics. If we highlight where the iterand magnitude monotonically diminishes over 3 values, revealing the Mandelbrot Butterfly, this spot is the boundary for a circular disc around a mini-M figure, extending all the way down to $(-2, 0i)$. This disc is the largest of a vast family of discs around the periphery of the Butterfly figure, near the Misiurewicz points—denoting basins of attraction. This particular basin of attraction appears to simulate gravity, denoting r_G the radius of gravitation (confluence in Fig. 7.2 and disc near base in Fig. 7.3).

The algorithmic nature of M and its associated figures is what allows us to show the wings and discs of the Mandelbrot Butterfly figure, but it also lets us suppress lower-order solutions to strip away layers of the Butterfly's form and see what is underneath. Ergo, we can observe the condensation process in more detail, by removing the largest disc and seeing what higher-order solutions are doing in that same region of M. One might ask what the Mandelbrot Set could possibly have to do with the Physics of gravity, black holes, and quantum condensation—apart from some nice analogies with physical processes to stimulate creative thought. Indeed, the author took a philosophical view of this work for many years, and had little communication with academics about it—apart from phone conversations with Benoit Mandelbrot some 30 years ago. But with the discovery in 1998 of the accelerating expansion of the universe (Perlmutter et al. 1998; Reiss et al. 1998), it became apparent to the author that what was being learned closely matched predictions based on features of the Mandelbrot Set. Ten years later, the author first presented these insights at the second crisis in Cosmology conference (Dickau

Fig. 7.2 M and its bifurcation diagram, where the confluence is a Misiurewicz point

Fig. 7.3 The Mandelbrot Butterfly figure highlights diminishing magnitude iterands, where the large disc near the base borders the same point

2009). To hammer out the connections with Physics has required quite a lot of work, including much additional study. Since that time, though, a large body of theoretical Physics has appeared that more closely mirrors the phenomenology I had worked out based on M. Only after returning from GR21 in 2016 did I learn how the ideas I presented there clearly resemble a combination of DGP gravity (as in Fig. 7.1) and the BEC formation—Event horizon model of Dvali and Gomez (shown in Fig. 7.4).

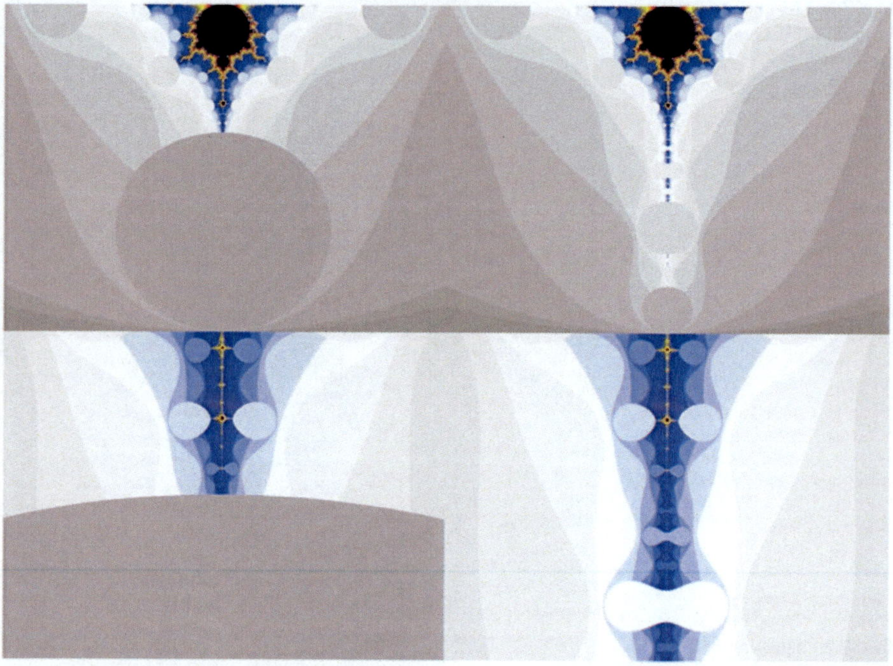

Fig. 7.4 Misiurewicz point at ~ $(-1.543689, 0i)$ illustrating an event horizon/quantum critical point (on left) and the condensation process (on right)

7.4 Conclusions

"Gravitation by Condensation" is offered as a way to explain the phenomenology suggested by the Mandelbrot Set and the remarkable way it fits to the ideas of top theorists. The advantages of this theoretical construction are that it resembles Newtonian or Einsteinian gravity at shorter scales but inherits the accelerating expansion at the largest scales that is seen in DGP gravity and cascading DGP, because it arises in a similar higher-dimensional framework. But a single context for all of this—condensation that proceeds from spaces with higher dimension, creates the current conditions, and explains the force of gravity—is provided by the Mandelbrot Set and its associated figures. Of course, we must assume that M is higher-dimensional, residing in the quaternions and octonions, but this is reasonable. We note that Kricker and Joshi (1995) used the Mandelbrot Set to map associative and non-associative regions in the octonions, in a paper on bifurcations in the octonionic quadratic. So we know there are natural correspondences to explore for brave souls prepared to deal with the complexity of octonion algebra. Thankfully, there are now tools available, helping to make the difficult calculations more routine to implement. Rick Lockyer has made available an octonion calculation package for NodeJS (Lockyer 2018) that I have been test driving, which offers some capabilities that will assist further research. However, the features of M seen along the real axis, such as the folded back edge at $(-0.75, 0i)$ and the Misiurewicz point at $(-1.543689012692, 0i)$, are present in both higher- and lower-dimensional cases.

Condensation appears to be a generic feature of a large variety of theories, including various flavors of emergent, entropic, thermodynamic, and holographic gravity theory. It is also obviously a feature of Cosmology if we envision a hot Big Bang preceded by some form of Inflation. Most of the concepts presented do not depend on an exotic higher-d framework, but they fit well with ideas that came about that way, like DGP gravity, as well as with more mainstream cosmologies. Conventional inflationary universe theory has its share of problems, as was pointed out by Professor Steinhardt at FFP11 (Steinhardt 2010), and which later became a feature article in Scientific American (twice) (Steinhardt 2011; Ijjas et al. 2017). But the theories cited above, and the new work presented here, suggest alternate roads to the present-day cosmos through cosmological transitions. Cosmologies based on M in the octonions show octonionic inflation ending in a 5-d volume, which flips into a 4-d spacetime at $(-0.75, 0i)$. The fact we now live in a 3-d space plus time, or a 4-d spacetime, may not mean that the current background dimensionality was always present, and it likely does not mean the dimensionality of space was always so well-defined. It is claimed that theories of emergent gravity like Verlinde's work as stated only if the fabric of spacetime is also emergent. In the context of gravitation by condensation, as a process of dimensional reduction, this happens automatically. In Mandelbrot gravity theory, we see represented desirable long-range features of higher-d theories combined with short-range behavior in line with the theories of Newton and Einstein that emerge from condensation, without having to put adjusting factors in by hand.

Acknowledgement The author wishes to thank the conference organizers, the MUDIC staff, the people and official representatives of Orihuela and Elche, and the Universidad Miguel Hernandez, for the warm reception and collegial atmosphere.

References

Anderson MH et al (1995) Observation of Bose-Einstein condensation in a dilute atomic vapor. Science 269(5221):198–201

Connes A (2000) Noncommutative Geometry Year 2000, arXiv:math/0011193

de Rham C et al (2008a) Cascading gravity and degravitation. J Cosmol Astropart Phys 2008. arXiv:0712.2821

de Rham C et al (2008b) Cascading gravity: extending the Dvali-Gabadadze-Porrati model to higher dimension. Phys Rev Lett 100:251603

Deffayet C (2001) Cosmology on a brane in Minkowski bulk. Phys Lett B 502(1-4):199–208. arXiv:hep-th/0010186

Dickau JJ (1987) Cosmology and the Mandelbrot set. Amygdala 1(7):2–3

Dickau JJ (2009) Does the Mandelbrot set offer clues to the cosmological evolution of form? In: 2nd Crisis in Cosmology Conference, CCC-2, ASP Conference series, vol 413, pp 116–121

Dickau JJ (2016a) Charting trends in the Mandelbrot set & showing their significance for cosmology. Prespacetime J 7(9):1319–1332

Dickau JJ (2016b) Entropic Gravity and the Mandelbrot Set, poster presented at GR21, Columbia University, NYC, July 2016
Dickau JJ (2018) Gravity is not fundamental; or is it? Essay for FQXi contest
Dvali G, Gomez C (2013) Black Hole's Quantum N-Portrait. Progr Phys 61(7–8):742–767. arXiv:1112.3359
Dvali G, Gomez C (2014) Black holes as critical point of quantum phase transition. Eur Phys J C 74:2752
Dvali G, Gabadadze G, Porrati M (2000) 4D gravity on a Brane in 5D Minkowski space. Phys Lett B 485:208–214. arXiv:hep-th/0005016
Eddington A (1920) Space, time, and gravitation—an outline of the general relativity theory. Cambridge University Press, Cambridge
Hänsel W et al (2001) Bose-Einstein condensation on a microelectronic chip. Nature 413:498–591
Hawking S, Ellis G (1973) The large-scale structure of space-time. Cambridge University Press, Cambridge, pp 149–156
Herdman CM et al (2017) Entanglement area law in superfluid 4He. Nat Phys 13:556
Ijjas A, Steinhardt P, Loeb A (2017) Pop Goes the Universe, Scientific American, January 2017. pp. 32–39
Jacobson T (1995) Thermodynamics of Spacetime: the Einstein equation of state. Phys Rev Lett 75:1260. arXiv:gr-qc/9504004
Kricker A, Joshi G (1995) Bifurcation phenomena of the non-associative octonionic quadratic. Chaos Solitons Fractals 5(5):761–782
Lockyer R (2018) Octonions for NodeJS, link to content at his FQXi essay forum. https://fqxi.org/community/forum/topic/3149#post_146884
Padmanabhan T (2010a) Thermodynamical aspects of gravity: new insights. Rep Prog Phys 73(4). arXiv:0911.5004
Padmanabhan T (2010b) Gravity as an emergent phenomenon. In: Gravitation—foundations and frontiers. Cambridge University Press, Cambridge, pp 670–688
Perlmutter S et al (1998) Discovery of a Supernova at half the age of the universe and its cosmological implications. Nature 391:51–54
Poplawski N (2010) Radial motion into an Einstein–Rosen bridge. Phys Lett B 687(2–3):110–113. arXiv:0902.1994
Pourhasan R, Afshordi N, Mann R (2014) Out of the White Hole: a holographic origin for the Big Bang. J Cosmol Astropart Phys 2014:1–18. arXiv:1309.1487
Reiss A et al (1998) Observational evidence from supernovae for an accelerating universe and a cosmological constant. Astron J 116(3):1009–1038
Sakharov A (1967) Vacuum quantum fluctuations in curved space and the theory of gravitation. Dokl Akad Nauk SSSR 177:70–71. reprinted (2000) General Relativity and Gravitation, 32, 2 (pp. 365–367)
Steinhardt P (2010) Inflationary Universe on Trial, Plenary Talk for FFP11, July 2010
Steinhardt P (2011) The inflation debate. Scientific American, April 2011. pp. 36–43
Steinhauer J (2016) Observation of quantum Hawking radiation and its entanglement in an analogue black hole. Nat Phys 12:959–965. arXiv:1510.00621
Verlinde E (2011) On the origin of gravity and the Laws of Newton. J High Energy Phys 2011:29. arXiv:1001.0785

Chapter 8
Dark Matter Anomaly

B. G. Sidharth and Abhishek Das

Abstract Latest observations by Riess and coworkers in 2017 have reconfirmed their earlier observation that the universe is accelerating some 8% faster than the currently accepted cosmological model described. In this chapter, we state again, in light of a recent paper, that this discrepancy can be eliminated by considering a universe consisting only of matter and dark energy.

8.1 Introduction

Very recently (Das and Sidharth 2019) Das and Sidharth have shown that the existence of dark matter is inconsistent with the observation of Riess et al. (2016). Now, Hubble's law is regarded as one of the major observational basis for the expansion of the universe. Later the existence of *dark matter* was hypothesized by Zwicky (1933, 1937) who inferred the existence some unseen matter based on his observations regarding the rotational velocity curves at the edge of galaxies. Although, Kapteyn (1922) and Oort (1932) had also had derived the same conclusions before Zwicky. Since then, various efforts have been made to prove the existence of *dark matter* (cf. ref. Bertone et al. (2005) for detailed review). Recently, after conducting experiments to detect weakly interacting massive particles (WIMPs) that interact only through gravity and the weak force and are hypothesized as the constituents of *dark matter*, it has been found that such hypotheses lead nowhere (Akerib et al. 2016).

Interestingly, authors such as Milgrom (1983), Bekenstein (2004) Sidharth (cf. Sect. 8.3) and Mannheim (2005) have endeavoured to find alternatives to the widely accepted *dark matter*. The author Sidharth (2000, 2006a, b) has also given a suitable alternative to the conventional *dark matter* paradigm.

Abhishek Das gave an invited talk at Physics and Astronomy Conference 2018, Hyderabad, India.

B. G. Sidharth · A. Das (✉)
B.M. Birla Science Centre, Hyderabad, India

B. G. Sidharth et al. (eds.), *Fundamental Physics and Physics Education Research*,
https://doi.org/10.1007/978-3-030-52923-9_8

The generally accepted ideas may have to be revisited in view of latest observations of Riess et al. which point to the fact that cosmic acceleration is some 5–8% greater than what the current cosmological model suggests.

Before proceeding, it may be mentioned that in 1997, the accepted model of the universe was that of a dark matter dominated decelerating universe. That year the author Sidharth put forward his contra model—an accelerating universe (Sidharth 1998), dominated by not dark matter but rather what is today being called dark energy.

8.2 Theory

We are well acquainted with the fact that the Friedman equations govern the expansion of space in homogeneous and isotropic models of the universe within the context of general relativity. Let us begin with the following equation

$$H^2 = \left(\frac{\dot{a}}{a}\right)^2 = \frac{8\pi G}{3}\rho - \frac{kc^2}{a^2} + \frac{\Lambda c^2}{3}$$

where H is the Hubble parameter, a is the scale factor, G is the gravitational constant, k is the normalized spatial curvature of the universe and Λ is the cosmological constant. Considering $k = 0$ (a flat universe) with the domination of both matter and dark energy, one can derive the Hubble parameter as

$$H(z) = H_0\left[\Omega_{\mathrm{M}}(1+z)^3 + \Omega_{\mathrm{DE}}(1+z)^{3(1+w)}\right]^{\frac{1}{2}} \tag{8.1}$$

where, z is the redshift value or the recessional velocity and the dimensionless parameter w is given by

$$P = w\rho c^2$$

P being the pressure and ρ being the density. Now, we would like to expand the function $H(z)$ using the Taylor expansion about the point z_0. This yields

$$H(z) = H(z_0) + \frac{H'(z_0)}{1!}(z - z_0) + \cdots$$

Neglecting terms consisting second and higher order derivatives of the Hubble parameter and considering that $H(z_0) = H_0$ we have using (8.1)

$$H(z) = H_0 + \frac{H_0}{2}\frac{3\Omega_{\mathrm{M}}(1+z_0)^2 + 3(1+w)\Omega_{\mathrm{DE}}(1+z_0)^{3(1+w)-1}}{\left[\Omega_{\mathrm{M}}(1+z_0)^3 + \Omega_{\mathrm{DE}}(1+z_0)^{3(1+w)}\right]^{\frac{1}{2}}}(z-z_0) \quad (8.2)$$

Now, we know that if the dark energy derives from a cosmological constant then

$$w = -1$$

Therefore, in such a case we have

$$H(z) = H_0 + \frac{3H_0}{2}\frac{\Omega_{\mathrm{M}}(1+z_0)^2}{\left[\Omega_{\mathrm{M}}(1+z_0)^3 + \Omega_{\mathrm{DE}}\right]^{\frac{1}{2}}}(z-z_0) \quad (8.3)$$

Now, since numerical values suggest that $\Omega_{\mathrm{DE}} > \Omega_{\mathrm{M}}$ we can use another series expansion for the denominator of the second term above to get

$$H(z) = H_0\left[1 + \frac{3}{2}\frac{1}{\sqrt{\Omega_{\mathrm{DE}}}}\left\{\Omega_{\mathrm{M}}(1+z_0)^2\right\}\left\{1 - \frac{\Omega_{\mathrm{M}}(1+z_0)^3}{2\Omega_{\mathrm{DE}}}\right\}\right](z-z_0) \quad (8.4)$$

Now, we would look at this equation at the point $z_0 = 0$ and for $z = 1$ to give

$$H = H_0\left[1 + \frac{3}{2}\frac{\Omega_{\mathrm{M}}}{\sqrt{\Omega_{\mathrm{DE}}}}\left\{1 - \frac{\Omega_{\mathrm{M}}}{2\Omega_{\mathrm{DE}}}\right\}\right] \quad (8.5)$$

Now, standard cosmological model suggests that the universe is comprised of baryonic matter, dark matter, dark energy and some other constituents. In a nutshell, we have (Knop et al. 2003).

$$\Omega_{\mathrm{Baryonic}} \approx 0.04$$

$$\Omega_{\mathrm{Darkmatter}} \approx 0.23$$

$$\Omega_{\mathrm{Darkenergy}} \approx 0.73$$

and

$$\Omega_{\mathrm{M}} = \Omega_{\mathrm{Baryonic}} + \Omega_{\mathrm{Darkmatter}}$$

Using all these values in (8.5) we have the Hubble parameter

$$H = H_0 + 0.39H_0$$

i.e. the acceleration of the universe should be approximately 39% greater than its value. But, due to recent observations it has been substantiated that the acceleration is about 5–8% greater than its value. So, in fact we should have

$$H = H_0 + 0.08H_0$$

If this is the case then doing some back calculations and using $\Omega_{DE} \approx 0.73$, we have the following two values for Ω_M.

$$\Omega_M \approx 1.41 \text{ or}, 0.044$$

Now, it is a fact that $\Omega_M < 1$ and so $\Omega_M \approx 1.41$ would be unphysical. Therefore we have the value of Ω_M as

$$\Omega_M \approx 0.044 \tag{8.6}$$

But, this is very nearly equal to the value of Baryonic matter, i.e. Ω_{Baryonic}. This suggests ostensibly that

$$\Omega_{\text{Darkmatter}} \approx 0 \tag{8.7}$$

In other words, the existence of dark matter is itself inconsistent according to the latest observations of Riess et al. In such a case, the total density of the universe is given by

$$\Omega = \Omega_{\text{Baryonic}} + \Omega_{\text{Darkenergy}} \approx 0.77 \tag{8.8}$$

which is less than the critical density. This suggests that the universe will be expanding in an accelerating manner.

8.3 Alternative to the *Dark Matter* Paradigm

Very recently the LUX detector in South Dakota has concluded (Akerib et al. 2016) that it has not found any traces of dark matter. So far this has been the most delicate detector. It will be recalled that dark matter was introduced in the 1930s by Zwicky to explain the flattening of the galactic rotational curves: With Newtonian gravity the speeds of these galactic curves at the edges should tend to zero according to the Keplerian law, $v \propto 1/\sqrt{r}$. Here r is the distance to the edge from the galactic centre. However velocity v remains more or less constant. Zwicky explained this by saying that there is a lot more of unseen matters concealed in the galaxies, causing this discrepancy. The fact is that even after nearly 90 years dark matter has not been detected.

The modified Newtonian dynamics approach of Milgrom (1983, 1986, 1989, 1994, 1997, 2001) was an interesting alternative to the *dark matter* paradigm. The objection of this fix has been that it is too ad hoc, without any underlying theory.

The author himself has been arguing over the years (Sidharth 2000, 2006a, b) (cf. ref. Sidharth (2008) for a summary) that the gravitational constant G is not fixed but varies slowly with time in a specific way. In fact this variation of the gravitational constant has been postulated by Dirac, Hoyle and others from a different point of view (cf. ref. Sidharth 2008; Narlikar 1993) which for various reasons including inconsistencies have in the author's scheme, exactly accounts for the galactic rotation anomaly without resorting to dark matter or without contradictions.

Our starting point is the rather well known relation (Narlikar 1993).

$$G = G_o\left(1 - \frac{t}{t_o}\right) \tag{8.9}$$

where G_o is the present value of G and t_o is the present age of the Universe, while t is the relatively small time elapsed from the present epoch. On this basis one could correctly explain the gravitational bending of light, the precession of the equinoxes of mercury, the shortening of the orbits of binary pulsars and even the anomalous acceleration of the pioneer spacecrafts (cf. references given above).

Returning to the problem of the rotational velocities at the edges of galaxies, one would expect these to fall off according to

$$v^2 \approx \frac{GM}{r} \tag{8.10}$$

However it is found that the velocities tend to a constant value,

$$v \sim 300\ \mathrm{km/s} \tag{8.11}$$

This, as noted, has led to the postulation of the as yet undetected additional matter alluded to, the so called dark matter.(However for an alternative view point cf. Sivaram and de Sabbata (1993).) We observe that from (8.9) it can be easily deduced that (Sidharth 2001).

$$a \equiv (\ddot{r}_o - \ddot{r}) \approx \frac{1}{t_o}(t\ddot{r}_o + 2\dot{r}_o) \approx -2\frac{r_o}{t_o^2} \tag{8.12}$$

as we are considering infinitesimal intervals t and nearly circular orbits. Equation (8.12) shows (cf. ref. Sidharth (2006a) also) that there is an anomalous inward acceleration, as if there is an extra attractive force, or an additional central mass, a la Zwicky's dark matter.

So,

$$\frac{GMm}{r^2} + \frac{2mr}{t_o^2} \approx \frac{mv^2}{r} \tag{8.13}$$

From (8.13) it follows that

$$v \approx \left(\frac{2r^2}{t_o^2} + \frac{GM}{r}\right)^{1/2} \tag{8.14}$$

From (8.14) it is easily seen that at distances within the edge of a typical galaxy, that is $r < 10^{23}$ cm, the Eq. (8.10) holds but as we reach the edge and beyond, that is for $r \geq 10^{24} cms$ we have $v \sim 10^7$ cm per second, in agreement with (8.11). In fact as can be seen from (8.14), the first term in the square root has an extra contribution (due to the varying G) which is roughly some three to four times the second term, as if there is an extra mass, roughly that much more.

Thus the time variation of G explains observation without invoking dark matter.

8.4 Conclusion

We have seen that the discrepancy in the acceleration value of the universe, as reconfirmed multiple times by careful studies of Riess and coworkers can be removed by considering a universe consisting only of matter and dark energy.

Recently, Swinbank (2017) and Genzel et al. (2017) have concluded that nearly ten billion years ago dark matter concentration was very small and the universe was dominated by baryonic matter. It is possible that this negligible concentration of dark matter boiled down to zero in due course of evolution of the universe.

References

Akerib DS et al (2016) Improved limits on scattering of weakly interacting massive particles from reanalysis of 2013 LUX data. Phys Rev Lett 116:161301

Bekenstein JD (2004) Relativistic gravitation theory for the modified Newtonian dynamics paradigm. Phys Rev D 70(8):083509

Bertone G, Hooper D, Silk J (2005) Particle dark matter: evidence, candidates and constraints. Phys Rep 405(5–6):279–390

Das A, Sidharth BG (2019) Anomaly of dark matter. MPLA 34(19):1950143

Genzel R et al (2017) Strongly baryon-dominated disk galaxies at the peak of galaxy formation ten billion years ago. Nature 543:397–401

Kapteyn JC (1922) First attempt at a theory of the arrangement and motion of the sidereal system. Astrophys J 55:302–327

Knop R et al (2003) New constraints on OmegaM, OmegaLambda, and w from an independent set of 11 high-redshift supernovae observed with the Hubble space telescope. Astrophys J 598:102

Mannheim PD (2005) Alternatives to dark matter and dark energy. Prog Part Nucl Phys 56(2):340

Milgrom M (1983) A modification of the Newtonian dynamics as a possible alternative to the hidden mass hypothesis. Astrophys J 270:365–370
Milgrom M (1986) Solutions for the modified Newtonian dynamics field equation. Astrophys J 302:617
Milgrom M (1989) Alternatives to dark matter. Comm Astrophys 13(4):215
Milgrom M (1994) Dynamics with a nonstandard inertia-acceleration relation: an alternative to dark matter in galactic systems. Ann Phys 229:384
Milgrom M (1997) Nonlinear conformally invariant generalization of the poisson equation to D>2 dimensions. Phys Rev E 56:1148
Milgrom M (2001) MOND – a pedagogical review. In: Presented at the XXV International School of Theoretical Physics particles and astrophysics-standard models and beyond. Ustron, Poland
Narlikar JV (1993) Introduction to cosmology. Cambridge University Press, Cambridge, p 57
Oort JH (1932) The force exerted by the stellar system in the direction perpendicular to the galactic plane and some related problems. Bull Astron Inst Netherlands 6:249–287
Riess AG et al (2016) A 2.4% determination of the local value of the Hubble constant. Astrophys J 826(1)
Sidharth BG (1998) The universe of fluctuations. IJMPA 13(15):2599–2612
Sidharth BG (2000) Effects of varying G. Nuovo Cimento 115B 12(2):151ff
Sidharth BG (2001) Chaotic universe: from the Planck to the Hubble scale. Nova Science, New York
Sidharth BG (2006a) Tests for varying G. Found Phys Lett 19(6):611–617
Sidharth BG (2006b) The puzzle of gravitation. Int J Mod Phys A 21(31):6315
Sidharth BG (2008) The thermodynamic universe. World Scientific, Singapore
Sivaram C, de Sabbata V (1993) On limiting field strengths in gravitation. Found Phys Lett 6(6)
Swinbank M (2017) Astrophysics: distant galaxies lack dark matter. Nature 543:318–319
Zwicky F (1933) Die Rotverschiebung von extragalaktischen Nebeln. Helv Phys Acta 6:110–127
Zwicky F (1937) On the masses of nebulae and of clusters of nebulae. Astrophys J 86:217

Chapter 9
Before the Big Bang

E. R. Siegel

Abstract The Big Bang has been typically identified as the beginning of the Universe: $t = 0$ if you extrapolate back our expanding Universe to a state of arbitrarily high temperatures and densities. While you can naively do exactly that in an expanding Universe under the rules of General Relativity, those conditions lead to observable effects that run contrary to what we see. Instead, the Universe is better described by cutting off the matter-and-radiation dominated Universe at some early time, patching on an inflationary epoch where space is expanding exponentially and dominated by some sort of vacuum energy. This is not mere theory nearly 40 years on, but is supported by a vast suite of observable evidence. The case for this conclusion, even in the absence of B-mode polarization in the CMB, is laid out here.

9.1 Introduction

There is, perhaps, no greater question in all of cosmology than the one concerning the origin of the Universe. When Albert Einstein first put forth his General Theory of Relativity, it was quickly recognized that the only static solutions containing matter required an inordinate amount of fine tuning and were inherently unstable. One of the earliest sets of solutions discovered were of isotropic, homogeneous Universes filled either with a cosmological constant or with a mix of matter, radiation, and spatial curvature. In both cases, it was found that the Universe cannot be static, but, depending on the initial conditions, will either expand or contract over time.

The revelation that, unlike in Newtonian physics, Einstein's equations admitted an evolving Universe solution, meant that spacetime was not necessarily static over time. In fact, the Universe could be full of many different types of matter and energy and still expand or contract. This was a big deal, because not only was our Universe full of stars, but galaxies as well. Making use of Leavitt's Law, Edwin Hubble became the first to measure the distance to a galaxy other than our own, by discovering Cepheid variables in the great Andromeda nebula, M31. Leavitt's Law

E. R. Siegel (✉)
Lewis & Clark College, Portland, OR, USA

B. G. Sidharth et al. (eds.), *Fundamental Physics and Physics Education Research*,
https://doi.org/10.1007/978-3-030-52923-9_9

provides a period/luminosity relationship between Cepheid variable stars, allowing an inference of the intrinsic brightness—and hence the distance—to any such star observed. Since Hubble could measure the apparent brightness and then the period of these Cepheids, a distance could be calculated.

By continuing on to measure Cepheids in a slew of spiral nebulae, Hubble and his assistant, Humason, derived the distance to a large number of galaxies. Complementary to that data set was Vesto Slipher's work, which noted the large redshifts of the spectral lines in a myriad of spiral and elliptical nebulae, corresponding to large recession velocities. These two data sets were first combined by Georges Lemaître to derive a redshift-distance relation, and then more robustly and independently by Hubble himself. This relationship is now known as Hubble's Law and has an astounding implication in the context of Einstein's relativity: that the fabric of space itself is expanding over time. The reason light gets redshifted is because the space between the emitting source and us, the observer, is stretching.

The rate of expansion is determined, on the observational side, by the measured Hubble constant, while on the theoretical side, it's a combination of the Universe's initial conditions along with the ratio of the various energy components present in the Universe. Although there are many possible cosmologies that admit expanding space like this, one of the most straightforward was put forth by George Gamow in what would grow into the idea of the Big Bang. If space was expanding and stretching, that implied that any gravitationally bound objects within it, like individual, isolated galaxies, would be expanding away from one another. Similarly, any radiation within this space would be stretched as is traveled through the Universe, causing its wavelength to lengthen and its energy to decrease. As you moved farther and farther forward in time, the Universe would become more dilute, and the temperature of the Universe would become cooler.

But as you extrapolated in the opposite direction—backwards in time—you'd move towards a hotter, denser, more uniform state. If the Universe becomes cooler, sparser, and gravitationally clumpier as we move forward in time, then the converse would be true as we moved backwards. Since the volume of our three-dimensional space scales is a^3, where a is the scale factor of the Universe, and the energy of an individual photon scales is a^{-1}, this implies that the matter density will shrink as a^{-3}, while the radiation density will shrink as a^{-4}. If we define the scale factor today, a_0, to equal 1, then we can extrapolate back to earlier times where the Universe was denser, hotter, and more uniform, and we can do this in a quantitative fashion, noting the cosmological implications along the way.

The remainder of these proceedings is laid out as follows. Section 9.2 focuses on the classical steps that take place as we move through time in the standard hot Big Bang. Section 9.3 focuses on the puzzles that arise if we extrapolate back to arbitrarily high temperatures, energies, and densities in such a model. Section 9.4 presents the solution to these puzzles in the form of cosmological inflation and includes the novel, generic predictions that arise from inflation and the status of where, whether, and how well they've been measured and tested. Finally, Sect. 9.5 presents an in-depth discussion of the implications for what occurred in our Universe before the Big Bang.

9.2 The Big Bang

If we begin at the present day, we can measure a whole slew of properties of the Universe as it is right now. There are specific temperature fluctuations that exist in the cold (2.725 K) cosmic microwave background (Planck Collaboration et al. 2015a), there are patterns to the clustering of galaxies on the largest scales in the Universe (Alam et al. 2017), and there are measurements of the expansion rate of the Universe out to many billions of light years (Suzuki et al. 2012). Combined, this gives us a concordance picture of the Universe, teaching us that it's composed of a specific mix of dark energy (68%), dark matter (27%), baryons (4.9%), neutrinos (0.1%), and photons (~0.01%), where these and other various methods of inquiry all produce measurements which point to the same, consistent cosmic picture (Freese 2017).

We also, when we look at the Universe nearby, find galaxies that are—on average—intrinsically redder in color, with high levels of metallicity (heavy element content), that are quite tightly clustered together, and that display relatively large masses and an advanced, evolved morphology (Baillard et al. 2011). The Universe shouldn't have always been this way, however. Gravitational attraction is a runaway process, pulling matter into structures slowly but increasingly when overdense regions are only slightly greater than the mean density (Mészáros 1974), and growing non-linearly once a certain density threshold is reached (Peacock and Dodds 1994). Applying this to the Universe, where looking back to great distances equates to looking back large amounts in time, we can reconstruct how more distant galaxies ought to be different in the past. Specifically, they should appear less clustered, with less power on large scales. For individual galaxies, they should be intrinsically bluer in color, less evolved in morphology, smaller in size and mass, with lower metallicities, and with different luminosity functions for their constituent stellar populations (Kawamata et al. 2018).

This should progress more and more severely as we go back in time, culminating in pristine populations of gas which have never yet formed any stars. This prediction of the Big Bang dates back 70 years (Alpher et al. 1948) and was at last verified just a few years ago (Fumagalli et al. 2011). As we continue to go back in time, we should reach an epoch where there was so little starlight that the atoms in intergalactic space remained neutral, since there was no ultraviolet radiation to ionize them (Gunn and Peterson 1965). This light-blocking matter was detected in high-redshift quasars; the Gunn-Peterson trough is now a solid part of observational cosmology (Becker et al. 2001). Prior even to this, there must be a time where no stars at all had formed, and the Universe was entirely dark, save for the leftover radiation from even earlier times. These cosmic "dark ages" are consistent with the timescale for reionization measured in the optical depth of the cosmic microwave background (Bennett et al. 2013).

But we can go back even farther, to when the Universe was more uniform, less clumped, and even smaller in size. Any radiation that existed would have its wavelength severely compressed in comparison to what exists today, implying a much hotter temperature under these early conditions. At some sufficiently early time, the wavelength of the most energetic photons would be short enough to enable the spontaneous ionization of neutral atoms: the photoelectric effect of this ultraviolet radiation would have prevented the stable formation of bound states of electrons and nuclei. In the face of an ionized plasma, this radiation would experience rapid Thomson scattering off of the free electrons, while after sufficient stretching and cooling, neutral atoms will form and the radiation will free-stream, creating a leftover radiation bath—a primeval fireball—whose remnants should exist with a low temperature and blackbody spectrum today (Dicke et al. 1965). The detection of this radiation (Penzias and Wilson 1965) and the subsequent measurement of its spectral properties (Fixsen et al. 1996) have thoroughly verified this cornerstone of the Big Bang. We've even come so far as to measure the minuscule temperature fluctuations in the spectrum of this radiation (Planck Collaboration et al. 2015a), corresponding to the density fluctuations that gave rise to the large-scale structure of the Universe today in great detail (Alam et al. 2017).

Even before that, the wavelength of photons would have been so short that atomic nuclei themselves would have been blasted apart. High-energy radiation can dissociate protons and neutrons from one another, creating a sea of unbound particles. The first step towards binding protons and neutrons together towards the formation of heavier nuclei is to create deuterium: a heavy isotope of hydrogen with a binding energy of just 2.2 MeV. When there are sufficient densities of photons in excess of those energies, no nucleosynthesis of heavier elements can proceed; the "deuterium bottleneck" delays those interactions due to the fragility of the products of this first, necessary step (Peacock 1999). Only when photons cool below this critical threshold can the synthesis of the first elements proceed. In this framework of the Big Bang, the only parameter that determines the relative element abundances is the baryon-to-photon ratio (Steigman 2006). The direct measurement of that ratio (Bennett et al. 2013) aligns spectacularly with the observations of the abundances of helium-4, helium-3, deuterium, and lithium-7, many of which are seen in distant quasar absorption lines (Riemer-Sørensen and Jennsen 2017).

So far, all of these predictions of the Big Bang have been spectacularly verified, but we can continue to extrapolate even farther back, to early times and high temperatures where no verifiable signatures remain. Before nucleosynthesis, particles should have high enough energies to spontaneously produce matter-antimatter pairs. Electrons and positrons remain the longest, but at even earlier times, any-and-all standard model particles and antiparticles are produced in great abundance, as defined by their fermionic or bosonic statistics. After the Universe cools and the particle/antiparticle pairs annihilate away, a tiny fraction of leftover matter remains: this baryon asymmetry is one of the greatest unsolved mysteries in all of physics.

When temperatures are high enough and the density of the Universe exceeds certain limits, protons, neutrons, and any type of baryon cease to exist, as they become replaced with a quark-gluon plasma. At even higher temperatures, exotic physics is expected to exist, particularly as we approach the theorized grand unification (GUT) scales. Particles mediating proton decay may appear; leptoquarks may exist; the couplings of the fundamental forces may run, and possibly run together; a slew of relic particles may be created at this time, including heavy Dirac neutrinos, potential dark matter candidates, and t'Hooft-Polyakov monopoles (t'Hooft 1974; Polakov 1974) may all exist under these extraordinary conditions. At the limit of our extrapolations, we exceed the Planck energy, achieve infinite (or arbitrarily high) densities, and run into the limits of our fundamental physics theories. We achieve a singularity, from where space and time themselves are believed to emerge.

9.3 Consequences

A singular beginning to the Universe, and its steady evolution from a set of hot, dense, uniform, expanding initial conditions, has tremendous implications for what we'd observe today. If the Universe began with arbitrarily high temperatures and energies, it would immediately create all the particles, antiparticles, and quanta of radiation that the quantum laws of nature—and energy constraints given by Einstein's $E = mc^2$—admitted. Beginning from the birth of the Universe, at a time $t = 0$, we would only be able to evolve the Universe forward, even in our imaginations. The first day in the Universe after the Big Bang occurred, as cosmologists sometimes put it, would be "a day without a yesterday."

After the first major verification of the predictions of the Big Bang came in, with the discovery of the cosmic microwave background radiation, the extrapolation back to a singularity seemed an inevitability. The singular beginning of the Universe would lead to a hot, dense state full of matter and energy, which would then evolve through the early stages in the Universe as outlined in Sect. 9.2. It's true that many of the stages we've passed through in the aftermath of the Big Bang have extraordinary amounts of evidence supporting them. But extrapolating back to arbitrarily high temperatures and energies not only lacks that evidence and remained a speculative extrapolation, but brought with it a significant number of unexplained puzzles (Siegel 2015) that the assumption of an initial singularity provided no answers for.

The first of these is the *horizon problem*. Even in an expanding Universe, there's a physical limit to how far, in a given amount of time, information (and temperature) can be exchanged between two disparate locations. Since the discovery of the cosmic microwave background, we've seen that the Universe has the same temperature in all directions in space, yet there's no physical mechanism that would have caused this to be the case. In the absence of such a cause, this simply needs to be an initial condition that the Universe was born with in order for it to be the case in the standard Big Bang, if it arises from a singular beginning. Without such conditions, we would expect the temperature (and density) to vary in different directions by roughly the order of the temperature itself.

The second is the related *isotropy problem*. When we look out at the Universe in all directions, today, we see the same types of large-scale structure everywhere. The types, sizes, populations, and clustering patterns of galaxies are direction-independent, illustrating the tremendous isotropy of the Universe. This also implies that the seed fluctuations from which today's large-scale structure arose were the same in all directions. But why should this be the case? Unless there were a mechanism in place to create these seed fluctuations, and unless that mechanism created fluctuations with the same properties in causally disconnected locations, this is simply another initial condition that arises without motivation.

The third is the *monopole problem*. In practically all grand unified theories, as well as many other extensions to the standard model, as the Universe cools through certain temperatures and conditions, it undergoes a phase transition. These transitions can create ultra-heavy relic particles, such as magnetic monopoles. In most scenarios of beyond-the-standard-model physics, these relics should not only be created, but they ought to be stable, persisting to the present day. Motivated by these calculations, there were a series of dedicated searches for isolated magnetic monopoles traveling through the Universe. Although there was a famously controversial detection of one monopole candidate (Cabrera 1982), follow-up experiments failed to confirm the existence of these particles. Their absence is a mystery that the hot Big Bang, if it were to have reached arbitrarily high temperatures and energies, has no explanation for.

The fourth puzzle is the *flatness problem*. Today, via a number of methods, we can measure the spatial curvature of the Universe. Particularly by looking at the patterns of fluctuations in the cosmic microwave background radiation (Planck Collaboration 2015a), we have determined that the Universe is spatially flat to approximately 1%. But flatness, rather than positive or negative curvature, which would imply a closed (sphere-like) or open (saddle-like) Universe, tells us that the balance between the expansion rate and the amount of matter and energy in the Universe must have been tremendously precise. If we extrapolate back to a time where the Universe was filled with a quark-gluon plasma, the energy density and expansion rate must have been balanced to better than one part in 10^{25}. If the Universe were just the tiniest bit denser, it would have already recollapsed; if it were the tiniest bit less dense, it would be more than double its present size. Without a mechanism to explain why these independent quantities are balanced, this too must be an initial condition that the Universe is simply born with.

Finally, the fifth puzzle can be called the *small fluctuation problem*. There are tiny imperfections in temperature (and therefore, in density) in the cosmic microwave background: on the order of $\pm 0.003\%$, compared to the mean value. This is an independent problem from the horizon problem, which offers no explanation for why the mean temperature value is the same everywhere. Even if you allow for that, there's no mechanism to generate temperature fluctuations that are not only the same magnitude everywhere, but merely one-part-in-30,000 times the average value. Why would these temperature fluctuations be so small, given that other out-of-equilibrium thermal systems where causal contact between disparate regions is disallowed display fluctuations that are thousands of times greater than what we seen in the cosmic microwave background? Again, this must be an initial condition that's added without any driving motivation, other than it's what we see when we examine the Universe in detail.

In theory, there's no reason why these puzzles couldn't simply be resolved by putting in the initial conditions we require. But if we solve these puzzles in that fashion, we give up on the very idea that physics is important for our Universe! We seek to apply laws, rules, and dynamical solutions to the phenomena we observe and measure; that's the key to scientifically examining and understanding the world and Universe around us. Given that these puzzles all suffer from the same problem—a set of initial conditions must be put in by an ad hoc method—it makes sense to attempt to concoct a physically well-motivated explanation for these initial conditions. Ideally, whatever scenario you cook up will not only explain these puzzles, but will make novel, additional predictions that can then be sought out and put to the test.

Historically, every new scientific idea has had to meet three criteria in order to be universally accepted as superior to the previously prevailing idea:

- It must reproduce the full suite of successes of the prevailing theory, meaning any replacement for the hot Big Bang with a singular beginning must successfully give the light element abundances, the cosmic microwave background, the large-scale structure of the Universe, and match the observed Hubble expansion.
- It must succeed where the prevailing theory has failed, which means it needs to provide an explanation for these five puzzles that the Big Bang cannot provide on its own.
- And it must make new, testable predictions for phenomena and properties that can be measured within this Universe. These predictions must be fundamentally, veritably, and quantifiably different from the predictions made by the alternative theory it's seeking to supplant.

If a scenario could be concocted that would satisfy all of those conditions, perhaps it could either replace or augment the hot Big Bang scenario, as it was originally formulated, to better explain the cosmic origins of our Universe.

9.4 The Inflationary Universe

By the late 1970s, many physicists and cosmologists were thinking about these problems, with a special emphasis on the horizon and flatness problems, owing to the robustness of the data indicating a uniform temperature and spatial flatness to the Universe as a whole. On December 7, 1979, a young postdoc named Alan Guth wrote the following in his notebook:

9.4.1 Spectacular Realization

This kind of supercooling can explain why the universe today is so incredibly flat—and therefore resolve the fine-tuning paradox pointed out by Bob Dicke in his Einstein day lectures.

This spectacular realization would very quickly lead to one of the most important papers in the history of modern cosmology, which detailed a new scenario for the Universe's origin: cosmic inflation (Guth 1981).

Unlike the standard hot Big Bang, where arbitrarily high temperatures, energies, and densities are achieved at a time $t = 0$ when you extrapolate backwards from today, cosmic inflation provided a cutoff to how hot, dense, and energetic the Universe was in the earliest stages. Rather than continuing back to a singularity, inflation postulated that there was a time period prior to the hot, dense, expanding state that the Big Bang describes. In the inflationary scenario, this prior phase contained no matter, antimatter, or radiation, but rather the Universe existed in a de Sitter-like state, dominated by energy in the inflaton field, which behaves similarly to vacuum energy or a cosmological constant.

During this inflationary phase, space expands exponentially, obeying the relation that $a(t) = a_0\, e^{Ht}$, where the value of H is determined by the energy scale of inflation. This exponential expansion doesn't necessarily need to occur at a rapid rate, but the exponential property is special because of how relentless it is. For inflation occurring at the hypothesized GUT scale, for example, it can turn a Planck-scale region ($\sim 10^{-35}$ m) into the size of the observable Universe today ($\sim 10^{27}$ m) in a timespan that's merely 10^{-33} s in duration.

This type of exponential expansion has a number of physical effects that are vital for both setting up the initial conditions of the Big Bang and ameliorating a number of cosmic puzzles. The inflationary phase, for example, stretches the Universe flat, by taking whatever amount and type of spatial curvature that may have pre-existed and expanding it so thoroughly that, to an observer that can only see the extent of the Universe that's accessible to us, it's indistinguishable from flat. It takes whatever properties existed in a tiny, causally connected region of space, and stretches that region across a volume far bigger than that of our observable Universe, giving it the same properties everywhere. Given that whatever field causes inflation is likely to be quantum in nature, the field ought to exhibit quantum fluctuations at all times; hence, these fluctuations should be stretched by the expansion of space to exist, with practically the same magnitude and spectrum, across all scales and in all locations in space.

Finally, when inflation ends, the energy that's inherent to the inflaton field—and hence, that's behaving as vacuum energy during the inflationary phase—gets converted into matter, antimatter, and radiation (Linde 1982; Albrecht and Steinhardt 1982). Inflation goes on so long as the field that drives it remains out of the ground state (such as at the top of a potential "hill"), but comes to an end when it reaches the lowest-energy, most stable state (such as rolling down into a potential "valley"). The temperature that the Universe reaches, post-inflation, is bounded from above by the energy scale of inflation, and may be lower depending on the particulars of the reheating process. In the aftermath of this conversion of energy, from the inflaton field to the matter, antimatter, and radiation permeating all of space, the conditions at the end of inflation are imprinted on the forms of energy that fill the Universe as it enters the familiar hot, dense, expanding-and-cooling state. Finally, given the size of our observable Universe, it's only the last few dozen *e*-foldings of inflation that have any observable impact on our Universe today.

Inflation, therefore, can reproduce the successes of the hot Big Bang by giving you the exact state you're seeking, beginning well above the earliest temperatures and energies we're currently capable of probing. In addition to that, cosmic inflation solves all five of the Big Bang's puzzles that we mentioned in Sect. 9.3.

- The horizon problem, that the Universe is the same temperature in causally disconnected regions, is explained by those different regions having been in causal contact in the past. Inflation stretched those regions apart to much larger distant scales, but they originated from the same Planck-scale volume, where they had time to achieve the same properties, like temperature and density.
- The isotropy problem, that the Universe has the same large-scale structure properties in all directions, is solved because the seed overdensities and underdensities that inflation creates, due to the quantum fluctuations that get stretched across the Universe, were created by the same process everywhere within our causal horizon.
- The monopole problem, that the Universe should be filled with these relic particles from a high-energy epoch in the distant past, is solved by a combination of two factors. First, any monopoles that existed during the inflationary phase are inflated away by the exponential expansion of space, meaning that there ought to be, at most, one magnetic monopole in the observable Universe. Second, so long as the maximum temperature we achieve after the end of inflation is below the scale that produces new monopoles, we wouldn't expect our Universe to contain any.
- The flatness problem, which notes that the Universe appears spatially flat, is now explained because no matter what conditions the Universe begins with, inflation will stretch it so that it appears indistinguishable from flat to an observer that can only see tens of billions of light years in any direction.
- And the small fluctuation problem is resolved, so long as inflation occurs well below (by a factor of ~10^3 or so) the Planck scale.

With these resolutions in place, we can see where inflation has succeeded in the realm where the original formulation of the hot Big Bang could not.

In addition, inflationary theories, including models of new inflation and chaotic inflation (Linde 1983), make six generic predictions that we can look to the Universe for confirmation, validation, or falsification. Inflation predicts a nearly-perfectly scale-invariant set of density fluctuations, parametrized by the scalar spectral index, n_s. Inflation predicts that n_s should be close to but not quite equal to 1; we measure it (Planck Collaboration 2015b) to be approximately 0.97. Inflation also predicts that there should be curvature fluctuations in the Universe, creating a spatial curvature that ought to be somewhere between $10^{-4} \geq \Omega_k \geq 10^{-6}$, consistent with our best constraints that $\Omega_k < 0.003$ (Alam et al. 2017).

In principle, there are two types of density fluctuations that could have existed and imprinted themselves on the Universe: adiabatic and isocurvature fluctuations. Inflation predicts that the fluctuations in our Universe should be 100% adiabatic, at least for all single-field models, both new and chaotic (Linde 1983). Thanks to the results from WMAP, that's exactly what we've observed (Bennett et al. 2013). Inflation also, owing to its stretching of quantum fluctuations across extremely large scales, predicts the existence of superhorizon fluctuations. These, theoretically, would be visible in the TE cross-correlation (temperature/polarization) spectrum of the microwave background, a prediction that has since been verified (Planck Collaboration 2015a). Inflation also predicts that there should be an upper limit to the maximum reheat temperature achieved at the hot Big Bang that follows the end of inflation; from the fluctuations in the cosmic microwave background, we find it's more than 100 times lower than the Planck scale (Bassett et al. 2005).

Finally, inflation also predicts a generic spectrum of tensor fluctuations—gravitational wave fluctuations—that ought to imprint themselves on the B-mode polarization signal in the cosmic microwave background. Although the spectrum is model-independent in inflationary spacetimes, the magnitude and amplitude of the signal is highly model-dependent (de Bernardis et al. 2009). Despite the purported and incorrect claims of detection by the BICEP2 collaboration a few years ago (BICEP2 Collaboration 2014), this remains a promising and active area of research that could further validate or challenge inflation.

9.5 Discussion

The Big Bang remains one of the greatest theoretical frameworks and achievements of modern cosmology. Its four cornerstone predictions, of the abundances of the light elements from Big Bang nucleosynthesis, of the existence and spectrum of the cosmic microwave background, of the Universe's expansion rate and its evolution over time, and the formation of large-scale structure, have been validated and verified in glorious detail. However, the inference that you can extrapolate the expanding, cooling Universe all the way back in time to $t = 0$ and the existence of a singularity is problematic at best, leading to the assumption of a slew of ill-motivated initial conditions.

The idea of cosmic inflation, however, removes the singular beginning to the Big Bang and instead replaces the earliest stages with an exponentially expanding phase to the early Universe. During this phase, the energy in the Universe exists in the form of the inflaton field, which drives the de Sitter-like behavior of the Universe during this time. Inflation is not only compatible with the four cornerstones of the Big Bang, but it dynamically sets up the initial conditions required to make the Big Bang compatible with our observations. In addition, cosmic inflation makes a number of generic predictions that are relatively model-independent; of the six presented in Sect. 9.4, four have been verified and two are consistent with inflation's predictions to the limits of our best observations.

Inflation, therefore, occurs before the Big Bang, setting up the initial conditions for what we see in our observable Universe during its final moments. It is only the final $\sim 10^{-33}$ s of inflation that imprint itself on what we can observe; although there are many fascinating predictions that arise from inflation, we have no way of knowing how long inflation lasted, how it arose, or whether it was eternal to the past or had, at some point, a singular beginning of its own. Although there is a theorem that demonstrates inflating spacetimes are past-timelike-incomplete (Borde et al. 2001), this theorem only demonstrates that particle trajectories separated by a finite distance in an inflating spacetime will have a common point-of-origin in the past. This does not necessitate the existence of a singularity, however, as numerous scenarios that avoid a singularity in inflating spacetimes have been found.

One remarkable consequence of inflation, even though it does not lead to any observables, is its prediction of the existence of a multiverse. The inflaton field, while slowly rolling towards the most stable, equilibrium field value, will spread out, being quantum in nature. If the rate of slow-roll is slower than the rate of field-spreading, which it is in practically all models that provide a sufficient amount of inflation, then there will inevitably exist regions of space where inflation continues indefinitely. No matter how many regions see inflation come to an end and give rise to a hot Big Bang, there will be inflating space between these regions to drive them apart from one another, continuing to generate new, inflating space, and—over time—to give rise to new, hot Big Bangs that will forever remain causally separated from the one that created our observable Universe. The generic phenomenon of eternal inflation (Vilenkin 1983) demands that once inflation begins, there are regions of it that continue inflating forever into the future. These regions of spacetime will continuously grow, exit inflation and reheat, creating an ever-increasing number of Universes similar to our own. This is why we expect our Universe to exist as part of a multiverse.

Back when it was first formulated, it was seen as an inevitability that the Big Bang was something you could extrapolate backwards to arbitrary times, energies, and temperatures, eventually reaching a singularity. Although the Big Bang can be accurately described as a hot, dense, expanding initial state, the Big Bang itself can no longer be said to imply a singularity. There may be a singularity at some early stage, but it is not a certainty, and if it does exist, it occurs before cosmic inflation does, which itself precedes and sets up the conditions for the Big Bang. Inflation makes many other predictions that are unique to it, and a great many of those predictions have been observationally confirmed. In addition, it predicts the existence of a multiverse, but sadly, the multiverse, as well as any properties of inflation that existed prior to the final 10^{-33} s (or so) of inflation, have no observable impact on our Universe. The Big Bang still represents an important moment in our cosmic history, but it is no longer the beginning of it all. Our matter-and-radiation-filled Universe started with a bang, but cosmic inflation takes us farther back, and does so more accurately, than the Big Bang on its own ever could.

Acknowledgement E. R. Siegel thanks B. G. Sidharth and FFP15 for support and direction; J. N. Fry for the education, training, and skills that will stay with me forever; my Patreon supporters for helping make this trip possible; my wife for all that she is; Will Kinney for helpful advice on the measurement technique for superhorizon fluctuations; and all the attendees of this program for enlightening discussions.

References

Alam S et al (2017) The clustering of galaxies in the completed SDSS-III Baryon Oscillation Spectroscopic Survey: cosmological analysis of the DR12 galaxy sample. Mon Not R Astron Soc 470(3):2617–2652

Albrecht A, Steinhardt PJ (1982) Cosmology for grand unified theories with radiatively induced symmetry breaking. Phys Rev Lett 48:1220–1223

Alpher RA, Bethe H, Gamow G (1948) The origin of chemical elements. Phys Ther Rev 73 (7):803–804

Baillard A et al (2011) The EFIGI catalogue of 4458 nearby galaxies with detailed morphology. Astron Astrophys 532:A74. 27pp

Bassett BA, Tsujikawa S, Wands D (2005) Inflation dynamics and reheating. Rev Mod Phys 78 (2):537–589

Becker RH et al (2001) Evidence for reionization at $z \sim 6$: detection of a Gunn-Peterson trough in a $z{=}6.28$ quasar. Astron J 122(6):2850–2857

Bennett CL et al (2013) Nine-year Wilkinson microwave anisotropy probe (WMAP) observations: final maps and results. Astrophys J Suppl 208(2):1–54. 54pp

BICEP2 Collaboration (2014) Detection of B-mode polarization at degree angular scales by BICEP2. Phys Rev Lett 112(24):241101

Borde A, Guth AH, Vilenkin A (2001) Inflationary spacetimes are incomplete in past directions. Phys Rev Lett 90(15):151301

Cabrera B (1982) First results from a superconductive detector for moving magnetic monopoles. Phys Rev Lett 48(20):1378–1381

de Bernardis P et al (2009) B-Pol: detecting primordial gravitational waves generated during inflation. Exp Astron 23(1):5–16

Dicke RH, Peebles PJE, Roll PJ, Wilkinson DT (1965) Cosmic black-body radiation. Astrophys J Lett 142:414–419

Fixsen DJ et al (1996) The cosmic microwave background spectrum from the full COBE FIRAS data set. Astrophys J 473:576–587

Freese K (2017) Status of dark matter in the universe. Int J Mod Phys D 26(6):31pp

Fumagalli M, O'Meara JM, Prochaska JX (2011) Detection of pristine gas two billion years after the big bang. Science 334(6060):1245–1249

Gunn JE, Peterson BA (1965) On the density of neutral hydrogen in intergalactic space. Astrophys J 122:1633–1641

Guth AH (1981) Inflationary universe: a possible solution to the horizon and flatness problems. Phys Rev D 23(2):347–356

Kawamata R et al (2018) Size⍰luminosity relations and UV luminosity functions at z = 6⍰9 simultaneously derived from the complete Hubble frontier fields data. Astrophys J 855(1):4, 47pp

Linde AD (1982) A new inflationary universe scenario: a possible solution of the horizon, flatness, homogeneity, isotropy and primordial monopole problems. Phys Lett B 108(6):389–393

Linde AD (1983) Chaotic inflation. Phys Lett B 129(3–4):177–181

Mészáros P (1974) The behaviour of point masses in an expanding cosmological substratum. Astron Astrophys 37(2):225–228

Peacock JA (1999) Cosmological physics. Cambridge University Press, Cambridge

Peacock JA, Dodds SJ (1994) Reconstructing the linear power spectrum of cosmological mass fluctuations. Mon Not R Astron Soc 267(4):1020–1034

Penzias AA, Wilson RW (1965) A measurement of excess antenna temperature at 4080 mc/s. Astrophys J Lett 142:419–421

Planck Collaboration et al (2015a) Planck 2015 results. XIII Cosmological parameters. Astron Astrophys 594(A13):63pp

Planck Collaboration et al (2015b) Planck 2015 results. XX Constraints on inflation. Astron Astrophys 594(A20):65pp

Polakov AM (1974) Particle spectrum in the quantum field theory. JETP Lett 20(6):194–195
Riemer-Sørensen S, Jennsen ES (2017) Nucleosynthesis predictions and high-precision deuterium measurements. Universe 3(4):44
Siegel ER (2015) Beyond the galaxy: how humanity looked beyond the milky way and discovered the entire universe. World Scientific, Singapore
Steigman G (2006) Primordial nucleosynthesis: successes and challenges. Int J Mod Phys E 15 (01):1–35
Suzuki N et al (2012) The Hubble space telescope cluster supernova survey. V. Improving the dark-energy constraints above z > 1 and building an early-type-hosted supernova sample. Astrophys J 746(1):85
t'Hooft G (1974) Magnetic monopoles in unified gauge theories. Nuclear Phys B 79(2):276–284
Vilenkin A (1983) Birth of inflationary universes. Phys Rev D 27(12):2848–2855

Part II
Physics Education Research

Chapter 10
Innovation of Curriculum and Frontiers of Fundamental Physics in Secondary School: Research-Based Proposals

Marisa Michelini

Abstract Conceptual knots in classical physics are often cited as motivation for the exclusion of modern physics in secondary school, but innovation in secondary school physics curriculum is a need for many aspects. The new ways of learning require new methods and active learning able to produce competences and not only knowledge; the laboratory work has to be integrated in every day school activities, as well as the contribution of ICT in the teaching/learning process; the new topics related to the physics of the last century is the main content need. Research is an urgent need in finding ways for integrating modern physics in vertical perspective in the curriculum, by means of new active methods to avoid the reductionist attitude to add not more than notions on special relativity and quantum physics as final topic in the curriculum in informative or narrative way. Modern physics in secondary school is a challenge which involves the possibility to transfer to the future generations a culture in which physics is an integrated part, not a marginal one, involving curricula innovation in a way that allows the students to manage them in moments of organized analysis, in everyday life, in social decisions. In the theoretical framework of the Model of Educational Reconstruction, we developed research-based educational proposals focused on contributions to the practice, developing coherent learning paths in vertical perspective. Design Based Research integrated with research and developments of new tools and studies on conceptual change are carried out to produce learning progression and finding ways to offer opportunities for understanding and experiencing what physics is, what it deals with, and how it works in operative way. Empirical data analysis of student reasoning in intervention modules supports proposed strategies. In this chapter, we present our research methods, we list the main path proposals prepared and implemented, and we describe in detail two of them: cross section and phenomenological approach to the superconductivity.

M. Michelini (✉)
Physics Education Research Unit, DMIF, University of Udine, Udine, Italy
e-mail: marisa.michelini@uniud.it

B. G. Sidharth et al. (eds.), *Fundamental Physics and Physics Education Research*,
https://doi.org/10.1007/978-3-030-52923-9_10

10.1 Introduction

Innovation in Physics curricula in Secondary School is a need for many different reasons and aspects: (1) methodological aspects and strategies related to the need of student's active learning promotion (McDermott 1993, 2008; Başer 2006; Mazur 2009; Deslauriers et al. 2011; Maloney 2011; Pizzolato et al. 2014), (2) multimedia role and relative contribution in learning process: it can be a simple support in teaching/learning in many ways integrated in school activities as learning tools (Wagner et al. 2006; Hennessy et al. 2006; Debowska et al. 2013), (3) experimental explorations or lab experiments to be integrated in an active learning process (Sokolowska and Michelini 2018), as well as (4) work activities integrated in the curriculum (Buongiorno et al. 2017) and (5) new contents to update curriculum contents (Constantinou and Papadouris 2008; Pavlin et al. 2013; Michelini et al. 2014a, b, c, d, e, f).

In the last 30 years, one of the main discussed content area is those of Modern Physics (Aubrecht 1989; Gil and Solbes 1993; Hake 2000; Ostermann and Moreira 2000; Silva 2015; Michelini et al. 2017). Modern Physics is now a chapter in the secondary school textbooks of many countries in the world, but it appears often as a final topic, added to the usual contents of a physics manual. Related contents are often organized in a traditional way, treated as information on the main ideas and research results eventually accompanied by a story telling of the main steps in the discoveries. The main topics treated are special relativity and physics of quanta story.

The different research areas aren't clearly identified.

The frontiers of fundamental physics topics are proposed in informative way as structured notions, in a static framework, without an in-depth discussion of the new concepts and of the problematic issue. There is a lack of discussion of the crucial aspects, of the formal aspects, of the motivated theoretical choices, and of evidence-based results.

Material science physics, for example, find some role in the proposed interpretative models of electrical and thermal conduction, but updated research topics, research questions, methods, and characteristics of the research do not appear.

High energy physics, when not omitted, is treated in even more reductionist and informative way. Cosmology is sometimes introduced in narrative way.

Additional available resources for teachers are the materials prepared for outreach or the presentation of research projects for popularization goals, which do not contain proper examples for educational approaches. For not expert in the specific field is very difficult to adapt this material in formative way for secondary students, so the episodic informative style remains the prevalent employ.

Physics Education Research literature offers studies on specific topics for what concern learning difficulties and new approaches. Special thematic issues on the main global topics under discussion as quantum physics and special relativity are published by some Journal (i.e., Am J Phys 2002, Special Issues 70(3) and Phys Educ 2000, Special Issues 35(6)). Each contribution is focused on a specific approach, but the integration of modern physics in the curriculum within a vertical perspective, selecting relevant topics and adopting conceptual based approaches is up to now an open question in physics education research.

To gain the needed appropriation for an active learning approach, a particular research-oriented study is required. The analysis of conceptual referents and learning difficulties in each specific approach offered in literature and the comparison of approaches in planning the teaching proposal, together with the selection and preparation of learning materials remain a preliminary step for the educational employ. Teachers don't have the resources and the time for that work. There is a need of research-based pilot implementation of educational proposals in the perspective of their integration in school curriculum for instructional scope and in addition support have to be guarantee to teachers.

We worked in that direction and we studied research-based proposals for new topics integration in secondary school curriculum looking in global perspective for a coherent framework and vertical paths in curriculum development and we start analyzing disciplinary, social, and educational relevance of topical areas in the different fields of updated research activities, looking to the relative educational power in gaining physics identity. We call here physics identity those conceptual competence consisting in the gained approach by students to the problems in phenomena interpretation, producing the attitude to identify the needed knowledge, instruments, and methods with the awareness of the relative power and limits for finding solutions and discuss evidence-based results by means of the treatment of formal aspects in a coherent way using basic math competences.

For this scope we selected topics and we studied vertical proposal in the research framework of the Model of Educational Reconstruction—MER (Duit et al. 2005). We carried out Design Based Research (DBR) (Anderson and Shattuck 2012) intervention modules to calibrate the proposal and to prepare educational materials useful for practice. The dimensions of such kind of research are many and include the study of relevance and of the educational reconstruction of each topic in accordance with MER, curricular studies (Constantinou and Papadouris 2008), comparative research for curricular aspects (Sokolowska et al. 2014), and empirical research on learning processes (Erickson 1998). Starting from learning difficulties and students' reasoning we study educational paths and individual strategies in relationship with conceptual stimuli (Vosniadou 2013), the role of different representations and of personal/cooperative work on specific aspects as ideal experiments. Research and development in producing new multimedia tools and experiments accompanied our work (Michelini 2010). We developed up to now a series of proposals that we do not consider exhaustive.

After a discussion of the adopted approach for this goals, we list here the proposals developed up to now, that we don't consider exhaustive, but examples of coherent paths in vertical perspective for modern physics in school curriculum. We leave the discussion of learning processes activated in classroom intervention modules and relative outcomes to the specific papers published during the research-based implementation of the paths development (Michelini et al. 2017, 2008; Greczyło et al. 2010; Vercellati 2010; Michelini and Stefanel 2008, 2010). Here we describe in detail two of them: cross section and phenomenologic approach to superconductivity.

10.2 The Research Approach in Building Modern Physics Proposals

Our perspective for modern physics is in the framework of content research (Meheut and Psillos 2004), oriented in building research-based proposals in a cultural perspective focusing on foundation of basic concepts as well as methods and applications in physics research, offering experience of what modern physics is in active research. Vertical paths are identified as learning corridor (diSessa 2004; Meheut and Psillos 2004; Michelini 2010) for individual learning trajectories and steps by steps concept appropriation modalities (Fedele 2005; Bradamante 2006; Vercellati and Michelini 2014).

The first step in our research approaches is to rethink the scientific contents as a problematic issue for the educational reconstruction, according to the Model of Educational Reconstruction (MER) (Duit et al. 2005). This task is integrated with empirical (Erickson 1998) research on student reasoning and research-based Teaching/Learning (T/L) intervention modules to individuate approaches able to favorite the building of scientific way of thinking on the spontaneous vision of students in phenomena description and interpretation. Cicles of Design Based Research (DBR) (Anderson and Shattuck 2012) intervention modules accompanied with action research in a collaborative dialectic between school and university produce the proposals for school activities, to contribute to classroom practice. The process includes empirical data analysis of each intervention module (Erickson 1998), which extension is growing with the path building. Data are collected in different ways. The main instruments are tutorials characterized by stimuli questions posed to promote attention in key questions; in particular students have to explain, interpret, or suggest further exploration on specific case studies or situations. Test-in and test-out on the way of interpret phenomena produce a global vision on the change in the reasoning of students as a community and for what concern individual profiles. Of course, it is often necessary to complete the data by means of semi-structured or Rogersians' interviews. In data analysis, attention is paid to identify strategic angles and critical details (Viennot 1996) used by common knowledge to interpret phenomena (Michelini 2010) to study spontaneous reasoning paths, to find new approaches to physics knowledge (Vosniadou 2013).

The foundation of basic concepts in the context of the different physics research fields is the referent goal in selecting the contents. The cultural perspective in offering experience of what modern physics is in active research orient the selection of topics. We introduce the foundation of theoretical physics by means of Dirac approach to quantum mechanics. We privilege the new Einstein relativistic dynamic with respect to the Lorenz transformations. We focus on the foundation of new theories as quantum mechanics and on the discussion of the change in basic concepts ideas as those of state, measure, mass, energy, cross section, as well as research methods in active research and on the adopted relative basic principles applied, as energy and momentum conservation in optical and Rutherford backscattering spectroscopy (RBS).

For the advanced topics, interpreted by high formalized difficult theories, as electrical transport properties and superconductivity, we chose phenomena exploration approaches to offer methodological education by means of active learning strategies and lab work.

The basic idea is to avoid reductionism to offer opportunities of learning and not only understanding of information, to gain competences of instruments and methods, to build interpretative solutions and results to become able to manage fundamental concepts.

Modern physics is proposed along the whole curriculum, integrated in physics curriculum and not as a final appendix. Dynamic of Einstein relativity is proposed with Newtonian dynamic to discuss the work energy theorem on the light of Bertozzi experiment (Bertozzi experiment n.d., MIT). RBS and cross section concepts are proposed when collisions are treated in the curriculum. The discussion of the polarization as quantum property of light to introduce basic concepts of quantum mechanics is proposed when optical physics is treated, preferably not after electromagnetism, to avoid the automatic assumption that light is an electromagnetic wave. Optical spectroscopy becomes an extension of optics in the same context, looking on the meaning of the colors of light and the processes in light emission. A deep discussion of the concepts of mass and energy and a reflection on the meaning of state of a system as well as those of measure starts with the first steps in physics curriculum and remains a warm attention for the whole curriculum to promote a comparison of that concepts in different theories.

The main coherent vertical planned and implemented paths in school are the following: (1) Phenomena bridging theories: diffraction (Michelini et al. 2014a, b, c, d, e, f); (2) Optical spectroscopy and physics of quanta foundation (Buongiorno and Michelini 2018); (3) The physics in modern research analysis technics: RBS, TRR, R&H (Fera et al. 2014; Corni and Michelini 2018; Michelini et al. 2017); (4) Explorative approach to superconductivity (a coherent path) (Michelini et al. 2014a, b, c, d, e, f); (5) Discussion of some crucial/transversal concepts both in classical and quantum physics: state, measure, cross section (Michelini et al. 2017); (6) Foundation of theoretical thinking: quantum mechanics according to a Dirac approach (Ghirardi et al. 1996; Michelini and Stefanel 2008; Michelini et al. 2014a, b, c, d, e, f); (7) Mass-Energy discussion for a path to understand $E = mc^2$ (Michelini et al. 2014a, b, c, d, e, f).

In the following, we describe in detail two paths very different: those on cross section and the explorative approach to superconductivity.

10.3 The Cross Section Educational Proposal

Cross section is an important concept in classical physics, it is central in the study of the interactions between elementary particles in nuclear physics at both low and high energies in atomic collisions and becomes essential in quantum mechanics where it is not possible to define the trajectory of a particle but only the probability of finding it in a certain space. We start in preparing this approach long time ago (Corni et al. 1996), starting from one-dimensional collisions and the basic elements about plane collisions for rigid bodies, using the conservation principles in classical physics to analyze the different cases and showing at the same time which associations can be derived from it.

We suggest to begin and go over a geometrical interpretation of the cross section, which is consistent only with the classical case of rigid spheres, to introduce its probabilistic meaning highlighting general aspects and showing how to relate measurements for interpretations, independently from the traditional force/equation model of motion/trajectory scheme.

In the elastic collision of two rigid spheres (1 and 2), with respect to the center of mass reference frame, in the case of infinite rigidity normal collision of identical spheres, we derive that the rebound of the two spheres is symmetrical with respect to the line through their centers. The interaction during the collision is characterized by the impulse vector $\boldsymbol{I}$ acting between the two spheres (i.e., from 2 to 1) and the comparison between the initial and final states of the system gives us all possible information about what happened during the collision. The scattering angle, θ, is connected to the distance b between the lines on which the centers of the two spheres were moving before the collision (impact parameter), as derived from the following equation: $b = (R_1 + R_2) \cos(\theta/2)$. In more complex cases, this straightforward procedure fails. The dependence of the scattering angle θ from the impact parameter b becomes more difficult, when we replace sphere 2 with an irregular shaped object. Another cause of unpredictability is the finite precision with which the impact parameter is determined. Final states obtained with many collisions changing initial conditions can be significantly different.

To understand how the scattering angle can offer information about the collision or the geometry of the two bodies, we discuss the asymmetry of results of the experiment of Fig. 10.1, where data on the scattering angles distributed over the range $(-160°, 160°)$ are collected for a certain number of impacts (400 events). This relation between *distribution asymmetry* and *asymmetrical shape* gives a qualitative example of how to extract the basic properties of the collision phenomenon from the scattering angle distribution. In general, a complete characterization of the collision requires the knowledge of the probability P_i for any given measurement outcome S_i. For N_{tot} observations carried out, these probabilities P_i can be obtained from the averages $P_i = \langle N_i/N_{tot} \rangle$, where N_i is the number of outcomes S_i.

In this way, it is clear that the probability, which is the element which is more related to the interaction, depends on the details of the measuring procedure. As in

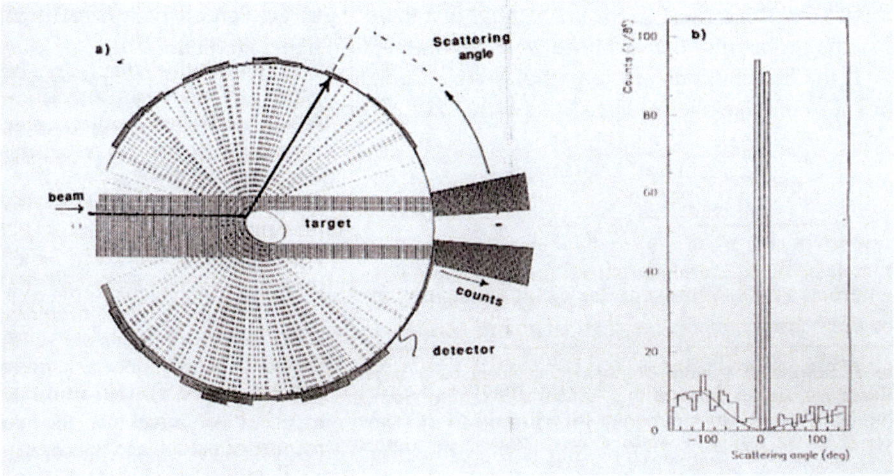

Fig. 10.1 (**a**) Results of 400 scattering events of rigid spheres colliding an ellipsoidal target disk at angles distributed in the interval (−160°, 160°). (**b**) Number of balls scattered for each channel (channel width 8°)

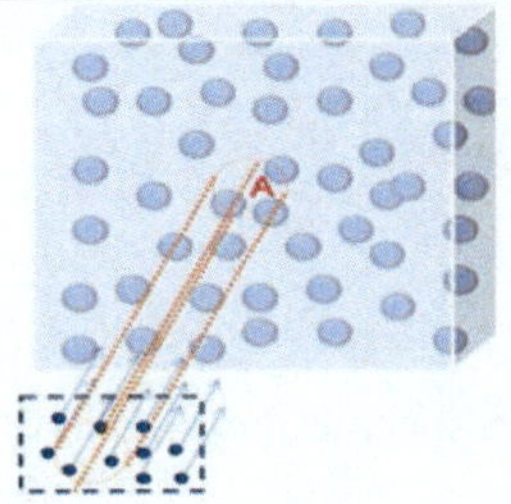	The average number of interactions $\langle N_i \rangle$ with outcome S_i can be obtained summarizing all the small squares A in which the section A_{tot} effectively crossed: $\langle N_i \rangle = N_A\, P_1\, P_2\, P_{int} = (A_{tot}/A)\,(n_1\, A)\,(n_2\, A)$ $P_{int} = A_{tot}\, n_1\, n_2\, (A\, P_{int})$. The factor $\sigma_I = (A\, P_{int})$ represent the characteristics more closely related to the observed interaction. This quantity is called cross section (for the reaction channel considered): it is clearly related to the probability that the specific interaction occurs and its dimension is that of an area.
	The cross section in the first considered example of the process in which the two particles are effectively scattered in the final state of the first case considered is called total as it includes all the possible results in which the final state of the system does not coincide with the initial state.

Fig. 10.2 An uniform beam of (identical) particles colliding with a target of evenly distributed (identical) particles

the considered case, the short range character of the interaction has to be considered explicitly.

To obtain a more general result, we suggest the discussion of the situation of Fig. 10.2 where an uniform beam of (identical) particles collide with a target of evenly distributed (identical) particles. The probability that an incident particle interacts with a target particle within the area A and give rise to a certain result S_i, will be given by: $P = P_1\, P_2\, P_{\text{int}}$, where P_1 is the probability that the incident particle impacts on the area A, P_2 the probability that there is a target particle in region A, and P_{int} the probability that occur the specific interaction being studied.

If the distributions of the particles are sufficiently sparse, P_1 and P_2 are proportional to the average number N'_i $(i = 1, 2)$ of particles in this area:

$$P_1 = \langle N'_1 \rangle / N1 = n_1 A / N1 \quad (1 : \text{incident})$$

$$P_2 = \langle N'_2 \rangle / N2 = n_2 A / N2 \quad (2 : \text{target particle})$$

where n_1 and n_2 represent the surface densities in a projection transversal to the axis of the beam.

The two particle will certainly collide when $b \leq R_1 + R_2$ and $\sigma_{tot} = AP_{int} = \pi(R_1 + R_2)^2$. This result is valid for any reference system and can be used for a geometrical interpretation of the concept of cross section. When $R_1 \ll R_2$, then $\sigma_{tot} \approx \pi R_2{}^2$, that is the projection of body 2 on a plane transversal to the direction of the collision, independently by the shape of body 2.

The histogram in Fig. 10.1b represents the angular distribution of the results of collisions in the second example we discussed; the ith bin of the histogram contains the number N_i of the outcomes for which the scattering angle is included between the extremes $(\theta, \theta + \Delta\theta)$ of the bin itself. Thus the cross section for the collisions with a scattering angle at such an interval is: $\sigma_i = AP_i \approx AN_i/(N_i A) = N_i/n_i$. This value depends on the bin width $\Delta\theta$.

To avoid such dependency, the differential cross section is introduced:

$$\frac{d\sigma}{d\theta} = \lim_{\Delta\theta \to 0} \frac{\sigma}{\Delta\theta} = \lim_{\Delta\theta \to 0} \frac{\langle N_i \rangle}{n_i} \frac{1}{\Delta\theta}$$

It is convenient to define differential cross section with respect to the solid angle:

$$\frac{d\sigma}{d\Omega} = \lim_{\Delta\Omega \to 0} \frac{\langle N_i \rangle}{n_i} \frac{1}{\Delta\Omega} \tag{3}$$

where $\Delta\Omega = \Delta\cos\theta\Delta\phi$ and ϕ is the azimuthal scattering angle, with respect to the axis of the beam.

The differential section for the scattering of the material point on a sphere with radius R can be calculated considering the number dN_1 of incident bodies which have a collision parameter included between b and $b + db$: $dN_1 = n_1\,(2\pi b db)$, all scattered at the same angle θ, which is linked to b by: $b = R\cos(\theta/2)$. Substituting in the equation of the cross section, we derive:

$$\frac{d\sigma}{d\Omega} = \frac{R^2}{4}$$

which is independent from θ and obviously coincides with the area of the section of the sphere.

The geometric nature of cross section we find considering the collisions of point particles with other bodies derived from the model used to describe the dynamic part of the process.

The cross section calculus in the case of Coulomb interactions of point-like bodies with Z, Z' charge is then suggested in the case in which one of the two bodies (scatterer) is much bigger than the other (incident), so that its position does not vary in the collision process and we can therefore consider it still at rest as in the case of scattering experiments to explore a solid system and in particular Rutherford Backscattering Spectroscopy (RBS) (Corni and Michelini 2018). The calculation of the differential cross section for this process is done in the same way as for the previous section: calculation of the relationship between the collision parameter and scattering angle and subsequent application of the definition. The detailed proposal is published in the path dedicated to RBS (Corni and Michelini 2018).

The discussion of the nuclear cross section and relative difficulties conclude the proposal for the last part of the curriculum. Due to the high number of nucleons involved in a nuclear reaction, we are not able to write down a simple expression for the nuclear force, then a simple expression for the cross section cannot be simply derived, but taking into account that we could have, at the same time, Coulombian and nuclear interactions for different nuclei, we can write: $\sigma_{NUC} = A\sigma_R N_{NU}/N_R$, where σ_R the Rutherford cross section, N_{NUC} the number of nuclear events, N_R the number of Rutherford events, and A a parameter depending on detectors' solid angles and intrinsic efficiencies. The N_{NU} number depends on the observed nuclear process we observe. In the case of a nuclear reaction where the two nuclei come close enough to fuse, one of these nuclei dissipate the extra energy and angular momentum removing nucleons from high energy levels to lower energy ones and emitting γ photons with energy equal to the difference of the two levels. For each nuclear fusion event, we will have a γ ray cascade from the first excited state to the ground state. The fusion cross section σ_{FUS} depend by the total number of transitions N_γ and by Rutherford scattering events as follows: $\sigma_{FUS} = A\sigma_R N_\gamma/N_R$.

Students learn how in a similar way it is possible to measure the cross section for other nuclear processes and understand why there are research fields on cross section.

10.4 A Conceptual Explorative Path to Superconductivity

Superconductivity is one of the most relevant topics of the twentieth-century physics. It is an important research field in material science for different kind of research, its application changes almost all the way in which magnetism is employed in technologies and produces new technologies as for the case of Maglev train, its theoretical interpretation founded a new theory (Bardeen et al. 1957). It can be explained at different level and in different interpretive frames (González-Jorge and Domarco 2004; Essén and Fiolhais 2012). From educational point of view, we offer to secondary school students the opportunity to experience how to perform a phenomenologic exploration, analyzing how to build explanations on observed phenomena and how to found interpretative aspects in a coherent and rigorous way. Students have to reflect on their own knowledge in electromagnetism and to put in the field the competence of using their electromagnetic knowledge in a new context. They are stimulated in the production of interpretative models and in the connections between science and in meaningful connections between classical and quantum physics technology. They analyze the history of the superconductivity discoveries and technological applications. They use kits for educational experiments and multimedia materials developed within the framework of European projects MOSEM 1–2 (Kedzierska et al. 2010; Greczylo et al. 2010; MOSEM2 Group 2011).

For the educational laboratory, we developed new systems for temperature dependence of resistivity measurements in a wide range of resistivity values (Gervasio and Michelini 2009), that we integrate in path proposal. Our educational proposal for a phenomenological exploration of superconductivity in secondary school is integrated in the electromagnetism curricula and for the feasibility study we support a group of teachers for research-oriented school implementation (Michelini and Viola 2011). From 2010 research experimentations were conducted in 20 different classes of eight different Italian schools with 393 students. Results documented the systematic interest of the students, which was not limited to the simple observation of phenomena unusual and surprising, but mostly focused on the exploration of interpretative hypotheses. Results of research-based school implementation showed that the personal involvement in the exploration of the collection of the many problematic proposed situations activates the planning of further explorations, aimed to test hypotheses able to explain the superconductor behavior. Starting from the ordinary electrical and magnetic properties of matter, students recognize the peculiar characteristics of the superconductivity behavior at liquid nitrogen (LN) temperature, comparing it to those of an ideal conductors and of diamagnetic materials (Stefanel et al. 2014).

The strategy adopted includes the following phases: presentation of a situation-problem, experimental exploration of it, student individual hypothesis for explanation and/or planning of a further exploration, discussion in little group to reach a common decision, discussion in large group at the end of each problematic issue. The following explorative steps are the core of the path proposal:

(S1) Interaction between a little strong magnet (M_1) and an YBCO disk at room temperature (T_{room}) and at liquid nitrogen temperature (T_{NL}). An YBCO disk at room temperature (T_{room}) does not present magnetic properties. When it reaches the temperature of T_{NL}, the levitation of the magnet occurs. This phenomenon called Meissner effect is analyzed by students as well as the stability of the levitation.

(S2) Comparison of the Meissner Levitation and the cases of magnetic suspensions. Interaction between two free magnets shows the rotation of one of them to attract the opposite pole of the other and suspension occurs only when we constrain magnets to face the same pole for example putting them in a tube. Levitation occur when YBCO and the magnet are free and we have to admit a fundamental difference in the two considered situations. They conclude that a phase transition can be involved, because when $T = T_{room}$ the superconductor does not evidence magnetic properties, but when $T \sim T_{NL}$, suddenly, magnetic properties emerge.

(S3) Students design and perform free experiments to explore the phenomenology and to characterize the YBCO behavior. To individuate the kind of magnetic property acquired by YBCO at T_{NL}, a systematic exploration of its interaction with different magnets and different materials (ferromagnetic materials in primis), in different configurations, is carried out. The superconductor is not becoming ferromagnetic, it is not a permanent magnet, it does not become like the mirror image of the levitated magnet. It always shows repulsive effects close to a magnet.

(S4) Study of the behavior of different materials interacting with a magnet to individuate which kind of material properties are assumed by YBCO at T_{NL}. The levitation of pyrolytic graphite on a quadrupole of magnet is observed. Students studied the interaction of a strong neodymium magnet with paramagnetic and diamagnetic systems suspended on a wire or on a yoke in order to make evident even very small repulsive/attractive forces. Comparing the behavior of diamagnetic materials with those of the YBCO at $T \sim T_{NL}$ it has to be classified diamagnetic.

(S5) Intensity of the diamagnetic interaction. The strength of the interaction between a superconductor and a magnet is several orders of magnitude greater than those observed with ordinary diamagnetic materials, suggesting to search for a more detailed characterization of the nature of the diamagnetism of the superconductor. Starting from the evidence that the superconductor shows magnetic interaction only when a magnet is close to it and that the YBCO do not interact with a ferromagnetic object, students recognize that the interaction with a magnet does not depend on the pole put close to the surface of the magnet, the equilibrium position is always the same. Changing magnet, the equilibrium position changes, but it is always the same, for the same magnet.

(S6) Discussion on the magnetic field inside to the Superconductor and search for an explanation. Analysis of the situation: a sandwich composed by the magnet M_1/YBCO/ferromagnetic ring at $T = T_{room}$ is lifted, pulling the magnet $M_{1.}$ At $T \sim T_{NL}$ the ferromagnetic ring is no more pulled. We have to admit that when $T \sim T_{NL}$ the magnetic field inside the superconductor sample can be zero or very little (in this condition the magnet and the iron ring do not interact when the YBCO disk is in between). Moreover, when $T_{NL} < T < T_{room}$ the magnetic field B can exist and be different of zero inside the YBCO sample, but the magnetization of the superconductor is always adjusted to react to the external magnetic field, tending to preserve the initial situation. In particular, if $B = 0$ when the superconductivity state is created, the system tends to react to an external magnetic field creating a counter field that tends to maintain $B = 0$ inside the superconductor (Meissner effect).

(S7) Search for an analogy able to explain the Meissner effect. Eddy currents produced in the electromagnetic induction have a similar behavior. A magnet is evidently braked when falling down on a thick copper layer or inside a copper tube, but never it can be stopped/maintained in levitation, as in the case of the YBCO. If the magnet stops, the induced current stops due to the Joule effect. The magnet would be stopped just falling over a conductor with resistance $R = 0\ \Omega$ (an ideal perfect conductor!). Supposing that an electromagnetic induction process will be at the base of the superconductivity levitation, the resistivity of the YBCO sample could be suddenly changed when $T \sim T_{NL}$ to have a levitation. This suggest the measurement of the resistivity of a superconductor.

(S8) The experimental measurement of the breakdown of the resistivity of a YBCO gives quantitative evidence to the phase transition for what concern the changes from the ordinary conductor state to the superconductor state of YBCO. The phenomenological exploration leads to the conclusion that $R = 0$ as well as $B = 0$ inside the superconductor. These two aspects characterize the Meissner effect.

(S9) Pinning effect in II type of superconductors is observed, when the superconductivity state is created in presence of an external field. The pinning effect, due to the penetration of the magnetic field inside the superconductor sample as vortexes created by supercurrents, emerges to explain the anchoring of the magnet to the superconductor at $T \sim T_{NL}$.

(S10) Students build a simple MAGLEV train, where the effect is clear because never derail and look for different applications of superconductivity.

During each intervention module, these steps were systematically monitored, using tutorial worksheets, audio recording of the student dialogues, and notes by the researchers/teachers conducting the activities.

For gifted students or advanced class groups having as a prerequisite the energy level model for the electrical conduction interpretation, we often discuss how to interpret the superconductivity and not only to explain by means of an analogy. Starting from the analysis of the energy of the electron system inside of a crystal lattice, we give into account how this superconductive state can be created. Cooper pairs, or in other words the formation of pair of correlated electrons is shown to be a process energetically favored. The collapse of these Cooper pairs on the same ground state is responsible of the $R = 0$ property at the base of the superconductive behavior.

10.5 Concluding Remarks

The recent demanding for innovation in secondary school has an important goal in introducing modern physics topics. Our contribution for modern physics, here discussed and exemplified for the cases of cross section and superconductivity, may be divided into four kinds of researches: (a) Research and development for new systems and tools; (b) Design based research to support practice with coherent paths; (c) Empirical research to study learning trajectories, appropriation, and kind of reasoning; (d) Teacher education and professional development. Our content research theoretical framework is the Model of Educational Reconstruction and Design Based Research guides the implementation of the studied proposals by means of conceptual change analysis of student reasoning. Instruments and methods are test-in/out, tutorials, and interviews using the standard methods of qualitative research. Research is carried out in the context of IDIFO Project (Innovation in Physics Education and Guidance Project led by our research unit) of the National Plan PLS-Fisica (Plan for improving Scientific degrees—section physics), with a peer review and cooperation between 20 Italian universities for different physics education actions and characterized by a strong cooperation between school and university. It follows different learning outcomes: (1) the physics in research analysis technics; (2) explorative approach to research edge topics; (3) discussion of crucial/transversal concepts both in classical physics and in modern physics; (4) foundation of theoretical thinking; (5) contexts for personal involvement in inquiry base learning to gain critical approach and competences in experimental physics work. The perspectives in developing proposals are focused in building in young people: physics identity, physics as a cultural issue, the idea of physical epistemic nature. Avoiding the reductionism our aim is to offer opportunities to: (1) experience of quantitative exploration of crucial phenomena (diffraction and optical spectroscopy), individuating laws, fitting data, and testing basic principal ideas and results with experimental data; (2) understand the crucial role of classical physics in modern research techniques (RBS, Resistivity and Hall coefficient measurements), manipulating data and interpretation like in a research laboratory; (3) focusing on reasoning to conduct a phenomena exploration (superconductivity) understanding the role of analogies for finding explanations; (4) reflect on physics meaning of basic concepts in different theories (state, measure, cross section) revising meanings in classical physics and understanding the different perspectives of new theories; (5) approach to the new ideas of QM theory: the first step toward a coherent interpretation with a supporting formalism experiencing aspects, cardinal concepts, elements peculiar to QM; (6) privilege dynamic aspects in special relativity to revise the concepts of mass and energy and to understand the law $E = mc^2$.

The vertical perspective with respect to the curriculum gives us the opportunity to integrate the different perspectives in different ways so each represents a specific methodological proposal. The two proposals for cross section and for superconductivity here presented offer a way to identify the different methods.

The implementations with secondary school students evidence positive learning progression concerning the crucial knots of the treated topics. It suggests to: (a) focus on the coherence of reasoning to create a reference framework, integrating hand-on and explorative work, mind-on interpretation of results, by means of real and ideal experiments and modelling; (b) use iconographic representations as conceptual tool; (c) use analogies for phenomena interpretation; (d) introduce formalism, using it to reinterpret explored situations; (e) analyze students' ideas in the framework of different interpretative schema; (f) integrate modern physics in classical physics developing coherent paths of conceptual understanding.

Acknowledgement I express my gratitude for the cooperation in the research work carried out to the colleagues of my physics education research group and in particular to Lorenzo Santi and

Alberto Stefanel, as well as to the PhD student Daniele Buongiorno and to the teacher Alessandra Mossenta.

We thank PLS-Fisica Italian Projects and in particular IDIFO project for the support.

References

Anderson T, Shattuck J (2012) Design based research. Am Educ Res 41(1):16–25

Aubrecht G (1989) Redesigning courses for the 21st century. Am J Phys 57:352–359

Bardeen J, Cooper LN, Schrieffer JR (1957) Theory of superconductivity. Phys Rev 108:1175–1204

Başer M (2006) Fostering conceptual change by cognitive conflict based instruction on students' understanding of heat and temperature concepts. Eurasia J Math Sci Technol Educ 2(2):96–114

Bertozzi experiment description (n.d.). https://www.youtube.com/watch?v=tmdp3jd8rig

Bradamante F, Michelini M (2006) Cognitive laboratory: gravity and free fall from local to global situations. In: Planinsic G, Mohoric A (eds) Informal learning and public understanding of physics. Selected papers in Girep Book, Ljubijana (SLO), pp 359–365. ISBN: 961-6619-00-4

Buongiorno D, Michelini M (2018) Research-based proposal on optical spectroscopy in secondary school, this book FFP15

Buongiorno D, Longo A, Michelini M, Pagotto S, Ricci D, Santi L (2017) APPs in sound measurements to gain a school-work experience. In: Proceedings of MPTL conference 2017, Milton Keynes, UK, 13–15 Sept 2017

Constantinou CP, Papadouris N (2008) Physics curriculum design, development and validation. In: GIREP proceedings. The Learning in Science Group, Cyprus

Corni F, Michelini M (2018) A didactic proposal about Rutherford backscattering spectrometry with theoretic, experimental, simulation and application activities. Eur J Phys 39:015501. (22pp). https://doi.org/10.1088/1361-6404/aa9053

Corni F, Michelini M, Santi L, Soramel F, Stefanel A (1996) The concept of cross section. In: Teaching the science of condensed matter and new materials. FORUM-GIREP, Udine, p 192

diSessa AA (2004), Metarepresentation: native competence and targets for instruction, cognition and instruction. Available at http://www.tandfonline.com/doi/abs/10.1207/s1532690xci2203_2

Debowska E, Girwidz R, Greczyło T, Kohnle A, Mason B, Mathelitsch L et al (2013) Report and recommendations on multimedia materials for teaching and learning electricity and magnetism. Eur J Phys 34(3):L47–L54

Deslauriers L, Schelew E, Wieman C (2011) Improved learning in a large enrollment physics class. Science 332:862–864

Duit R, Gropengießer H, Kattmann U (2005) Towards science education research that is relevant for improving practice: the model of educational reconstruction. In: Fischer HE (ed) Developing standards in research on science education. Taylor & Francis, London, pp 1–9

Erickson F (1998) Qualitative research methods for science education. In: Fraser BJ, Tobin KG (eds) International handbook of science education, part 2. Kluwer, Dordrecht, pp 1155–1174

Essén H, Fiolhais MCN (2012) Meissner effect, diamagnetism, and classical physics—a review. Am J Phys 80:164–169

Fedele B, Michelini M, Stefanel A (2005) 5–10 years old pupils explore magnetic phenomena in cognitive laboratory (CLOE). In: Pitntò R, Couso D (eds) CRESILS. Selected papers in ESERA publication, Barcellona. ISBN: 689-1129-1

Fera G, Michelini M, Vercellati S (2014) Reasoning and models of talented students on electrical transport in solids. In: Tasar F (ed) Proceedings of the world conference on physics education 2012. Pegem Akademiel, Istambul, pp 155–162. ISBN: 978-605-364-658-7

Gervasio M, Michelini M (2009). http://www.fisica.uniud.it/URDF/mptl14/contents.htm

Ghirardi GC, Grassi R, Michelini M (1996) A fundamental concept in quantum theory: the superposition principle. In: Thinking physics for teaching. Plenum, Aster, p 329

Gil DP, Solbes J (1993) The introduction of modern physics. Int J Sci Educ 15:255–260

González-Jorge H, Domarco G (2004) Superconducting cylinders aid in an understanding of current induction. Phys Educ 39:234

Greczyło T, Michelini M, Santi L, Stefanel A (2010) Measuring and analyzing the resistivity break down of high temperature superconductors in a didactic laboratory. In: Michelini M, Lambourne R, Mathelisch L (eds) (eds) Multimedia in physics teaching and learning, SIF, Bologna and in Il Nuovo Cimento, 33 C, 3. NIFCAS 33(3):147–155. https://doi.org/10.1393/ncc/i2010-10620-3

Greczylo T et al (2010) High-tech kit. Il Nuovo Cimento C 33(3):221–229

Hake RR (2000) Is it finally time to implement curriculums? AAPT Announcer 30(4):103

Hennessy S, Deaney R, Ruthven K (2006) Situated expertise in integrating use of multimedia simulation into secondary science teaching. Int J Sci Educ 28(7):701–732

Kedzierska E, Esquembre F, Konicek L, Peeters W, Stefanel A, Farstad VS (2010) MOSEM 2 project: integration of data acquisition, modelling, simulation and animation for learning electromagnetism and superconductivity. In: Michelini M, Lambourne R, Mathelisch L (eds) Multimedia in physics teaching and Learning, SIF, Bologna and in Il Nuovo Cimento, 33 C, 3. NIFCAS 33(3):64–74. https://doi.org/10.1393/ncc/i2010-10616-y

Maloney D (2011) An overview on physics education research on problem solving. Rev PER 2:1. https://www.compadre.org/per/items/detail.cfm?ID=11457

Mazur E (2009) Farewell, lecture? Science 323:50–51

McDermott LC (1993) How we teach and how students learn – a mismatch. Am J Phys 61 (4):295–298

McDermott LC (2008) Physics education research: the key to improving student learning. In: Jurdana-Sepic R et al (eds) Frontiers of physics education. Zlatni, Rijeka

Meheut M, Psillos D (2004) Teaching–learning sequences. Int J Sci Educ 26(5):515–535

Michelini M (2010) Building bridges between common sense ideas and a physics description of phenomena. In: Menabue L, Santoro G (eds) New trends in STE. CLUEB, Bologna, pp 257–274

Michelini M, Stefanel A (2008) Learning paths of high school students in quantum mechanics. In: Jurdana-Sepic R et al (eds) Frontiers of physics education, selected papers in Girep-Epec book. Zlatni (CRO), Rijeka, pp 337–343. ISBN: 978-953-55066-1-4

Michelini M, Stefanel A (2010) High school students face QM basic concepts. In: Menabue L, Santoro G (eds) New trends in science and technology education, selected papers, vol 1. CLUEB, Bologna, pp 308–322. ISBN: 978-88-491-3392-9

Michelini M, Viola R (2011) Research-oriented training for Italian teachers involved in European MOSEM Project. Il Nuovo Cimento C 34(5):255–275

Michelini M, Santi L, Viola R, Corni F (2008) Superconductivity in Italian Secondary Schools: the experimentation carried out by Udine University with Supercomet2 (SC2) materials. In: Sidharth BG, Honsell F, Mansutti O, Sreenivasan K, De Angelis A (eds) Frontiers of fundamental and computational physics – FFP9, selected papers. American Institute of Physics – AIP 1018, Melville, NY, pp 225–226. ISBN: 978-0-7354-0539-4

Michelini M, Stefanel A, Zuccarini G (2014a) Exploring university student ideas on the relationship between formal aspects and physical meanings in quantum mechanics. In: Szarková D, Richtáriková D, Záhonová V (eds) 12th international conference APLIMAT 2013. Institute of Mathematics and Physics, Faculty of Mechanical Engineering, Slovak University of Technology in Bratislava, Curran Associates, Inc., Red Hook, NY.. ISBN: 978-1-63266-512-6, pp 435–446

Michelini M, Santi L, Stefanel A (2014b) Teaching modern physics in secondary school. In: Kajfasz E, Masson T, Triay R (eds) Proceedings FFP14 conference. Aix Marseille University (AMU), Saint-Charles Campus, Marseille, France. http://pos.sissa.it/archive/conferences/224/231/FFP14_231.pdf

Michelini M, Santi L, Stefanel A (2014c) Upper secondary students face optical diffraction using simple experiments and on-line measurements. In: Kajfasz E, Masson T, Triay R (eds) Proceedings FFP14, Marseille 2014, 15–18 July 2014. Aix Marseille University (AMU), Saint-Charles Campus, Marseille, France. http://pos.sissa.it/archive/conferences/224/240/FFP14_240.pdf

Michelini M, Pugliese E, Santi L (2014d) Mass from classical physics to special relativity: learning results. In: Tasar F (ed) Proceedings of The WCPE. Pegem Akademiel, Istambul, pp 141–154

Michelini M, Santi L, Stefanel A (2014e) Basic concept of superconductivity: a path for high school. In: Burra GS, Michelini M, Santi L (eds) FFP&PER. Springer, Heidelberg, pp 453–460

Michelini M, Santi L, Stefanel A (2014f) How students link quantum concept and cormalism. In: Szarková D, Richtáriková D, Záhonová V (eds) 12th international conference APLIMAT 2013. Institute of Mathematics and Physics, Faculty of Mechanical Engineering, Slovak University of Technology in Bratislava, Curran Associates, Inc., Red Hook, NY. ISBN: 978-1-63266-512-6, pp 423–434

Michelini M, Santi L, Stefanel A (2017) Research based proposals to build modern physics way of thinking in secondary students. In: Király A, Tél T (eds) Teaching physics innovatively, proceedings of TPI-15. Graduate School for Physics, Faculty of Science, Eötvös Loránd University, Budapest, pp 331–349. ISBN: 978-963-284-925-6

MOSEM2 Group (2011) Teacher guide. Simplicatus AS, Norway

Ostermann F, Moreira MA (2000) Física Contemporánea en la escuela secundaria: una experiencia en el aula involucrando formación de profesores. Rev Enseñ Cienc 18(3):391–404

Pavlin J, Vaupotic N, Čepic M (2013) Liquid crystals: a new topic in physics for undergraduates. Eur J Phys 34(3):745–761

Pizzolato N, Fazio C, Sperandeo-Mineo RM, Persano-Adorno D (2014) Open-inquiry driven overcoming of epistemological difficulties in engineering undergraduates: a case study in the context of thermal science. Phys Rev 10:010107. (25pp). https://doi.org/10.1103/PhysRevSTPER.10.010107

Silva RS (2015) Modern physics for school students. GJSFR-E, vol 15, issue 1. Global journals Inc., (USA). Online ISSN: 2249-4626 and print ISSN: 0975-5896

Sokolowska D, Michelini M (eds) (2018) The role of laboratory work in improving physics teaching and learning, selected paper book. Springer, Switzerland. ISBN: 978-3-319-96183-5, pp 1–278

Sokolowska D, Peeters W, De Meyere J, Mathelitsch L, Constantinou C, Pospiech G, Michelini M, Folmer E, Rovšek B, Nordström G, Ireson G (2014) The SECURE project research on science curricula and teachers' and learners' opinions on science education. In: Tasar F (ed) Proceedings of the world conference on physics education 2012. Pegem Akademiel, Istambul. ISBN: 978-605-364-658-7, pp 201–208

Stefanel A, Michelini M, Santi L (2014) High school students analyzing the phenomenology of superconductivity and constructing model of the Meissner effect. In: Tasar F (ed) Proceedings of the world conference on physics education 2012. Pegem Akademiel, Istambul. ISBN: 978-605-364-658-7, pp 1253–1266

Vercellati S (2010) A discussion of disciplinary knots on electromagnetism and superconductivity using a web environment in the context of an EU Project (MOSEM) for research-based. Il Nuovo Cimento C 33(3):189–193. https://doi.org/10.1393/ncc/i2010-10637-6

Vercellati S, Michelini M (2014) Magnetic field nature and magnetic flux changes in building formal thinking at secondary school level. In: Tasar F (ed) Proceedings of the world conference on physics education 2012. Pegem Akademiel, pp 559–566. ISBN: 978-605-364-658-7

Viennot L (1996) Raisonner en physique. De Boeck Université, Paris, Bruxelles

Vosniadou S (ed) (2013) International handbook of research on conceptual change, 2nd edn. Routledge, London

Wagner A, Altherr S, Eckert B, Jodl HJ (2006) Multimedia in physics education: a video for the quantitative analysis of the centrifugal and the Coriolis force. Eur J Phys 27:L27–L30

Chapter 11
Physics Education Research and the Foundations of Physics: A Case Study from Thermodynamics and Statistical Mechanics

David Sands

Abstract The connection between mathematics and physics poses problems for students, but even professional physicists do not always notice when mathematical treatments fail to reflect physical processes. By way of example, I draw on the foundations of thermodynamics, which is a highly mathematical but conceptually challenging subject. The failure of mathematics to connect to physics concepts, in particular energy conservation, raises a number of questions. First, how is it that generations of professional physicists have been seduced by the elegance of mathematics into overlooking underlying difficulties with the physics? Secondly, what can this tell us about the *how* of physics: how is it done, how do we justify a theory? Thirdly, should PER encourage us to look more critically at the fundamental foundations of physics, and if so, what does it reveal about the way we should be teaching these subjects? Not all of these questions can be answered yet, but in this chapter I illustrate some of the more profound mathematical difficulties in thermodynamics and briefly discuss what approach to teaching thermodynamics might fruitfully be taken.

11.1 Introduction

At the GIREP-ICPE-EPEC conference in Dublin, 2017, I presented an argument that the foundations of thermodynamics were flawed and asked how this should affect the approach to the teaching of thermodynamics. This chapter is on the same topic and inevitably draws on many of the same arguments, so some overlap is unavoidable. However, the arguments against the concept of thermodynamic entropy as a property of a body are numerous and in this work I shall attempt to present the case against entropy slightly differently.

D. Sands (✉)
Department of Physics and Mathematics, University of Hull, Hull, UK
e-mail: D.Sands@hull.ac.uk

B. G. Sidharth et al. (eds.), *Fundamental Physics and Physics Education Research*,
https://doi.org/10.1007/978-3-030-52923-9_11

Just what is it about entropy that I disagree with? Simply that entropy is not a property of a body and that Clausius erred in his derivation of the fundamental inequality of irreversible thermodynamics, namely;

$$TdS \geq dQ \tag{11.1}$$

I shall argue in this chapter that this expression is not consistent with energy conservation as expressed in the First Law and that its use as the basis of the commonly accepted form of the Second Law due to Clausius, which holds that the entropy of the universe can never decrease, is incorrect. To begin, I go back to foundations of thermodynamics to review the origin of the concept of entropy, starting with Carnot and Clausius. Carnot idealised the operation of a heat engine and paved the way for people like Clausius and Thompson (Kelvin) to incorporate the equivalence of heat and work into a theory of thermodynamics that has passed down the generations almost unchanged at its core.

Clausius reworked Carnot's theory in 1850 (Clausius 1898) to take into account the change in the theory of heat from a material substance, caloric, to idea that heat is motion and developed the modern notion of reversibility from the idea of a reversible cycle. It is perhaps not widely appreciated, however, that Clausius and Kelvin differed fundamentally in their view of thermodynamics (Magie 1899). Indeed, the very word, "thermodynamics" was coined by Kelvin to describe the idea of using heat to produce work and vice versa, generating heat from work. Kelvin was concerned with cyclic processes, without which a heat engine cannot operate. It was implicit in everything he wrote and is central to his statement of the Second Law: it is impossible to convert an amount of heat completely into work in a cyclic process in the absence of other effects. Kelvin's statement is simple and intuitive. Work can be done during the expansion stage of a cycle but the piston has to be returned to the starting position with the working fluid back at the starting state for the engine to continue working. This also requires work, which is derived from a portion of the heat taken in from the hot reservoir. Clausius' view of thermodynamics was much less intuitive.

Clausius phrased his statement of the Second Law as the impossibility of transferring heat from a cold to a hot body without some other effect also occurring at the same time. Although not explicitly stated, this also relies on the existence of a cyclic process and indeed the proof of the equivalence of Kelvin's and Clausius' different forms of the law usually considers two cyclic processes in tandem, with one operating as a refrigerator converting work into heat and the other as a heat engine converting heat into work. Violation of one form of the law leads to violation of the other (Zemansky 1968). The cyclic process is therefore central to the foundations of thermodynamics and it was Carnot's innovation to represent the ideal, reversible heat engine as an ideal cycle comprising alternate isothermal and adiabatic processes. Ironically, it was Kelvin who followed Carnot most closely in regarding the cycle itself as reversible, but it was Clausius' ideas that gained prominence and for him the separate stages of the cycle were themselves reversible processes.

The modern notion of a reversible process as quasi-static is derived directly from Clausius, but curiously, the notion of a quasi-static process as reversible has never, to my knowledge, been questioned. Quite possibly, this is because the Carnot cycle comprises four stages which can be represented as occurring along contours in thermodynamic phase space defined by either constant temperature or zero heat flow and the differential form of work, PdV, is integrable between any two points along such contours. Thus, mathematically at least, the work done by the gas between two states, denoted by, say, A and B, on any such contour is,

$$W = \int_A^B \mathrm{P}dV = -\int_B^A \mathrm{P}dV \tag{11.2}$$

The key question is whether there is any physical process that corresponds to this mathematical abstraction. If not, the separate stages of Carnot's cycle are not in themselves reversible and we should consider, in line with Carnot and later Kelvin, the reversibility of the cycle. Moreover, entropy cannot be considered a property of a body.

This last argument is quite subtle and not so easily comprehended, but it can be understood with reference to cyclic processes. It was Clausius himself (1898) who argued that within a cyclic process,

$$\oint \frac{\mathrm{d}Q}{T} \leq 0 \tag{11.3}$$

The equality applies to a reversible cycle, such as Carnot's, and the inequality applies to a cycle that contains an irreversible process. Although Clausius claimed that this was susceptible to mathematical proof, there is scant evidence in the literature that he actually did prove it, though it appears to hold in practice in as much as there are no reported violations within the literature. The irreversible process that Clausius considered and which demonstrates this theorem in practice was the Joule expansion, in which a gas expands freely into a vacuum. There is no work done, no heat flow, and therefore no change in internal energy during such a process. In order to restore the initial state following such an expansion it is necessary to compress the gas, which requires work. If there were no flow of heat out of the gas its temperature would rise and the initial state could not be restored. No matter whether it occurs during the compression or after, the internal energy of the gas must be reduced and this inevitably requires a flow of heat out of the gas. If a flow of heat out of a body is defined as negative, then the inequality in Eq. (11.3) holds. If entropy is a property of a body, then it is clear that it has decreased as a consequence of this net outflow of heat. If further, the entropy is considered to have a unique value in a given thermodynamic state, then it must have increased during the irreversible process and Eq. (11.1) in seen to hold. Moreover, as the body itself has returned to its initial state with no change in entropy, the inflow of heat into the environment represents a positive increase in entropy and Clausius' view of the

Second Law that the entropy of the universe can only increase or remain the same, is seen to operate.

At the heart of this view is the relationship between Eqs. (11.1) and (11.3) and the notion that entropy is property of a body with a unique value in a thermodynamic state. A body is defined here as simply any collection of atoms or molecules in whatever state, solid liquid or gas. This phrasing makes the distinction between entropy as a state function, which is essentially a mathematical notion, and entropy as a property of a body, which is a physical idea. The connection between the two is an area that has been neglected in the literature on entropy, but lies at the heart of the present discussion. In relation to physics education research, the link between mathematics and physics should be central to any physics education programme, but I will argue here that the history of entropy shows that even professional physicists can put mathematics first ahead of physics. In this chapter, the connection between the law of increasing entropy, as expressed by Eq. (11.1) and the law of conservation of energy will be examined with reference to particular examples. In addition, the relationship between the mathematical idea of a state function and the corresponding physical properties will also be examined and examples of students' confusion over entropy will be presented.

11.2 Entropy and Energy Conservation

The preceding example on the free expansion illustrates the fundamental problem of the notion of the entropy of a body. As there is no change in internal energy during the free expansion, there is also no change in temperature. Even if it is argued that during the expansion the state of the gas is not well defined, the quantity TS, which has the units of energy, must be defined for the initial and final states. The change in the Gibbs free energy is given by the difference, yet there is no change in U, no heat flow and no work done. We can meaningfully ask about the physical meaning of the Gibbs free energy in the light of this change.

This difficulty was built into the structure of thermodynamics by Clausius, as summarised in the 1898 (Clausius 1898) collection of his nine Memoirs. It is evident in all his early writings that Clausius was interested in what he referred to as, "internal work", which is the work associated with inter-particle forces when a gas is either compressed or expands. In treating internal work, Clausius borrowed from his earlier work on cyclic processes: "*… as there is no essential difference between interior and exterior work, we may assume with certainty that a theorem which is so generally applicable to exterior work cannot be restricted to this alone*". In fact, there were two theorems in Clausius' view of thermodynamics, which Clausius explained as the equivalence of heat and work, or Joule's principle, and the equivalence of transformations. The latter will be unfamiliar to the modern physicist, as it is an obscure concept not taken up by Clausius' contemporaries and which has subsequently disappeared altogether from the thermodynamics lexicon. Clausius regarded two transformations as being equivalent in some way: the conversion of heat into work, and *vice versa*, in a cyclic process and the conversion of "heat at one temperature to heat at another temperature". Mathematically, the theorem of the equivalence of transformations is expressed by Eq. (11.3), though originally heat was defined such that the integral is positive for an irreversible cycle. It was sometime later that Clausius adopted the modern convention that negative heat corresponds to a heat flowing out of a body.

In referring to exterior work, Clausius meant the work produced by a heat engine. In an ideal reversible engine of the kind considered by Carnot, all the transformations are, to use Clausius' terminology, compensated by equivalent transformations and the equality in Eq. (11.3) applies. For example, the heat taken in from the hot reservoir is converted to work, but in returning the piston to its starting state work is converted to heat which is ejected to the cold reservoir. Both processes have the same equivalence value, Q/T. In an irreversible cycle, at least one of the transformations is uncompensated, leading to the inequality as previously discussed. In comparing internal work to external work, Clausius believed that there must be a similarly uncompensated transformation and actively sought an equivalent inequality. In consequence, he derived Eq. (11.1).

The essential difficulty with Clausius' work is that he did not base it on conservation of energy. The concept was not fully developed at the time and this can be seen in his approach to irreversible, noncyclic processes. Equation (11.3) for cyclic processes is fully compatible with energy conservation whereas Eq. (11.1) for noncyclic processes is not. In a cyclic process, an irreversible stage can by offset by some other process within the cycle in which heat is extracted to restore the original state, but this cannot occur in a single, noncyclic process. Clausius overcame this incompatibility in his Sixth Memoir by disregarding the work that is actually done in a process, which of course is governed by energy conservation, and looking instead at the work that might be done: "The law does not speak of the work which the heat *does*, but of the work which it *can do* ..."; "...similarly, in the first form of the law, it is not of the resistances which the heat overcomes, but those of which it *can overcome* that mention is made". The emphasis is Clausius' and by this reasoning he introduced an inequality into the First Law of thermodynamics. In the following pages the consequence of this inequality are explored for a classical ideal gas subject to a change in the number of particles.

11.3 Extensivity, Entropy and Open Systems

Consider two systems at the same temperature and pressure. One contains N particles of a classical ideal gas in a volume V and the other $N + \delta N$ in a volume $V + \delta V$. Clearly, the difference in internal energy between the two systems is directly proportional to δN, so we have U, V and N all increasing by the same factor. Write,

$$\frac{\delta N}{N} = \frac{\delta V}{V} = \frac{\delta U}{U} = \alpha \tag{11.4}$$

Then,

$$N + \delta N = (1 + \alpha)N \tag{11.5}$$

If entropy is a homogeneous function of degree 1, then

$$S([1 + \alpha]U, [1 + \alpha]V, [1 + \alpha]N) = (1 + \alpha)S(U, V, N) \tag{11.6}$$

In other words, the entropy increases in proportion to the increase in the size of the system. This is standard and on the face of it presents no difficulties.

However, now consider what happens if we compress the second system isothermally through a volume change δV such that the work done is $P\delta V$. Energy conservation requires an outflow of heat corresponding to the work done, $P\delta V$, and the entropy of the systems decreases. However, we now have $N + \delta N$ particles in a volume V at temperature T and we would expect the entropy of this system to be greater than the entropy of N particles at V and T. If the entropy of the second system is now $(1 + \beta)S(U,V,N)$, where $\beta < \alpha$, then

$$(1 + \alpha)S(U, V, N) > (1 + \beta)S(U, V, N) > S(U, V, N) \tag{11.7}$$

Equation (11.7) appears to be consistent with known thermodynamics, but in fact there is a difficulty.

According to Landsberg [p128], the two systems considered above are closed, simple systems for which the equation,

$$T\mathrm{d}S = \mathrm{d}U + P\mathrm{d}V \tag{11.8}$$

holds. As closed systems, Eq. (11.8) cannot be used to describe the transformation from one to another at the same volume, otherwise we would have the simple result,

$$T\delta S = \delta U = \frac{3}{2}kT \cdot \delta N \tag{11.9}$$

This would lead to the difference in entropy being directly proportional to the number of additional particles, which is demonstrably not the case. In order to see this more clearly, consider the case when $\alpha = 1$. This corresponds to the famous Gibbs paradox in which there are two identical systems each containing N particles at temperature T and volume V separated by a partition. Removal of the partition creates a single, larger system with $2N$ particles at temperature T, and hence energy $2U$, in a volume $2V$. The entropy of this larger system is simply double that of each single system. If we were now to compress this larger system isothermally into half the volume, so that we had $2N$ particles in a volume V at temperature T, the entropy would have decreased and self-evidently the entropy would be less than twice the entropy of N particles in a volume V at temperature T. In short, the change in entropy cannot be given by the change in the number of particles and Eq. (11.9) is shown to be invalid.

Landsberg (1961) attempts to put the change in entropy of an open system on a firm mathematical footing and concludes on page 153 of his 1961 book that for a gas which is, "homogeneous in all its states of interest, and which contains only one type of molecule", Eq. (11.8) can be extended to an equation of the form

$$T\mathrm{d}S = \mathrm{d}U + P\mathrm{d}V + T\left(\frac{\partial S}{\partial N}\right)_{U,V}\mathrm{d}N \tag{11.10}$$

Then, one has, by the "laws of partial differentiation" [p153],

$$\mu = -T\left(\frac{\partial S}{\partial N}\right)_{U,V} \tag{11.11}$$

Applying Eqs. (11.10) and (11.11) to our two systems, we arrive at the conclusion,

$$T\delta S = \delta U - \mu\delta N = \left(\frac{3}{2}kT - \mu\right)\cdot\delta N \tag{11.12}$$

This is the desired result. We have shown that the entropy of $N + \delta N$ particles in a volume V at temperature T is greater than the entropy of N particles in a volume V at temperature T, but less than would be obtained if the entropy were simply proportional to the number of particles. This, then, accords with the thermodynamics of the simple systems we have so far developed.

11.4 Discussion and Conclusion

Having derived Eq. (11.12), it remains to show how this conflicts with energy conservation. If we have two systems at the same volume and temperature with the only difference between them being that one has δN more particles than the other, the difference in energy between the two systems is given by,

$$\delta U = \frac{3}{2}kT\cdot\delta N \tag{11.13}$$

Yet, Eq. (11.12) shows that there is some property of the body with the units of energy ($T\delta S$) that differs by an amount less than this. In other words, there is some energy contained in the quantity $-\mu\mathrm{d}N$ that offsets the increase in internal energy due to the increase in the number of particles. Thermodynamics gives us no clue as to the physical meaning of μ and the physics of the ideal classical gas in the form of kinetic theory gives no insight. Whilst the internal energy can be equated with the average energy of particles with a range of velocities given by the Maxwellian distribution, there is no quantity analogous to the chemical potential.

The physical meaning of μ and the validity of Eq. (11.11) can also be queried. It is not clear to this author that Eq. (11.11) is correct for a classical ideal gas. This equation is valid only if U and N are independent; that is, particle number N can be changed whilst holding both U and V constant. In a classical ideal gas N can be changed independently of V but not of U except at absolute zero, as shown by Eq. (11.13), so the partial differential in Eq. (11.10) cannot be applied.

It should be acknowledged that entropy might not be extensive, which would redefine the entropies of the different bodies we have discussed and alter the relationship between them. However, it is not clear that it would solve the problem, which fundamentally arises from the notion that a body in a particular state has a particular entropy. This can be illustrated by the following argument. It is reasonable to assume that entropy must in some way increase with the number of particles and that the entropy of $N + \delta N$ particles in a volume V at temperature T is greater than the entropy of N particles in a volume V at temperature T. We would expect some functional relationship between S and N which would permit partial differentiation, but in a classical ideal gas N can be varied independently of only T and V or, equivalently, P if V is allowed to vary with N. Expressing entropy as a function of T would make it difficult in general to combine the First and Second laws, but notwithstanding this difficulty let us suppose that we end up with something of the form,

$$\delta S(N) = \frac{\mu}{T}\delta N \tag{11.14}$$

Here, for complete generality μ can be positive or negative, though normally it is the latter. We are left with the difficulty that $T\delta S$ has the units of energy and the only change in energy of a classical ideal gas on changing the number of particles is given by Eq. (11.13). If the total change in entropy contains a term δU, the only value of μ that will satisfy energy conservation is zero, but if S does not explicitly depend on U but on T, μ must be equivalent to the average energy per particle in order to satisfy energy conservation. This is not the end of the difficulty, however, because the system can then undergo a Joule expansion to $V + \delta V$ in which there is no change in internal energy, but the quantity $T\delta S$ increases.

Whichever way we look at it, the concept of the entropy of a body would appear to be incompatible with energy conservation. The Joule expansion highlights the difficulty most clearly and illustrates the connection with Clausius' flawed reasoning. In a Joule expansion, $T\delta S$ is reckoned to increase by $P\delta V$ by comparison with the so-called equivalent process of a reversible isothermal expansion. The process is only equivalent in as much as the initial and final states are the same. However, in the isothermal expansion real work is done, but in the Joule expansion, no work is done and this is reflected in the fact that $P = 0$. This is not commonly appreciated, but follows from Newton's second law: if the molecules are expanding freely into a vacuum and do not change momentum through collisions with the wall of a container they are not themselves subject to a force and cannot exert a force. We therefore have an immediate conflict with the First Law that is built into the structure of thermodynamics via Clausius' assertion that, "The law does not speak of the work which the heat *does*, but of the work which it *can do* ...". The idea that heat can do work is a direct consequence of the view of the time that heat was somehow converted into work, but we now know that work is a consequence of repeated collisions of particles on a piston. Heat flows into a gas to replace the energy lost during work and in this sense there is no such thing as isothermal heat flow. However, borrowing from the terminology of heat engines prevalent at the time of Clausius, we can regard heat as capable of doing work $P\delta V$. If, following Clausius, we use this work term to give the increase in entropy during an irreversible expansion, we are adding a term in energy that does not in fact reflect the physical processes and violates energy conservation.

This failure of the mathematics to reflect the physics has been overlooked by the majority of physicists for well over 160 years and consideration of this alone shows that the process of interpreting mathematical formalisms in terms of physics is not straightforward. It is perhaps not surprising that students struggle. It is over 10 years since Rebello, working with Dean Zollmann and others (Rebello et al. 2005), looked at the transfer of mathematical knowledge from one domain to another, but little progress seems to have made since then. Authors such as Karam (2014) and Redish and Gupta (2010) stress that understanding mathematics in physics is not just about understanding mathematical operations, but how those operations connect to and describe physics concepts. I suggest that we have only just begun to understand some of the complex interactions in not just learning physics, but in *doing* it and that this process of examining critically the way physics is done should impact on our understanding of the fundamentals.

Revisiting fundamental concepts and their connection to mathematical formalisms means keeping an open mind and rejecting a utilitarian approach. That is, just because a mathematical approach appears to be useful does not mean it is correct and it has been argued in this chapter that Clausius' conception of entropy was flawed in so far as it was based around the concept of transformations rather than conservation of energy. Clearly, the difficulties engendered by this approach were not realised at the time and the failure to reflect on the disconnection between the mathematics and the physics has left its mark on thermodynamics today. That leads inevitably to the question of what might usefully be taught in thermodynamics.

It is my own personal view that a fundamental revision of the foundations of the subject is required. Carnot was concerned with the reversibility of the cycle itself, which was also the view of Kelvin: the cycle can be performed in one direction to convert heat into work or in the other to convert work into heat. This view of thermodynamics has been overlooked in favour of Clausius, but the implication of the work summarised here is that this view is actually correct. There is a powerful argument, therefore, for returning to the origins of the subject and basing thermodynamics education not on entropy as a driving force or a determinant of equilibrium, but on cyclic processes. What, though, of entropy itself? If entropy is not a property of a body it must instead be associated with the flow of heat and the inequality of Eq. (11.1) has no meaning. Extending the First law to include changes in the number of particles and combining it with the Second Law would yield;

$$TdS = dU + PdV - \eta dN \quad (11.15)$$

Here, η represents the average energy of the particles.

This equation looks remarkably like the outcome of combining Eqs. (11.10) and (11.11), but is actually quite different. For the two systems considered in this chapter, this would give $TdS = 0$ because entropy in this formulation is not associated with a property of a body but is solely the ratio of heat exchanged to the temperature at which it is exchanged. Given the widely accepted current view of entropy, this would appear to be a radical shift, but in fact it is consistent not only with energy conservation but the foundations of thermodynamics.

Acknowledgement The author would like to acknowledge the invaluable assistance of his friend and colleague, Dr Jeremy Dunning-Davies, for his insight into the mathematics of thermodynamics and the illuminating discussions over not just the topics in this chapter, but extending over many years.

References

Clausius R (1898) In: Hirst TA (ed) The mechanical theory of heat with its applications to the steam engine and to the physical properties of bodies. Taylor & Francis, London, p 215

Karam R (2014) Framing the structural role of mathematics in physics lectures: a case study on electromagnetism. Phys Rev Spec Top Phys Educ Res 10:010119

Landsberg PT (1961) Thermodynamics with quantum statistical illustrations. Interscience Publishers, New York

Magie WF (1899) The second law of thermodynamics: memoirs by Carnot, Clausius and Thomson. Harper Brothers, New York

Rebello NS, Zollman D, Allbough AR, Engelhardt PV, Gray KE, Hrepic Z, Itza-Ortiz SF (2005) Dynamic transfer. In: Mestre JP (ed) Transfer of learning from a modern multidisciplinary perspective. Information Age Publishing, Charlotte, NC

Redish EF, Gupta A (2010) Making meaning with math in physics: a semantic analysis. arXiv:1002.0472v1 [physics.ed-ph]

Zemansky MW (1968) Heat and thermodynamics, 5th edn. McGraw-Hill, New York

Chapter 12
Stem, Inquiry Practices and Technology in Physics Education

Ton Ellermeijer and Trinh-Ba Tran

Abstract How to make Physics Education more challenging, relevant and attractive for our school students? How to stimulate the development of creative thinking, problem solving, and other higher cognitive skills? What are the realistic conditions that must be fulfilled so that teachers can realise this kind of quality Physics Education? Governments in many countries stimulate science and technology in schools; the more recent Alphabet Soup acronyms are IBSE, STEM (or STEAM), and MINT (Germany). What is meant with that, and what are the main expectations? And, can technology applied in physics education bring us closer to the desired goals?

Clearly it has been demonstrated that technology can help to make physics education more relevant, more linked to real life and more authentic. And can increase the opportunities for own investigations by the students. So really has an added value, and not just provides another way of teaching the same. This is known for decades but still applied at a relatively small scale.

A major challenge is the preparation of teachers for using technology in this direction. The authors recently investigated the development of an effective and relatively short course for teachers to prepare them for the use of ICT/Technology in IBSE lessons. The final course setup is based on several rounds of tryouts and improvements, and has been applied in the Netherlands, Slovak Republic and Vietnam. The course will be presented, some attention will be given to the differences in application in different settings (pre-service and in-service, different

Most part of the text is taken from the same authors' work published in Pietrocola (ed.), Upgrading Physics Education to Meet the Needs of Society, https://doi.org/10.1007/978-3-319-96163-7_3.

T. Ellermeijer (✉)
CMA, Amsterdam, The Netherlands
e-mail: ton@cma-science.nl

T.-B. Tran
Faculty of Physics, Hanoi National University of Education, Hanoi, Vietnam
e-mail: trinhtb@hnue.edu.vn

B. G. Sidharth et al. (eds.), *Fundamental Physics and Physics Education Research*,
https://doi.org/10.1007/978-3-030-52923-9_12

cultures) and the learning effects on the participants. Interesting and important conclusion is that such a high-quality course design can be applied broadly.

12.1 Introduction

The knowledge society requires innovative products and services, and so workforce must be able to apply knowledge in innovative ways. It is challenging, because many students entering technical universities are not able to cope with open tasks; they are not trained in divergent thinking. About secondary school students, there has been a decline in interest for science (OECD 2006); many school students do not see science and technology attractive, relevant, and related to jobs. Therefore, the overall challenge in physics education today is probably attracting more students for Science and Technology; this results in driving questions in physics education: *How to make physics more challenging, relevant, and attractive for students? How to stimulate their development of creative thinking, problem solving, and other higher cognitive skills?*

In many countries, governments have been innovating the science curriculum, considering these questions. For example, to let students experience physics as a relevant topic for them, students' authentic projects is included in the school agenda, and the teacher is stimulated to link regular physics lessons to real life. To show students that physics itself is still very much alive, modern topics (e.g. quantum physics, elementary particles), and applied physics (medical imaging, biophysics) are introduced. More connections between schools and universities, research institutes, companies, outreach are established, and more facilities for improvements of school labs are provided. To prepare students also for higher cognitive skills, the curriculum includes students' own investigations (with minds-on) and design tasks (divergent thinking).

Since the 1980s, advances in technology and physics education research have stimulated intensive development of Information Communication Technology (ICT) for data logging with sensors, video measurement, and dynamical modelling. These tools resemble those of scientists and engineers but are designed for educational purposes and primarily aimed at classroom use. Can the technology applied in physics education bring us closer to the desired goals? Clearly it has been demonstrated that technology can help to make physics education more relevant, more linked to real life, and more authentic. And can increase the opportunities for own investigations by the students. So it really has an added value, and not just provides another way of teaching the same. Many initiatives such as STEM or STEAM (USA), MINT (Germany), ICT in IBSE (EC) are the more recent Alphabet Soup acronyms. In this chapter, we present the main contributions of technology: *showing students how physicists work today* (e.g. beats plus signal analysis, numerical model for cooling down); *enabling authentic projects by powerful tools for doing investigations* (e.g. reentry of a capsule in atmosphere, bungee jumping, bouncing balls);

and *facilitating a better link to real-life phenomena* (e.g. high jumping, parachute jump).

Educational benefits of these ICT tools are known for decades, but they are still applied in physics education at a relatively small scale. Factors involved are among others:

- Limited curriculum time and limited teacher preparation time.
- Lack of equipment, resources and technical support.
- Mismatch between assessment, examinations and curriculum objectives (e.g. inquiry).
- Pupil problems (e.g. high cognitive load of inquiry learning).
- Teacher problems (e.g. prescriptive instruction with ICT) and lack of continuous teacher education.

Needed are systemic changes in these factors and concerted, simultaneous actions of all stakeholders, especially teachers who should be well-prepared, be professionals and be treated like that! Such changes take time and there needs the right actions.

In our recent study, we focus on teacher preparation and training, which are driving forces for change in classroom practices regarding ICT incorporation. We investigated the development of an effective and relatively short course for pre- and in-service teachers to prepare them for the use of ICT in Inquiry-Based Science Education (IBSE). Several pedagogical principles, like the depth-first and one theory-practice cycle, were applied to (re)design, implement, evaluate and optimise this ICT in IBSE course. The course was aimed not only at learning ICT skills, but also at awareness of benefits and motivation. The final course setup is based on several rounds of tryouts and improvements, and has been applied in the Netherlands, Slovak Republic and Vietnam. The course is presented in this chapter as well; some attentions are given to the differences in application in different settings (pre-service and in-service, different educational systems, and different cultures) and the learning effects on the participants.

12.2 Inquiry Practices in Physics Education

Science educators have been aware of the potential benefits of an inquiry-based approach in science teaching and learning at both primary and secondary levels, and the term "Inquiry-Based Science Education"—IBSE has been popular for a long time. In an article published in 1910, Dewey (1910) remarked that science is not only a body of knowledge to be acquired, but it also includes inquiry methodologies to generate and validate knowledge. In this chapter, we consider inquiry as a process of generating and validating knowledge through moving back and forth between the theoretical world (ideas, concepts, relationships, theories and models) and the physical world (objects, phenomena, observations, measurements and experiments). According to Van den Berg (2013), ideally, IBSE will engage pupils in thinking back and forth between these two worlds like scientists; and "the phenomena and experiments serve as a source for validating ideas and theories and as a playground for generating new ideas and theories in a complex mix of inductive and deductive mind play" (p.75).

Inquiry as process of generating and validating knowledge fits into a view on learning as knowledge creation, discussed by Paavola et al. (2004). Inquiry, under the knowledge-creation perspective, is the process whereby new knowledge and understandings are (re)constructed. From the knowledge-creation perspective, knowledge is not always objectively true. Knowledge is not always given by teachers and scientists or in other knowledge containers (e.g. journal articles, textbooks). Knowledge and its representations (e.g. ideas, concepts, relationships, theories and models) can also be created, elaborated and restructured by learners and researchers. This is in line with Duschl et al. (2007) that the brain is filled with preconceptions from early life experiences; some of these preconceptions match with science, others do not. Therefore, much learning involves reconstruction of prior ideas, which are already in the learner's brain. In addition to the knowledge-creation model, Paavola et al. (2004) discussed two other metaphors of learning: acquisition and participation. The knowledge-acquisition metaphor focuses on learning within individuals' minds, whereas the participation metaphor emphasises learning as a process of participation in various practices and activities. The knowledge creation perspective encompasses both acquisition and participation.

In the book: "The scientist in the crib", Gopnik et al. (1999) implied that from young ages, children can create new knowledge by inquiry, and scientists make the most of this capacity, which lets "children learn so much so quickly" (p.9). Consequently, we indeed concur with Duschl et al. (2007, p.83) that pupils can "engage in and profit from instruction that incorporates relatively complex scientific practices from the very beginning of their schooling".

The science-education community has suggested making authentic inquiry of science more accessible to pupils (e.g. Gaskell 1992; Edelson 1998; Braund and Reiss 2006). Authenticity of inquiry in the school can be interpreted as resemblance of pupil activities to experimentation/modelling activities of practicing scientists in constructing new knowledge, considering the three following aspects (Heck 2009):

- A real-life context for learning that provides pupils with opportunities to investigate realistic science problems in history or present-day research and so pupils will appreciate the relevance of scientific knowledge in everyday life.
- Tools and techniques that enable pupils to carry out experiments/modelling and to analyse and process high-quality data in much the same way scientists do.
- Scientific attitudes of learning that stimulate pupils' pursuit of unanswered questions, commitment to challenging tasks, and social interactions (e.g. cooperation, argumentation).

Authentic inquiry is close to real science, so makes school science more attractive and relevant. Moreover, considering the "learning as participation" metaphor (Paavola et al. 2004), pupils can appreciate inquiry as a scientific method of generating and validating knowledge through being engaged in practices similar to those of scientists.

IBSE is an integral component of the intended science curricula in many countries (Jeskova et al. 2015). In such school science curricula, "inquiry" refers to learning goals (i.e. understanding of the methods of scientific inquiry and the ability to carry out scientific inquiry). "Inquiry" also refers to teaching strategies that stimulate and support pupils to exercise inquiry practices, including hands-on activities, minds-on discussions and meaning making (Hodson 2009; Minner et al. 2010; NRC 2012).

12.3 Technology for Inquiry Practices in Physics Education

12.3.1 An Open Computer Learning Environment

As known for a long time, ICT can enable physics education in a direction that brings students in a similar position as researchers in physics (Heck and Ellermeijer 2014). The development of ICT tools for physics education should be driven by a combination of educational research, curriculum development and innovative technology. We envisioned a scenario of teachers and students using a set of tools for inquiry-based study of natural and mathematical phenomena. This set of tools is integrated in one open environment designed for a broad educational setting. Openness means that it is:

- A flexible, customisable, multipurpose system.
- An environment for solving open problems that need definition, setup, exploration, data processing and analysis, mathematical modelling, and so on; that is primarily a cognitive tool.
- As much as possible free of didactic context or principles; that is less considered as a pedagogical tool, but more as a tool for doing physics.

This computer learning environment does not only exist in the minds of software designers, but it has already been realised to a large extent in the Coach learning and authoring environment, which is the result of three decades of sustained research and development work at CMA, Amsterdam, the Netherlands to improve STEM education. "Coach" refers to coaching and support of learning.

A one-sentence description of Coach is as follows: Coach is a single, activity-based, open computer learning environment that is designed for the educational setting and that offers a versatile set of integrated tools for the study of Science, Technology, Engineering, and Math (STEM) (Heck 2009). For more detail about the latest version of Coach, please visit http://cma-science.nl/coach-7-overview.

Coach activities are mostly based on the powerful tools for *data logging with sensors*, advanced *video measurement* up to *dynamical modelling*. Teachers can use readymade activities or author new activities to structure the lesson materials (i.e. experiments, models). Such activities may contain:

- Texts with activity explanations or instructions.
- Pictures illustrating experiments, equipment, and/or context situations.
- Video clips or digital images to illustrate phenomena or to use for measurement.
- Measurement data presented as graphs, tables, meters, or digital values.
- Models (textual, equations based, or graphical) to describe and simulate phenomena.
- Programmes to control devices and to make mathematical computations links to Internet sites and other external resources for students.

Coach activities are presented below, but please note that all kinds of activities are supported in a single computer environment and not in a suite of separate programmes. Having a single environment instead of a bunch of special purpose software packages brings many benefits. First, students and teachers only need to familiarise themselves with one environment, in which components are geared with each other. They can grow into their roles of skilled users of the system during their learning and teaching. A learn-once-use-often philosophy of educational tools is realisable. Additionally, students may experience the connections between different school subjects using a single environment instead of a grab bag of disconnected tools. Another advantage of a single environment compared to a software suite is the possibility to easily combine different tools in one activity.

12.3.2 Coach Tool for Data Logging with Sensors

12.3.2.1 Characteristics

The Coach tool for data logging with sensors enables students' experimentation activities in which the sensor, connected to an interface, measures a quantity (e.g. temperature, voltage, pH) in the physical world and transforms this quantity into a voltage or other signal(s), which is then read by the interface. The interface converts the signal into digital data that are transferred to, then interpreted and processed by the connected computer or other devices with dedicated software (Fig. 12.1). A computer equipped with an interface and ample sensors becomes a universal measuring instrument, which has a wide range of sampling frequencies from very low to very high (e.g. 10,000 samples per second). This computer-based instrument certainly can take the place of instruments such as thermometers, voltmeters and pH meters, used in conventional practical work. It enables automatic, accurate, conditional measurements and includes ample ways of storing, displaying and analysing data. During the measurement, real-time data can be represented in graphs, tables or displayed as digital values.

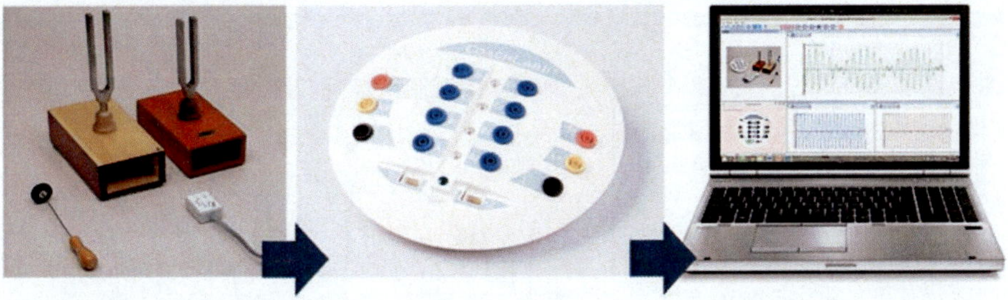

Fig. 12.1 Diagram of the tool for data logging with sensors (including sensor, interface and computer with dedicated software)

12.4 Challenges of Technology in Teaching Physics

Among different software packages, Coach has common measurement methods:

- Time-based measurement in which data are automatically gathered at regular time intervals according to the sampling frequency; with this setting, it is possible to specify a signal condition from which the computer automatically starts a measurement (i.e. triggering). For example, Fig. 12.1 shows time-based measurements of interfered sound and the obtained graph displays the sound signal.
- Event-based measurement in which measurements are taken each time a pulse (i.e. an event) is received on an interface input. For example, in an experiment with stroboscopic light, each light flash generates a pulse via a light sensor, and then a measurement is taken.
- Manually triggered measurement in which a single measurement is taken every time the user presses a button; this setting also allows to type in data (e.g. observational data, standards, given values) via the keyboard. For example, in the Boyle's law experiment, pressure of air, which is trapped inside a cylinder fitted with a piston, is measured by a pressure sensor each time a value of gas volume read by naked eyes is manually entered in the software.

Data logging with sensors is a generic experimental tool for physics, chemistry and biology.

12.4.1 Examples of Coach Data-Logging Activities

Figure 12.2 shows a screen shot of the measurement and signal analysis of the voice sound "eeh" recorded with the €Sense interface, which is mostly used at primary school or by beginners. The diagrams show that the sound signal is well described by a sinusoidal signal that consists of four frequencies. A visual representation of the €Sense interface with the built-in microphone is also present in the activity screen to make the experimental setup clear to students. A text window is used for explanation and description of tasks.

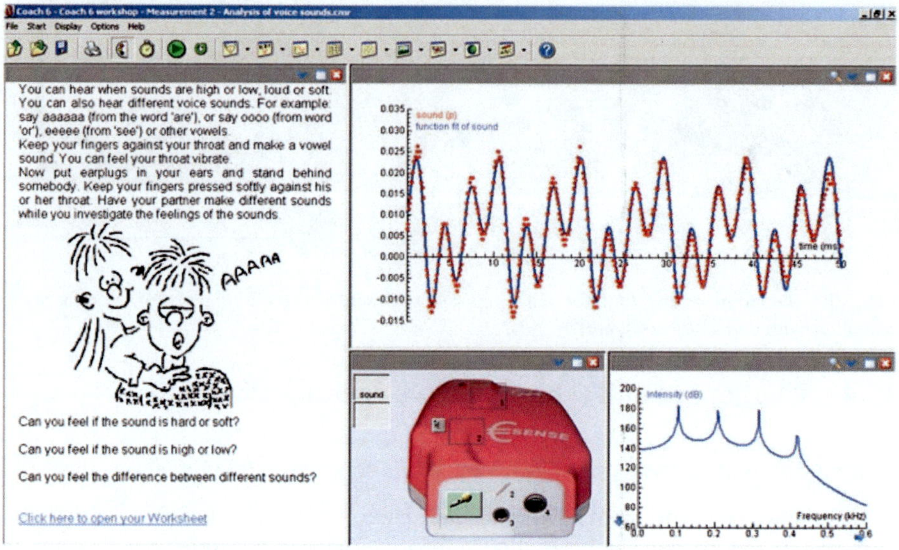

Fig. 12.2 Measurement and signal analysis of voice sounds with the €Sense interface

12.4.1.1 Educational Benefits

The tool for data logging with sensors has many educational benefits if it is properly integrated in the science lessons. First, this tool enhances new possibilities and contexts for physics experiments that might not be otherwise possible due to time constraints and technical difficulties (Barton 2004; Newton and Rogers 2001). This increases access to real-life phenomena, facilitates new classroom experiments, and allows measurements in the field. Second, the tool enables collecting, recording and representing of many data and even repeating this process several times in short time (physical world). Consequently, pupils will have time in the classroom to design the experiment, interpret data and/or explain relationships (theoretical world).

Third, the "real-time graphing" feature of the data-logging tool stimulates pupils to move back and forth between the physical world and the theoretical world. For example, a pupil walks in front of a motion sensor, and immediately the software shows in the graph her or his position and/or velocity in real time. By observing the pupil walking and the graph showing up at almost the same time, other pupils in the class can easily realise the connection between the motion of their classmate and the kinematics concepts. According to Brasell (1987), this immediacy between the phenomenon and real-time graphing of data stimulates pupils' conceptual understanding, and this feature is critical for both understanding and motivation. Sokoloff et al. (2007) showed research evidence that the use of the tool for data logging with sensors in a laboratory curriculum (i.e. RealTime Physics Mechanics) improved pupils' understanding of dynamics concepts, and the retention of the concepts by those pupils was excellent.

Finally, the incorporation of the data-logging tool enables pupils to participate in aspects of scientists' experimental inquiry, considering that the data-logging tool is like those used by scientists. According to Ellermeijer et al. (1996), once pupils get used to the data-logging tool, they can decide and reflect at any time about what to measure, how to calibrate and what readings should be taken. This shows that such participation in authentic inquiry with the data-logging tool will stimulate pupils to comprehend scientific inquiry.

12.4.2 Coach Tool for Video Measurement

12.4.2.1 Characteristics

The Coach tool for video measurement enables pupils to conduct experiments in which, for instance, position and time data of a moving object, registered in a digital video, are collected in the successive video frames by mouse clicking. Among different software, Coach includes common steps to gather real-life data from a video. First, the user must define the video scale, time calibration and coordinate system. The video clip is scaled by specifying which distance on the video screen corresponds to which actual distance (e.g. 1 m viewed in the video frame in Fig. 12.3). A video is a collection of rapidly displayed pictures called video frames. The time interval between two successive frames shown in the software is calibrated by entering the actual frame rate of the video (i.e. how many video frames were taken in a second as the video was recorded). Next, the user moves the cursor over the video screen to locate the point(s) of interest (e.g. a baseball) and then click to store the first video point (i.e. first coordinate and time data). The video clip automatically advances to the next frame, and then the user continues clicking on the reference

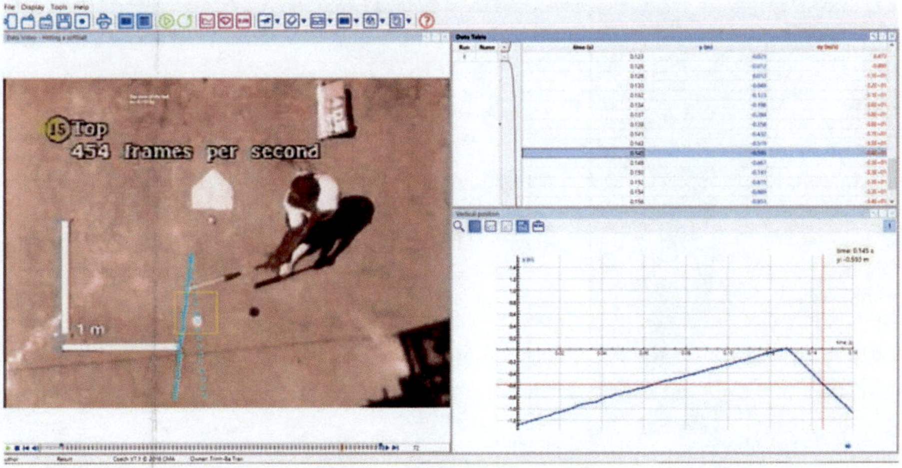

Fig. 12.3 Screenshot of an experimentation activity facilitated by video-measurement tool

point. This procedure with the software is repeated until the user obtains a desired number of data points.

Coach allows automated tracking of the movement of objects and enables collection of different video points in a single video frame. Like the data logging with sensors, during the manual measurement and automated tracking from a video, the collected data are simultaneously displayed in a diagram or table (real-time graphing) (Fig. 12.3). Other dynamics quantities such as velocity, acceleration, momentum, kinetic energy, force can be numerically computed based on the collected data. Finally, collected and computed data are analysed and processed further by the software. Video measurement is mostly limited to movement, so it is mostly used in physics.

12.4.2.2 Examples of Coach Video-Measurement Activities

Figure 12.3 illustrates an experimentation activity facilitated by the video-measurement tool. In this activity, position and time of a baseball is collected from a high-speed video and displayed in the graph and table by the software. The dotted cross in the graph indicates that the scan feature of the software is activated. In this illustration, the data point (– 0.593 m, 0.145 s) on the graph and the table is scanned, and the video advances to Frame 72, which shows the corresponding position of the baseball.

Another example is the video measurement of the motion of a self-made yoyo, which is winding up and down (Fig. 12.4). In this case, the position of the point near the rim of the disk and marked by a sticker (P1) is measured in a slightly moving coordinate frame whose origin is at the hand of the person holding the end of the

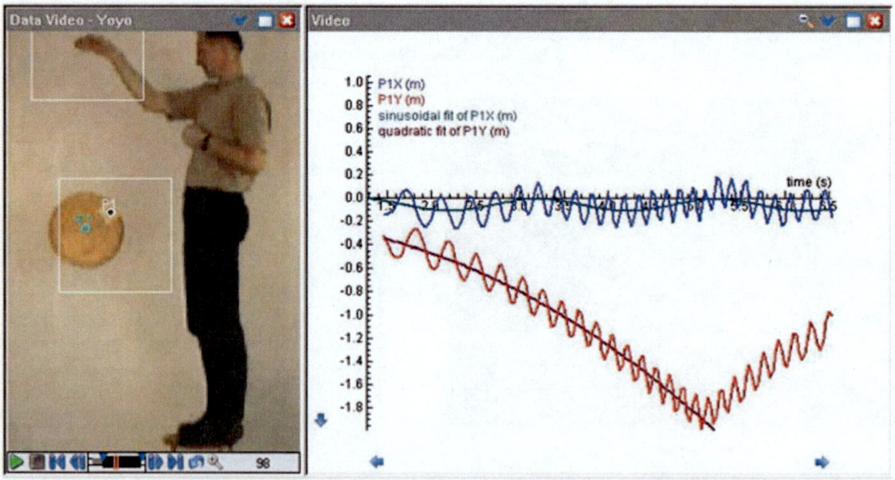

Fig. 12.4 Screen shot of a video analysis activity about the motion of yoyo

cord of the yoyo. But point-tracking makes the measurement at hand easier and less time-consuming: in the starting frame, the positions of the hand and of the sticker are specified and the shapes of the search areas (white boxes) are set, and then the coordinates of these points are automatically recorded in subsequent frames.

In the diagram to the right, the horizontal position and the vertical position of P1 are plotted against time. This is combined in the diagram with a sinusoidal fit of the horizontal displacement of the yoyo, due to an unintentional pendulum motion of the yoyo, and a quadratic function fit of the vertical position during the first phase in which the yoyo unwinds. These trend curves can be used as coordinate functions of a computed point that is displayed in the video clip (P2): It turns out to be close to the position of the axle during the unwinding phase of the yoyo. Please refer to Heck and Uylings (2005) for detailed modelling of the yoyo motion.

12.4.2.3 Educational Benefits

The Coach tool for video measurement has much added value if it is incorporated appropriately in school science. First, like the data-logging tool, video measurement creates new possibilities and contexts for experimentation activities. With the video-measurement tool, the teacher can bring real-life, attractive scenes of motion into classroom activities that show pupils the relevance of science concepts and theory in everyday life (Heck 2009; Zollman and Fuller 1994). Such realistic scenes of motion can be quite ordinary (e.g. basketball shots, amusement-park rides, dancing) or unusual (e.g. car crashes, jumps on the Moon, rocket launch). With high-speed videos (i.e. up to 1200 frames/s), the teacher and pupils can quantitatively explore many more situations of realistic motions (e.g. multidimensional collisions between billiard balls, gun recoil) that would be mostly impossible to investigate with traditional instruments and even with sensors for school science. Additionally, the video-measurement tool can serve as a cost and time effective instrument for the school laboratory, which might replace rulers, timers, photogates and motion sensors in motion-related experiments.

Second, the tool enables the collection and representation of many video data from different realistic situations in a short time (physical world). Consequently, pupils will have time in the classroom to interpret data and/or explain relationships (theoretical world). Third, the "real-time graphing" and "scan" features of the video-measurement tool stimulate pupils to think back and forth between the physical and theoretical worlds. This becomes more likely as images of these two worlds are shown in the same software interface (Fig. 12.3). When pupils scan a data point in one of the graphs, the corresponding video frame, where the data were collected, displays simultaneously. This feature enables pupils to identify events during the realistic situation (physical world) and connect them to abstract representations in the graph (theoretical world). This results in pupils' deeper understanding of the motion and related kinematic concepts (Beichner 1996; Gröber et al. 2014).

Finally, the incorporation of the video-measurement tool makes it possible for pupils to exercise experimental inquiry practices like those of biomechanics and movement-science scientists (Heck 2009; Kearney and Treagust 2001; Laws and Pfister 1998). Pupils can participate in many aspects of experimental inquiry using video measurement. For example, formulating problems; designing the scenario and setup for appropriate video recording by a webcam, a smartphone or a video camera; calibrating time and scale of the video; defining from which frames to get data and with which techniques to collect data; and processing and interpreting the collected video data.

12.4.3 Coach Tool for Dynamical Modelling

12.4.3.1 Characteristics

Modelling has different meanings for different communities, depending upon the context in which it is discussed. The term "modelling" will refer to computational, dynamical modelling that is a tool used by scientists in many different fields (e.g. science, technology, economics, sociology) to describe, explain and predict complex dynamical systems. It helps to understand a system's structure, the interaction between its objects, and the behaviour it can produce. Many of such systems can be built as models on the computer, which can carry out many more simultaneous calculations than human mental models and which can enable solution of differential equations. These differential equations cannot be solved with secondary school mathematics.

The Coach tool for dynamical modelling provides the teacher and pupils with possibilities to be engaged in the modelling process in science: "analyse a situation in a realistic context and reduce it to a manageable problem, translate this into a model, generate outcomes, interpret these outcomes, and test and evaluate the model" (van Buuren et al. 2010, p.112). First, a realistic context (e.g. a tennis ball bouncing on the floor) is analysed and simplified to be manageable by ignoring realistic effects or situational factors (e.g. the ball moving vertically without rotation, air resistance and aerodynamics effects); the stripped-down, mental model is then translated into a computational model. Next, the computational model is constructed by graphical elements: state variables (e.g. height, velocity); in- and out-flows of state variables (i.e. rates of change); auxiliary variables; constants (e.g. acceleration due to gravity); events (e.g. bounce) that provoke discrete, instantaneous changes of state variables; and relations that are visualised by connectors between variables, constants, events (Fig. 12.5) and are specified by simple mathematical formulas.

As the model is executed, differential equations behind the model are automatically solved by numerical iteration methods and so result in values of variables as a function of time. To interpret these modelling data, the modeller needs to choose relevant representations of the resulting values of variables such as (a) graphs that show more explicit, comprehensible relationship between variables; (b) animations that visualise behaviours of modelled objects. To validate the model (i.e. evaluating its descriptive, predictive and explanatory quality), the modeller compares modelling

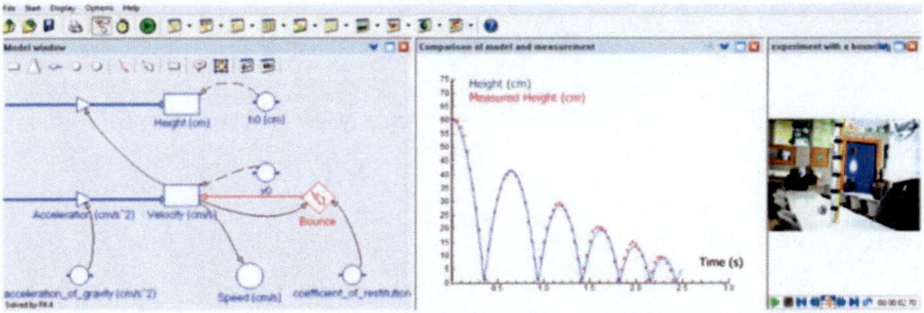

Fig. 12.5 Screenshot of a modelling activity facilitated by the modelling tool

outcomes with their counterparts in the physical world (i.e. standards, measured data, empirical graphs).

There are different ways to represent variables and relationships behind a dynamical model in Coach, including (a) the stock and flow mode (graphical representation) and (b) text-based modes, using equations or a textual representation. In this chapter, we confined ourselves to graphical modelling with a stock and flow representation of variables and relationships among these variables. This was because the stock and flow representations stimulate pupils to focus on qualitative relationships (theoretical world) and connections of these to the realistic situation (physical world) rather than mathematical equations or programming syntaxes.

12.4.3.2 Examples of Coach Modelling Activities

Figure 12.5 illustrates a modelling activity facilitated by the modelling tool. In this activity, bouncing of a solid, rubber ball is modelled, and the modelling result is compared with data obtained from video measurement of the bouncing ball. The graph shows the modelling result (solid curve) and the measurement (dots) for height versus time. Another example is a graphical model of the main span of Golden Gate bridge (Fig. 12.6), which is based on the approximation of the suspension cable by k_{max} straight line segments with horizontally equidistant joint.

Coach is in fact a hybrid system that combines a traditional system dynamics approach with event-based modelling. The left window of Fig. 12.7 shows a graphical model of a ball hanging on a vertical spring attached to the ceiling and that can also bounce against the ceiling; a special event-icon (with the thunderbolt symbol) is used to specify what should happen when the ball bounces. The window in the middle is an animation window that displays the simulation results as animations where model variables are presented as animated graphics objects. A student can interact with the animation through a slider bar, that is, select the value of the spring coefficient before the start of the simulation or change it while the simulation runs. Animation allows students to first concentrate on understanding a phenomenon with the help of simulations before going into the details of how the simulations have been implemented by means of computer models.

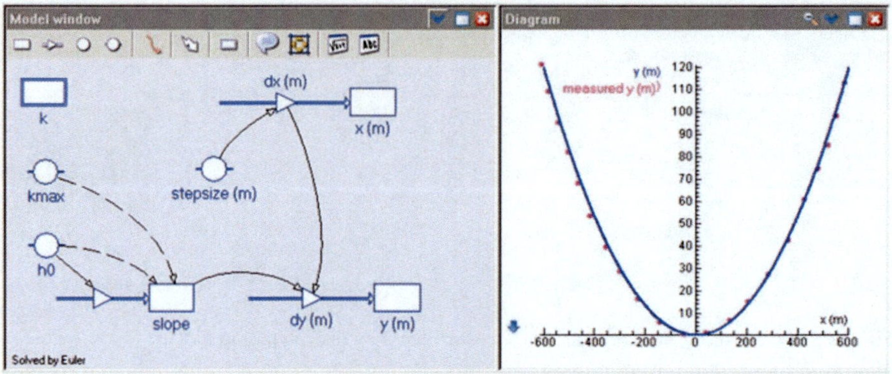

Fig. 12.6 A graphical model of the shape of the Golden Gate Bridge

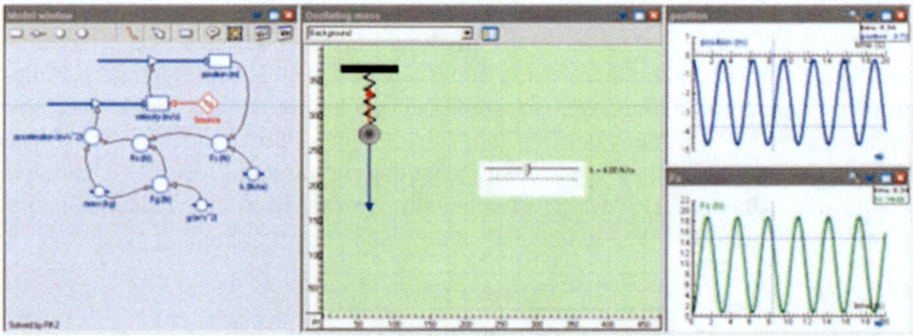

Fig. 12.7 A model of the harmonic motion of an oscillating ball hanging on a spring and an animation of the motion

12.4.3.3 Educational Benefits

First, the modelling tool holds the potential to enlarge possibilities for pupils' theoretical inquiry of realistic, dynamic phenomena (e.g. motion with air resistance, charging and discharging capacitors, combustion of carbon monoxide, and chemical equilibrium). These phenomena are difficult to describe with school mathematics but relatively easy to model with software (Heck 2009; Velanova et al. 2014). There are different patterns in which pupils can move back and forth between the theoretical and physical worlds and so learn with the modelling tool. For example, pupils run a given model (e.g. a parachute jump with air resistance) to understand a phenomenon and/or explore its structure to gain insight into interactions between the model elements. Based upon their understanding, pupils can also make a small change to a given model, try out various modelling ideas, and then evaluate if the revised model describes the phenomenon better. For example, "unfortunately, the parachute does not open right away. Therefore, there is first free fall for two minutes and then fall with air resistance while the parachute already opens". With a certain mastery of the modelling tool, pupils may construct a new model from their mental model of the realistic phenomenon and validate the model by comparing modelling outcomes with experimental results. Patterns for teachers to prepare a lesson, using the modelling tool are similar; the teacher might use a ready model, modify it a bit or develop a new model.

Second, the software allows importing measured data and graphs to the modelling activity. This enables simultaneous observations of the modelling graph (i.e. an outcome from the theoretical world) and the experimental graph (i.e. an outcome from the physical world) in the same diagram (Fig. 12.5). It is convenient for pupils to compare these outcomes of the two worlds. If the modelling result does not fit the real data, then pupils can adjust the model (e.g. changing parameters, adding variables, correcting relationships), execute it again, and compare new modelling results with the real data. The modelling tool enhances opportunities for many rounds of thinking back and forth between the theoretical and physical worlds.

Last but not least, the incorporation of the modelling tool enables pupils to (a) get used to modelling as a scientific tool in computational science (doing science with computer), (b) appreciate what modelling is as a way of thinking, (c) understand how important it is in science, and (d) develop a critical attitude by working with several models for one and the same phenomenon (i.e. modelling cycle). In their article, submitted in November 2008, Heck and Ellermeijer (2009) used Coach video measurements and models of runners to predict the possible time: 9.6 s for 100 m of Usain Bolt based on his Olympic run in Beijing 2008. This model accurately predicted or apparently affirmed his world record at the 2009 World Championships in Athletics in Berlin (9.58 s). This instance illustrates the power of the Coach tools in explaining and predicting real-world phenomena like sprinter's run. Additionally, Heck and colleagues showcased students' research projects (i.e. yoyos, alcohol metabolism, beer foam, bouncing balls) in which pupils could build models from simple to more complex (i.e. progressive modelling approach) by incorporating more factors aimed at better matching between the model and reality (Heck 2007, 2009; Heck et al. 2009a).

12.4.4 Data Processing and Analysis in Coach: Generic Components of the Coach Tools

Generated from the model or collected from the experiment or the video, numerical data can be then quickly transformed into more comprehensible, graphical representations: graphs, tables and animations. If such representation forms are arranged in advance, then pupils can see, for example, how empirical graphs appear (i.e. real-time graphing) or how animated objects move during the measurement or generation of data.

Additionally, just requiring simple manipulations, the software provides pupils with many possibilities for elementary analysis such as scan, slope, area and further processing such as (a) fitting or modelling the data with analytic functions (e.g. function fit); (b) integrating, differentiating data; and (c) displaying Fourier transforms of the data (Sokoloff et al. 2007; Heck et al. 2009b). Moreover, swift analysis and processing with three tools save time on labour intensive, repetitive tasks (e.g. drawing graphs) and so allow pupils to focus on inquiry skills like interpreting data, inferring relationships, and testing different assumptions, which are otherwise impossible due to time constraints. These features of data processing and analysis crucially add to specific characteristics of the three Coach tools mentioned above. This makes each of the ICT tools an authentic platform where pupils can easily move back and forth between the physical and theoretical worlds within the classroom time to generate or validate knowledge of science.

12.4.5 An Example of ICT in a Student Project: A Surprising Result

12.4.5.1 The Student Project with ICT: "Physics of Bungee Jumping"

The Dutch curricula require pupils to gain exposure to research projects in physics and other subjects where they must make their own choices with respect to topic, questions, and experiments/models; collect and analyse data; and compare outcomes with literature. A final investigation project is intended for 80 h outside of regular lessons and spread over a whole school year (about 2 h a week). In 2003, two Dutch students teamed up to investigate the physics of bungee jumping. In the first phase of bungee jumping, the bungee jumper falls, and the bungee rope is still slack. In instructional material, this phase is often considered a free fall. Considering the mass of the bungee rope, the students formulated the research question: *How large is the acceleration at a bungee jump and to what degree is this acceleration influenced by the relative mass of the rope and the jumper?*

The students collected position-time data through Coach video measurements on a dropped scale model (an Action Man toy figure) and on dropped wooden blocks of various weights attached to ropes of various stiffness. The velocity and acceleration of the dropped object were computed by numerical differentiation. Soon the students realised that the mass ratio between rope and objects was too low to see an outstanding result and they repeated the experiment with objects of larger mass ratio. The graph of the acceleration at the moment that the block has fallen a distance equal to the rest length of the elastic as a function of the mass ratio of elastic and block is shown in Fig. 12.8, together with the graph of the following theoretical result:

$$a = g\left(1 + \frac{\mu(4 + \mu)}{8}\right) \tag{12.1}$$

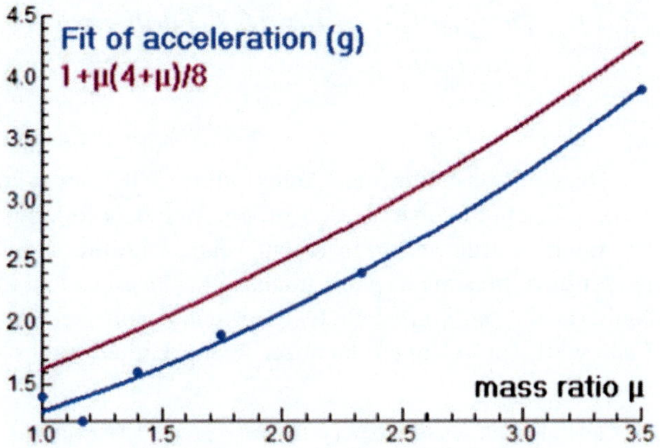

Fig. 12.8 Graphical display of experimental results (below) and computed values (above)

where μ is the mass ratio of the elastic and the wooden block. The students noted that the graphs obtained by measurement and theory are alike, with the theoretical values just a bit higher. They attributed the difference mainly to the development of heat during the motion.

12.4.5.2 The Surprising Result

Not knowing that a Dutch physics teacher had published around the same time about an experimental verification of the physics of bungee jumping, the students wrote an article about their work that was published in the journal of the Dutch Physics Society. The students' article claimed the result for acceleration during first phase "free fall" up to $a = 3.9\ g$ for mass ratio $m/M = 3.5$. It triggered quite several reactions in the journal and for almost a year on Internet. It seemed that a major part of the physics community, at all levels of education, was suddenly playing with ropes, chains, elastics, and so on. The result of the student project is contrary to the usual experience with free falling objects and therefore hard to believe by many a person, even by an experienced physicist.

It was a starting point for heated discussions about the quality of the experiments and the physics knowledge of the experimentalist, and it even prompted complaints about the quality of current physics education in the Netherlands. However, experiments did reveal the truth, and students could do this supported by ICT tools. Two theoretical physicists agreed with the findings of the students, and they explained that physics intuition is easily fooled, as everyone is taught the Galilean paradigm of the motion of constant masses, according to which acceleration must be produced by a force. A launched rocket and a falling chain or slinky are important counterexamples to this line of thought. As can be seen in the theoretical section, believing the statement $a > g$ means giving up or generalising the law $F =$ ma. For other bungee-jumping experiments, which investigate the phenomenon further and make more use Coach, we refer to Heck et al. (2010).

12.4.6 Integration of the ICT Tools in Recent Physics Curricula

Already for a long time, physics curricula have included laboratory activities. Together with a widespread integration of ICT in schools, there have been more and more school curricula that incorporate data logging with sensors in the science laboratory. Furthermore, in recent years, learning about modelling and learning to model have become explicit goals of science curricula in many countries (e.g. the Netherlands, the United States, Germany, and the United Kingdom). The current framework for science education in the United States (NRC 2012), for example, stated that

> Curricula will need to stress the role of models explicitly and provide students with modelling tools so that students come to value this core practice and develop a level of facility in constructing and applying appropriate models (p.59).

Modelling is now included in the Dutch Physics curriculum (and in Chemistry and Biology) for the pre-university track in secondary schools. Recently, a learning path to achieve modelling skills has been developed for the Dutch lower secondary physics curriculum. This learning path is completely integrated into the curriculum and has been tested in school practice (van Buuren 2014).

Looking back at thirty years of research on the use of ICT in education, curriculum development, and software/hardware development, it is fair to say that a lot has been achieved. Hardware and software development, including the development of the working environment—Coach, has been able up to now to meet more or less the requirements of trends in STEM education such as the change towards context-rich education, emphasis on scientific approaches, better preparation for higher education through a stronger focus on competencies, and emphasis on individual learning and provision of students' autonomy over the process of knowledge and skills acquisition. This work will undoubtedly continue, due to the very nature of technology and education, and new demands from society.

12.4.7 Conclusions About Technology for Inquiry Practices in Physics Education

It has been demonstrated that the Coach tools for data logging, video measurement, and modelling can stimulate inquiry by pupils. The proper use of these tools enhances opportunities and time for pupils' generation and validation of knowledge in the classroom. These meaningful opportunities and enough time enable pupils to move back and forth between the physical and theoretical world within the inquiry process. Students can work directly with high-quality, real-time data in much the same way professionals do. Physics learning with the ICT tools resembles practice in contact with current research work. With these ICT tools, investigations are characterised as being challenging, complex, and open-ended, cross-disciplinary, and requiring strong commitment and broad range of skills.

12.5 Integration of Technology into Inquiry: Challenges

12.5.1 Integration of ICT into IBSE in Teaching Practice

12.5.1.1 ICT and IBSE in Teaching Practice

ICT, IBSE and ICT in IBSE are still not implemented sufficiently and properly in the classrooms in most countries. Abrahams and Millar (2008) summarised results of observations in 25 typical laboratory lessons in the United Kingdom as follows:

> Practical work was generally effective in getting students to do what is intended with physical objects, but much less effective in getting them to use the intended scientific ideas to guide their actions and reflect upon the data they collect (p. 1945).

Teachers sometimes try to apply ICT tools (e.g. data logging with sensors) in the classroom, but mostly in traditional ways in which they provide a prescriptive list of tasks for pupils to follow ritualistically. Meanwhile, pupils know in advance from the textbook what the results of the practical work should be (Hofstein and Lunetta 2004). Prescriptive instruction enables pupils to operate the ICT tool on their own (manipulation of equipment), but such instruction limits inquiry opportunities for pupils (manipulation of ideas).

International ICT projects (e.g. KLiC and ICT for IST) highlighted possibilities and good practices of ICT tools for IBSE but were not of the scale needed to bring about the substantial impact on classroom implementation. At the national level, in many countries (e.g. Greece, Russia and Ireland), governmental education-reform projects have invested in ICT apparatus and software, delivered it to schools nationwide, and provided teachers with short training. Meanwhile, knowledge-oriented curriculum objectives and paper-and-pencil assessments were not reformed. Eventually, sufficient and proper implementation of the ICT tools was not realised. In this situation, projects with a huge funding failed to make the intended change in the classroom due to lack of consistent and concerted efforts to design, align, and implement curriculum innovation regarding ICT integration.

12.5.2 Integration of ICT in IBSE: Challenges to Pupils

Pupils' authentic inquiry in the classroom should be similar to, but cannot be the same as those of scientists, because interests, background knowledge and motivations of pupils are enormously different from those of scientists (Edelson 1998). "Real" authentic inquiry is open-ended and requires a solid discipline-knowledge base. As Ogborn (2014) claimed, real inquiry takes some years and requires scientists "full critical attention", whereas "replicated" inquiry in the classroom is often intended for very limited time (i.e. half an hour) and relies on pupils' "intuitive responses". These responses "will most often be wrong or misguided, yet seem good to them, and be difficult to counter" (p.42). Reviewing several studies that empirically examined the inquiry learning process, de Jong and van Joolingen (1998) identified "intrinsic problems" of pupils in inquiry learning. These problems are related to hypothesis generation, design of investigations, interpretation of data, and regulation of learning. Consequently, IBSE might generate a heavy cognitive load for pupils (Kirschner et al. 2006).

12.5.3 Integration of ICT in IBSE: Challenges to Teachers

Even without ICT integration, proper implementation of IBSE activities is still a problem for teachers. A number of studies have reported that teachers often find it difficult to elicit pupils' ideas about the research questions, to guide pupils in planning, executing investigations and analysing, interpreting data, and to manage pupils' independent learning at different paces (Davis et al. 2006; Hofstein and Lunetta 2004). ICT provides innovative tools for IBSE, but the use of these tools will "further complicate the complex web of overlapping factors, which characterise pedagogical thinking involved in planning and executing lessons" (Rogers and Twidle 2013, p. 229).

Effective ICT integration assumes that teachers have to learn possibilities of the ICT tools for their subject, acquire skills to operate the software and hardware, and get used to troubleshooting technical problems. More importantly, teachers need to adapt and improve their pedagogical knowledge to be able to design suitable ICT in IBSE activities and engage pupils in implementation of such activities in the classroom.

12.5.3.1 Inquiry Teaching Versus Prescriptive Instruction with ICT

Pupils need sufficient instruction and practice time in order to handle laboratory equipment in the classroom. Recipes like "do this, then do that" might be the fastest and most convenient way in helping pupils to get over hurdles in manipulating equipment within a limited time, but it hinders pupils' minds-on inquiring and meaning making. Research on practical work often uses the "cookbook" metaphor to describe this prescriptive instruction (Fig. 12.9).

Considering the ICT integration, extra instructions needed to handle the necessary software might further reinforce the prescriptive nature of ICT-enhanced experimentation/modelling activities. These instructions can unintendedly come to dominate the activity, although there are practical ways to get around this. For example, after a cookbook phase for learning to manipulate the tools should come

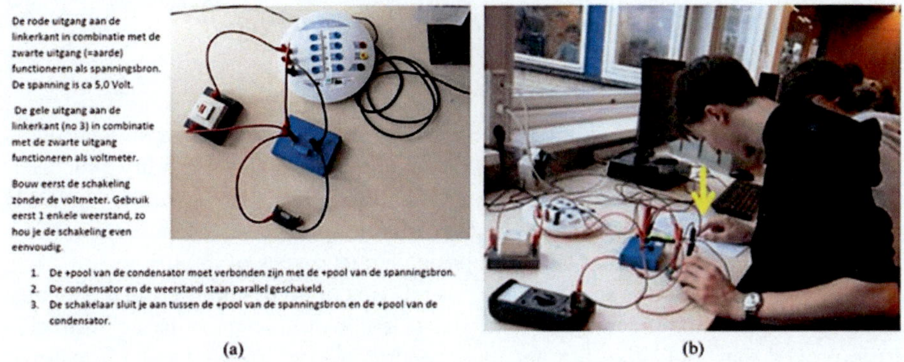

Fig. 12.9 In the photograph (**b**), a pupil was following a cookbook instruction (**a**) to set up the experiment using sensors and the computer

an inquiry phase in which pupils themselves have to make decisions on how to use the new tools in their investigations.

12.5.3.2 Limited Preparation Time and Limited Curriculum Time

In many countries including Slovakia and Vietnam, teachers are pressed to teach all the content standards within a tightly structured, explicitly expressed syllabus (Woolnough 2001). In the Netherlands, the school science curricula are rather overloaded with standardised requirements, too; except that the way to attain these requirements is mostly left open to the teacher. In such constrained circumstances, teachers tend to get through the content (Bencze and Hodson 1999) rather than engaging pupils in investigations with the ICT tools. This is because these investigations require ample time and aim at not only conceptual learning but also other goals. Incorporating the ICT tools in school science might add extra problems to time-constraint situations, considering that teachers need extra time, effort, and/or training to learn to handle the ICT tools and to practice their use. Logistics of organising the use of hardware and software cause extra preparation time for teachers. The constraints on preparation time and curriculum time can be a factor explaining for the fact that: although the ICT tools have many innovative features that stimulate pupils' authentic inquiry practices, really authentic inquiry learning is limited to a few special projects, for example in Dutch schools, once in junior secondary and once in senior secondary.

12.5.4 Conclusions About Challenges of Technology Integration in Inquiry Practices

The science education community mostly agrees about the relevance of using ICT, IBSE, and their integration: ICT in IBSE for pupils exercising inquiry practices, acquiring inquiry skills, and understanding scientific inquiry. However, ICT in IBSE is still very much under-used and applied at a relatively small scale in most countries. When it is used, the use often lacks the basic characteristics of inquiry.

Although integration of ICT into IBSE is relevant, it is cognitively and practically challenging for both students and teachers. ICT-enhanced inquiry-based strategies have proved their potential, but not their general efficacy in the hands of average teachers. Science teachers need effective training and practical guidelines to handle the complexity of the ICT-enhanced, inquiry-based activity without reducing it to simple "cookbook recipes". Furthermore, integration of ICT into IBSE needs sufficient time to be faithfully implemented. An actual impact of ICT in IBSE teaching on pupils needs both a longer period and consistent incorporation of ICT in IBSE in regular teaching.

12.6 Development of a Short and Effective Course for Teachers on Technology in Inquiry-Based Teaching of Physics

12.6.1 Aim and Research Questions

Factors involved in the under use of ICT and IBSE in teaching practice are among others: (a) limited curriculum time and limited teacher preparation time; (b) mismatches of the IBSE goals with commonly used lesson materials; teaching methods; and assessment and examination (e.g. prescriptive nature of materials and methods, predominance of content over inquiry goals), and (c) insufficient teacher preparation and training on integrating ICT into IBSE. All these factors need to be changed consistently and in concert to realise proper incorporation of ICT and IBSE into a classroom where *manipulation of equipment and software* is turned to *manipulation of ideas and concepts* for knowledge generation and validation.

Within our recent research project (Tran 2016), we focussed on preparation and training of science teachers on ICT in IBSE teaching and developed an effective and relatively short course for student teachers and teachers with diverse teaching experience. The present research confined the ICT in IBSE teaching to (a) three Coach tools: data logging with sensors, video measurement, and dynamical modelling and (b) the use of these tools to support inquiry by pupils. In an ICT in IBSE activity, the pupils should have some role not only in executing the experiment/model but also to some extent in formulating research questions, designing the experiment/model, and interpreting the results. We developed a short course so that it can be accommodated within typical overloaded teacher-education programmes or adopted as an in-service course. Furthermore, educational theories and products, such as our ICT in IBSE course, do not always travel well as educational and cultural contexts in different schools and countries can be very different. That is why the present research included three case studies in The Netherlands, Slovakia and Vietnam and in pre- and in-service teacher education. That way we could test the transferability of our course design and the generalisability of the pedagogical principles at the basis of this design.

The aim of the present research was twofold. First, the objective was to design a short course, which—with some adaptations—will be effective in widely different educational settings. Second, this research was to investigate the validity of pedagogical principles, which were used to guide (a) the design, implementation, evaluation, and optimisation of the course and (b) the extent to which the course can be adjusted to the different settings in the Netherlands, Slovakia and Vietnam. The pedagogical principles are at a higher level of abstraction and intended to be generalisable across educational and cultural contexts.

The design research approach was applied as it can provide guidelines and scientific reasoning for such a research and design process, which was guided by the two research questions.

First, *what are characteristics of an effective, short course for Dutch student teachers to learn to apply the ICT tools in IBSE?*

Second, *to what extent is the course applicable in different educational and cultural contexts of pre- and in-service teacher education in different countries (i.e. the Netherlands, Slovakia and Vietnam)?*

12.6.2 Course Design and Research Design

12.6.2.1 Objectives of the ICT in IBSE Course

The general aim of the course was elaborated into the four objectives as follows:

1. Awareness objective: participants become aware of educational benefits of the ICT tools in science education.
2. ICT-mastery objective: participants master skills to operate the ICT tool.
3. ICT in IBSE objective: participants can design, implement, and evaluate an ICT in IBSE lesson.
4. Motivation objective: participants are motivated to continue studying the ICT tools and trying out ICT in IBSE lessons with pupils.

The ICT in IBSE objective (3) was considered as the main objective of the course. In order to reach this objective, participants had to achieve a certain minimum level of mastery of the ICT tools (2). The awareness objective (1) and motivation objective (4) were aimed at the course's long-term effects on participants' teaching practice.

The Coach platform for data logging, video measurement, and modelling was used together with available support materials (i.e. Coach introductory, tutorial and exemplary activities) and materials that we developed (i.e. forms for designing and self-evaluating the ICT in IBSE lesson).

12.6.2.2 Pedagogical Principles Underlying the ICT in IBSE Course

The literature on design research and on professional development of teachers led us to the following pedagogical principles as the basis for (re)designing, evaluating, and optimising the ICT in IBSE course:

1. *One theory-practice cycle*: participants are required to go through at least one complete cycle of designing, implementing and evaluating an ICT in IBSE lesson within the course. Participants will apply the IBSE theory in a design for an ICT in IBSE lesson, which they will also try out in the classroom, self-evaluate, and report in the final session of the course.
2. *Distributed learning*: participants study in live sessions and carry out individual assignments in between the sessions with the support materials and in consultation with the course instructor. Learning time is distributed between live sessions and individual assignments but is also carefully distributed over a longer period to provide opportunities for a well-planned tryout in a real classroom.
3. *Depth-first*: participants are introduced to the possibilities of the three tools after which they specialise in only one ICT tool. Learning time is prioritised for an in-depth study and application of one tool (one-tool specialisation) rather than broad study of all three tools at a more superficial level, so depth-first—breadth-later.
4. *Ownership of learning*: participants have freedom to select what to learn and how to learn it, using the course scenario and support materials in order to achieve the course objectives. The individual participants pursue their self-tailored learning process in which they make their own choices regarding the tool, the grade level, topic and activity for their ICT in IBSE tryout with pupils.

The four objectives, four pedagogical principles, and support materials together form the general design of the ICT in IBSE course.

12.6.2.3 Evaluation of the ICT in IBSE Course Through Three Case Studies

To examine the effectiveness of the ICT in IBSE course; validity and generalisability of the pedagogical principle; and transferability of the course design in different education contexts, we conducted the Dutch, Slovak and Vietnamese case studies (Fig. 12.10). These three case studies were related; the Dutch case study was the earliest and most extensive, followed by the Slovak case study, and then the Vietnamese case study. All three case studies (a) concerned the same questions

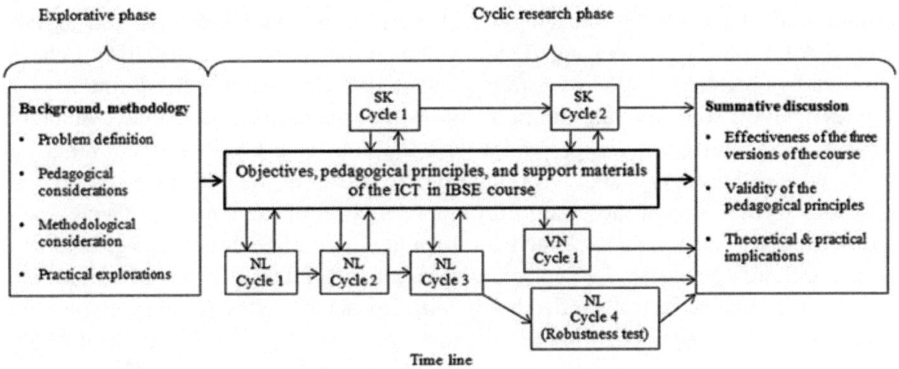

Fig. 12.10 Design and research process including an explorative phase and a cyclic-research phase with three case studies in the Netherlands (NL), Slovakia (SK) and Vietnam (VN)

about implementation of the pedagogical principles and course design, usefulness of the support materials, and attainment of the course objectives, (b) applied the same evaluation framework, and (c) used the same instruments for data collection and analysis. The course evaluation was guided by two main questions:

A. *To what extent were the four pedagogical principles implemented as intended?*
B. *To what extent did the ICT in IBSE course achieve its four objectives?*

Question B involves the evaluation of the effects of the course on participants, which resulted from actual implementation of the pedagogical principles, the course design, and the support materials. The evaluation of this actual implementation was guided by Question A and based on a comparison between (a) the intended course programme and (b) the actual activities of participants during the course. To evaluate attainment of the course objectives, we first operationalised performance levels for each objective. The definition of these levels was based on theoretical considerations and aligned with time-constraint conditions of the course. After that, we collected data, compared the data-analysis outcomes with the pre-defined levels of the course objectives, and concluded which level(s) of each objective the participants achieved.

This evaluation framework includes instruments for data collection and analysis (i.e. pre-course, post-course, and follow-up questionnaires; observations and video recording of live sessions and classroom tryouts; participants' ICT in IBSE lesson plans and self-evaluation reports of the classroom tryouts; computer performance test for each tool; the inquiry-analysis inventory; and the communication records). With these instruments, data were collected from a variety of sources and by different data collectors (i.e. the researcher, the course instructor, course participants). Accordingly, we could record both intended and possibly unintended outcomes as the course was implemented. Most outcomes were evaluated by more than one instrument thus allowing for data triangulation.

In the Dutch case study, we further operationalised the pedagogical principles in the initial scenario of the ICT in IBSE course. With "scenario", we mean the programme of the course and all instructor and participant activities and assignments. After that, we implemented and evaluated the course with 40 physics/chemistry student teachers spread over four sequential cycles. Among these four cycles, Cycles 1 and 2 were for fine-tuning of the course scenario. The course evaluation (Questions A and B) and experiences with the course in Cycle 1 (incl. What did work, what did not work, and why) suggested revisions of the initial scenario. These revisions were aimed at more faithful implementation of the course in Cycle 2 and with respect to many factors such as diversity of participants' background and ability; school schedules; and curriculum time for ICT in IBSE tryouts. Likewise, the Cycle 2 evaluation was guided by the objectives and pedagogical principles and resulted in further optimisation of the course scenario. We achieved faithful implementation of the four principles in Cycle 3. Consequently, in this cycle, the summative effects of the Dutch version of the ICT in IBSE course were evaluated, and only minor suggestions were made for further optimisation. The robustness of the course design and the ecological validity of the pedagogical principles were tested in Cycle 4 under routine implementation conditions without the extra support of the researcher.

The new understanding of how the course was developed and why it was effective (Dutch context) together with the basic course design (including course objectives, pedagogical principles and support materials) enabled the tailoring of local versions of the ICT in IBSE course in different contexts. The ICT in IBSE course was adapted and tested in (a) two cycles with 66 physics/biology/chemistry teachers with diverse teaching experience (1–33 years of teaching) in Slovakia. The two cycles of the Slovak course were already in routine implementation conditions without the direct participation of the researcher. The ICT in IBSE course was adapted and tested in one cycle in Vietnam with 22 master students in physics education, who either had taught for 2–9 years or came straight from a Bachelor teacher-education programme. Evaluations of the three local versions of the course enabled us to draw conclusions about (a) the extent to which the four objectives can be attained, (b) the validity and generalisability of pedagogical principles, (c) the transferability of the course design, and d) the practical relevance of the course. These evaluations led us to new understanding of the extent to which the pedagogical principles can guide the fine-tuning of the basic design of the ICT in IBSE course to varying boundary conditions.

12.6.3 Findings, Discussion and Conclusions

The Dutch case study resulted in an improved and successful course scenario in which the Dutch participants achieved the course objectives also when the course was taught under routine conditions (Cycle 4). A typical reaction of one of the participants was:

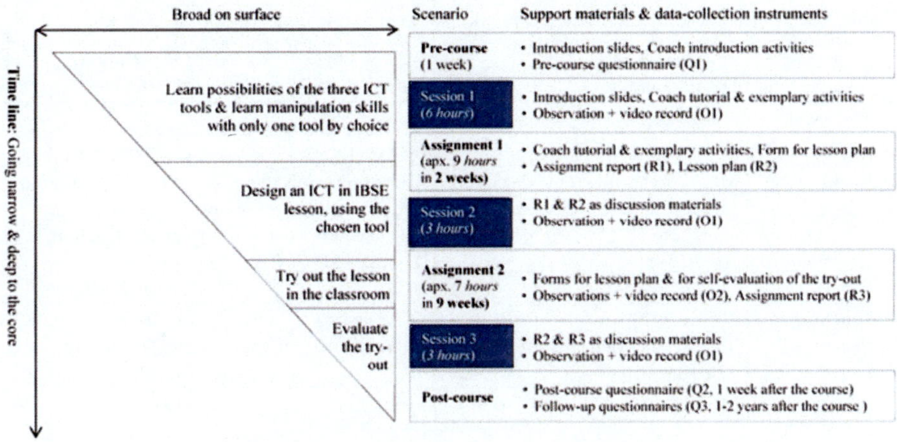

Fig. 12.11 The optimised scenario of the ICT in IBSE course in the Dutch context

> Very relevant, I am happy that the course was offered and that I took it. In the teacher-education programme, we learn a lot of theory but for the implementation of it, I feel rather on my own, with very little guidance. The ICT in IBSE course was very hands-on (though the theory was also presented), gave me the opportunity to learn with and from others (knowledgeable, friendly experts and other (beginning) teachers). I got the guidance I needed to develop an activity and implement it, and the feedback after completion. The duration was fine. I strongly advise the training to other student teachers

Figure 12.11 visualises the optimised scenario, illustrates the specific operationalisation of the pedagogical principles in the Dutch context, and describes the time distribution for the three live sessions and two assignments and the plan on when to use which support materials and data-collection instruments. For more detail about the programme for the live sessions, the requirements for the individual assignments, and the support materials, see Tran (2016).

The iterative evaluation and refinements of the Dutch course confirmed the validity of the pedagogical principles in the Dutch context. The support materials proved necessary and useful for the sufficient implementation of the pedagogical principles and the satisfactory attainment of the course objectives. To conclude, the four course objectives, the four pedagogical principles, and the optimised scenario with the support materials establish the core characteristics and basic design of an effective short ICT in IBSE course for Dutch student teachers. Findings from the iterative refinement of the course show that fine-tuning the distribution of time and individual assignments is crucial as far as distributed learning is concerned. Direct, personalised support (in live meetings and/or via emails) and sense of direction (via explicit support framework plus assignment tracking and stimulation) are crucial factors to ensure effectiveness of independent learning, especially the quality of the ICT in IBSE lesson plan. These factors create a balance between much freedom of choice and appropriate guidance, which is essential to establish ownership of learning.

Table 12.1 Differences among Dutch, Slovak and Vietnamese contexts for the course

	Dutch context	Slovak context	Vietnamese context
Participants			
• Number	40 student teachers	66 experienced teachers	22 master students
• Background and age	Mix of fresh-master and second career graduates; physics & chemistry Age: 23–55	Mix of physics, chemistry, biology and geography teachers Age: 23–55	Mix of fresh graduates and experienced teachers, Physics Age: 23–31
• Teaching experience	1–5 years 83% at first-year teaching	1–33 years 19 years on average	0–9 years 23% with no teaching
• Entrance level of ICT skills	Moderate	Low	Low
Scheduling requirements			
• Programme	Postgraduate teacher education	Accredited professional development	Master in physics education
• Total study time of the course	28 h	40 h	60 h
• "Spread" of the course	11 weeks	15 weeks	5 weeks
Teaching conditions			
• Availability of the ICT tools	Sufficient *Available in most schools*	Limited	None
• Pupil experience with ICT	Sufficient *Pupils have ever used Coach/similar software*	Insufficient *ICT is starting to be introduced*	None
• Pupil experience with IBSE	Insufficient *Certain experience with laboratory but less with IBSE*	Poor *A little experience with laboratory but very limited with IBSE*	None
• IBSE in curriculum	Explicit and required	Starting to be required	In new 2016 curriculum
• Teacher autonomy	High	Moderate	Low

The Dutch, Slovak and Vietnamese contexts for the ICT in IBSE course were different in many aspects (e.g. scheduling requirements, school conditions and characteristics of participants) (Table 12.1). First, the Dutch course was limited to 12 contact hours out of 28 h of total study time, but it was spread over 11 weeks. The Vietnamese course was compressed in 5 weeks, but 30 h out of total 60 study hours were scheduled for live activities. The Slovak case had the least constraints, regarding both contact hours (25 out of total 40 study hours) and "spread" of the course (15 weeks). Second, the Dutch school conditions (e.g. curriculum time, teacher preparation time, national examinations, pupils' experience with ICT and IBSE, availability of equipment and software) were not excellent but sufficient. Meanwhile, the Slovak school conditions were insufficient, and the Vietnamese conditions were very poor. Third, the Vietnamese and Slovak participants were experienced teachers, but their ICT-mastery entrance level was low. The Dutch participants had more experience with the ICT tools and felt freer to decide their own lesson objectives and teaching methods. However, they lacked teaching experience, especially classroom management skills. Vietnamese teachers work in an education system with a strong hierarchical culture and much less autonomy than in the Dutch system. Lessons are teacher-centred and there is no tradition of open learner investigations in secondary school and teacher education. All three groups of participants lacked practical experience with inquiry teaching with or without ICT, so ICT in IBSE teaching was challenging for them. For all three versions of the course, diversity of participants and time constraints were challenging contextual factors.

Across the three case studies, the *awareness* and *motivation objectives* of the ICT in IBSE course were achieved as expected. The participants could enumerate relevant benefits of the ICT tools. They devised plans and continued studying the ICT tools and teaching ICT in IBSE lessons after the course. About the *ICT-mastery objective*, all three groups of participants were able to operate the Coach tool fluently after the course. Compared with the Dutch participants, the Vietnamese participants attained a higher mastery level for the chosen tool, and the Slovak participants achieved a similar ICT mastery but with all three ICT tools. This shows effectiveness of the many more contact hours with direct, personalised support scheduled for the ICT-mastery objective to compensate for the low ICT entrance of the Slovak and Vietnamese participants.

About the *ICT in IBSE objective*, all three groups of participants were able to design and realise acceptable ICT in IBSE lessons considering their teaching conditions and their inexperience with inquiry teaching with ICT. The Dutch participants could design and realise better ICT in IBSE lessons than the Slovak and Vietnamese participants. Many Dutch participants were able to engage pupils in designing experiments or models and predicting and interpreting results as expected. Meanwhile, the Slovak and Vietnamese participants focused too much on pupils' execution of experiments or models (manipulation of equipment and software) and did not sufficiently involve pupils in moving back and forth between the physical and theoretical worlds (manipulation of ideas and concepts). Most Slovak and Vietnamese participants intended to take control over the entire classroom activity through plenary systematic explanations and/or prescriptive worksheets for the group work. In contrast, in half of the Dutch ICT in IBSE lesson plans, pupils were required to take a larger role in conception, planning and interpretation of the experiment/model in more-open inquiry patterns. This shows a clear difference in teacher/pupil centeredness and education culture among the three countries.

Although familiar with theory of IBSE, all three groups of participants had trouble to operationalise real inquiry in lesson plans and even more so in the classroom. There were many deviations between intended and actual ICT in IBSE lessons, and these resulted from reasons such as shortcut of intended inquiry opportunities; tasks that were too demanding; over ambitious timing; and ineffective communication with pupils. However, Dutch, Slovak and Vietnamese participants were able to identify the shortcomings in their ICT in IBSE lessons and suggest relevant revisions of their lesson plans for future use. To conclude, the basic design of the ICT in IBSE course was *effective, practical and transferable* in the different educational and cultural contexts of pre- and in-service teacher education in different countries. The course can cater to diverse groups of teachers and teacher-education programmes, and it fits into time-constraint conditions. For all three cases, the ICT in IBSE course achieved its objectives to the pre-determined acceptable level, except that for the ICT in IBSE part there was still much room for improvement.

Considering the issue of teachers learning to teach by inquiry, we prepared and expected our course participants to get their first experience with inquiry teaching with ICT. The theory-practice cycle was valuable to make them *more aware* of what IBSE involves, of what are differences between guided versus open inquiry, and of how to involve pupils in planning and interpretation of an experiment. It was concluded that the educational and cultural system influences teachers' perception and implementation of inquiry-based teaching with ICT. This results in different typical patterns of ICT in IBSE in different countries. The analysis of the lesson plans and classroom tryouts using the inquiry-analysis inventory revealed considerable *inconsistency between inquiry objectives and activity specifications* and noticeable *deviations between intended and actual IBSE lessons*. These are persistent problems, which have been reported worldwide (Abrahams and Millar 2008; Abrahams and Reiss 2012; Tamir and Lunetta 1981). Many teachers do have problems to operationalise inquiry in the classroom, even in countries like the UK and the US where inquiry has been emphasised in the curriculum for a long time. Research findings from the Vietnamese case study shed light on challenges of and potential solutions to the application of IBSE in a hierarchical education culture. Obviously, the ICT in IBSE course under the time constraints does not push its participants far enough yet in the direction of inquiry teaching with ICT. Participants' achievement through the course is a starting point; more theory-practice cycles are needed to bring them further in such ICT in IBSE direction.

In the present research, the applied pedagogical principles proved to be valid in providing not only the framework for implementing, evaluating, and optimising the course in a specific context but also guidelines for effective adaptation of the course to varying boundary conditions. When adapting the course to a different context, the "one theory-practice cycle" principle should not be changed. Instead, the "depth-first" and "distributed learning" principles can be adjusted by the course instructor to some extent to the specific context, considering the entrance level and other characteristics of the participants and the scheduling requirements. The "ownership of learning" principle has to be enabled to provide a dial for participants to self-tune the course to their own interest and ability. The adjustment with distributed learning and depth-first makes the first flexible phase: ICT mastery, which can be lengthened (Slovak case) and compressed (Vietnamese case) in order to compensate for the low ICT entrance, accommodate diversity of the participants, and align their activities, assignments, and efforts with the intended attainment of the ICT-mastery objective. Such ICT-mastery attainment is necessary for the participants to be able (a) to design and teach the ICT in IBSE lesson and (b) to continue studying and using the ICT tools after the course. Among the four course objectives, the ICT-mastery objective can be achieved in a compressed course with sufficient contact hours, whereas the learning with respect to the ICT in IBSE objective needs to be distributed sufficiently to allow for a well-planned and mature lesson plan and curriculum time for classroom tryout. The support materials proved necessary, useful and robust in different contexts. This finding suggests that it is not always necessary to develop materials locally to have effective educational innovations. Instead, with certain adaptations, one can use existing materials.

12.6.4 Reflections on the Findings and Methods

Based on our positive experiences with one theory-practice cycle, we think that this principle should be wider applied in teacher education. Would it be possible to identify a small number of core practices and have student teachers go through one theory-practice cycle for each? For example, the study of formative assessment could be followed by classroom practice with embedded formative assessment and feedback. Regarding the depth-first principle, deeper understanding of one ICT tool has surplus value compared to partial understanding of all three tools, and it leads to better transfer to the whole ICT environment (breadth-later). This further suggests the application of the depth-first principle as part of a solution for content overload in teacher-education programmes.

Regarding the learning of ICT skills, collective practice of ICT skills in small groups is more effective than either individual practice at home or the practice under plenary step-by-step instructions to the whole class. Personalised, direct support from the course instructor and peers is essential for participants to get over initial hurdles of learning a new tool and to troubleshoot "Technological Content Knowledge" (TCK) problems. Beyond the basic manipulation skills, TCK problems involve:

- Advanced features of the ICT tool (e.g. how to add the empirical graph to the modelling activity and compare it with the modelling graph; how to make control bars to change constants and initial values of variables of the model).
- Participants' understanding of the phenomenon (i.e. concepts, laws and events) to be experimented or modelled with the ICT tool and general experimentation/modelling skills.
- Participants' knowledge of mental model behind the digital tool that enables (a) collection and modelling of information about the phenomenon (sampling, digitalisation, numerical iteration) and (b) representation, analysis, and processing of the collected/modelled data.

To troubleshoot TCK problems independently, participants need to understand how the ICT system works and to be confident and committed in searching and trying out solutions. Taking the computer performance test can be a first step to learn such troubleshooting skills.

At the beginning of the present research, we defined the unique objectives of the course to be developed and a clear view about design criteria. Based on these objectives and criteria, we chose the design research approach to develop, evaluate and optimise the course as an educational product through research, and this approach worked well. We started this research and design project with the pedagogical principles and concluded it with these principles as the core of the basic design of the ICT in IBSE course. These principles can be considered as independent, validated educational products, which teacher educators can "buy into" and use for broader aims than only ICT in IBSE integration. Pedagogical principles establish the theoretical model underlying the course design, provide guidelines and structure to the (re)design, implementation, evaluation and optimisation process, and help to communicate the design research to others. The role of pedagogical principles in design research is indeed essential. Moreover, in our design research, we incorporated (a) a "*robustness test*" step to try out the course under routine conditions and (b) a "*generalizability and transferability testing*" step to tryout the course in different programmes or even countries. We achieved successful outcomes with these steps. Consequently, we strongly recommend robustness and generalisability/transferability tests as part of design research.

Main limitations of the present research were that it only measured the quality of one theory-practice cycle, that it did not have an opportunity to measure the further development of participants in later ICT in IBSE activities in their classrooms, and that it did not measure pupil results. Learning to teach the IBSE way takes a longer trajectory than this course, and so our measurements only show the start of the participants' development. For the same reason, there was no point in measuring pupil achievement, as improvement of their inquiry skills would only become visible after a series of lessons rather than one lesson.

It is common practice that each teacher-education programme invents its own wheels. Our research outcomes through the development/adaptation, implementation, and evaluation of the ICT in IBSE course in the Netherlands, Slovakia, and Vietnam indicate that with careful design and well-chosen pedagogical principles, courses and other educational products could be fine-tuned and shared. The fourth and fifth Dutch course and the second Vietnamese course were implemented by the local course instructors without any involvement of the researcher. The third Slovak course is planned within a new large-scale national project, which is aimed at implementation of ICT tools across science subjects in Slovakia. The Dutch ICT in IBSE course (implemented with five batches already) is unique, since it is the only teacher-education course offered by several Dutch universities together, as far as we know. The Vietnamese ICT in IBSE course is unique as it is the only Master course of which design and materials were entirely developed outside the university. These institutionalisations do not happen often for general educational projects. These show not only the *practical relevance of the basic design* of the ICT in IBSE course for different educational and cultural contexts of pre- and in-service teacher education, but also suggest the possibility to have *more-productive standardisation* among teacher-education courses. Education should be less political-driven and more expert-driven!

References

Abrahams I, Millar R (2008) Does practical work really work? A study of the effectiveness of practical work as a teaching and learning method in school science. Int J Sci Educ 30 (14):1945–1969

Abrahams I, Reiss MJ (2012) Practical work: Its effectiveness in primary and secondary schools in England. J Res Sci Teach 49(8):1035–1055

Barton R (2004) Teaching secondary science with ICT. McGraw-Hill Education, Maidenhead

Beichner RJ (1996) The impact of video motion analysis on kinematics graph interpretation skills. Am J Phys 64(10):1272–1277

Bencze L, Hodson D (1999) Changing practice by changing practice: toward more authentic science and science curriculum development. J Res Sci Teach 36(5):521–539

Brasell H (1987) The effect of real-time laboratory graphing on learning graphic representations of distance and velocity. J Res Sci Teach 24(4):385–395

Braund M, Reiss M (2006) Towards a more authentic science curriculum: the contribution of out-of-school learning. Int J Sci Educ 28(12):1373–1388

Davis EA, Petish D, Smithey J (2006) Challenges new science teachers face. Rev Educ Res 76 (4):607–651

De Jong T, van Joolingen WR (1998) Scientific discovery learning with computer simulations of conceptual domains. Rev Educ Res 68(2):179–201

Dewey J (1910) Science as subject-matter and as method. Science 31:121–127

Duschl RA, Schweingruber HA, Shouse AW (2007) Taking science to school: learning and teaching science in grades K-8. The National Academies Press, Washington, DC

Edelson DC (1998) Realising authentic science learning through the adaptation of scientific practice. In: Fraser BJ, Tobin KG (eds) International handbook of science education, vol 1. Kluwer, Harrogate, pp 317–331

Ellermeijer AL, Landheer B, Molenaar PPM (1996) Teaching mechanics through interactive video and a microcomputer-based laboratory. In: Tinker RF (ed) Microcomputer–based labs: educational research and standards. Springer-Verlag, Amsterdam, pp 281–290

Gaskell PJ (1992) Authentic science and school science. Int J Sci Educ 14(3):265–272

Gopnik A, Meltzoff AN, Kuhl PK (1999) The scientist in the crib: what early learning tells us about the mind. Perennial, New York, NY

Gröber S, Klein P, Kuhn J (2014) Video-based problems in introductory mechanics physics courses. Eur J Phys 35(5):055019

Heck A (2007) Modelling intake and clearance of alcohol in humans. Paper presented at the International Conference on Technology in Mathematics Teaching (ICTMT8), Hradec Králov

Heck A (2009) Bringing reality into the classroom. Teach Math Its Appl 28(4):164–179

Heck A, Ellermeijer AL (2009) Giving students the run of sprinting models. Am J Phys 77 (11):1028–1038

Heck A, Ellermeijer AL (2014) Realizing authentic inquiry activities with ICT. In: Tasar MF (ed) Proceedings of the world conference on physics education. Pegem Akameni, Istanbul, pp 775–786

Heck A, Uylings P (2005) Yoyo joy. In: Olivero F, Sutherland R (eds) Technology in mathematics teaching, proceedings of ICTMT7, vol 2. University of Bristol, Bristol, pp 380–387

Heck A, Ellermeijer AL, Kedzierska E (2009a) Striking results with bouncing balls. Paper presented at the physics curriculum design development and validation, Nicosia

Heck A, Kedzierska E, Ellermeijer AL (2009b) Design and implementation of an integrated computer working environment for doing mathematics and science. J Comput Math Sci Teach 28(2):147–161

Heck, A., Uylings, P., Kędzierska, E., Ellermeijer, A. L. (2010). Cross-disciplinary, authentic student research projects. Paper presented at the SMEC 2010 conference: inquiry-based learning: facilitating authentic learning experiences in science and mathematics, Dublin, pp. 40–45

Hodson D (2009) Teaching and learning about science: language, theories, methods, history, traditions and values. Sense Publishers, Rotterdam

Hofstein A, Lunetta VN (2004) The laboratory in science education: foundations for the twenty-first century. Sci Educ 88(1):28–54

Jeskova Z, Kires M, McLoughlin E, Finlayson O, Ottander C, Ekborg M (2015) In-service and pre-service teacher education in IBSE – the ESTABLISH approach. In: Fazio C, Sperandeo Mineo RM (eds) Proceedings of the GIREP-MPTL international conference on physics education. Università degli Studi di Palermo, Palermo, pp 811–818

Kearney M, Treagust DF (2001) Constructivism as a referent in the design and development of a computer program using interactive digital video to enhance learning in physics. Australas J Educ Technol 17(1):64–79

Kirschner PA, Sweller J, Clark RE (2006) Why minimal guidance during instruction does not work: an analysis of the failure of constructivist, discovery, problem-based, experiential, and inquiry-based teaching. Educ Psychol 41(2):75–86

Laws P, Pfister H (1998) Using digital video analysis in introductory mechanics projects. Phys Teach 36(5):282–287

Minner DD, Levy AJ, Century J (2010) Inquiry-based science instruction—what is it and does it matter? Results from a research synthesis years 1984 to 2002. J Res Sci Teach 47(4):474–496

Newton L, Rogers L (2001) Teaching science with ICT. Continuum, London

NRC (National Research Council) (2012) A framework for K-12 science education: practices, crosscutting concepts, and core ideas. Committee on a Conceptual Framework for New K-12 Science Education Standards. Board on Science Education, Division of Behavioural and Social Sciences and Education. The National Academies Press, Washington, DC

OECD (Organisation for Economic Co-operation and Development) (2006) Evolution of student interest in science and technology studies: policy report. OECD, Paris. http://www.oecd.org/science/sci-tech/36645825.pdf

Ogborn J (2014) Curriculum development in physics: not quite so fast. In: Tasar MF (ed) Proceedings of The World Conference on Physics Education. Pegem Akademi, Istanbul, pp 39–48

Paavola S, Lipponen L, Hakkarainen K (2004) Models of innovative knowledge communities and three metaphors of learning. Rev Educ Res 74(4):557–576

Rogers L, Twidle J (2013) A pedagogical framework for developing innovative science teachers with ICT. Res Sci Technol Educ 31(3):227–251

Sokoloff DR, Laws PW, Thornton RK (2007) RealTime physics: active learning labs transforming the introductory laboratory. Eur J Phys 28(3):S83–S94

Tamir P, Lunetta VN (1981) Inquiry-related tasks in high school science laboratory handbooks. Sci Educ 65(5):477–484

Tran TB (2016) Development of a course on integrating ICT into inquiry-based science education. VU University Amsterdam, Doctoral thesis

van Buuren O (2014) Development of a modelling learning path. Doctoral thesis. University of Amsterdam, Amsterdam

van Buuren O, Uylings P, Ellermeijer AL (2010) Towards a learning path on computer modelling. In: Raine D, Hurkett C, Rogers L (eds) Proceedings of the GIREP-EPEC & PHEC 2009 international conference, vol 1. Lulu/The Centre for Interdisciplinary Science, Leicester, pp 110–125 van den Berg E (2013) The PCK of laboratory teaching: turning manipulation of equipment into manipulation of ideas. Scientia in Educatione 4(2):74–92

Velanova M, Demkanin P, Gergelova B, Demkaninova D (2014) The experienced physics teacher and her first experience with data-logger. In: Proceedings of INNODOCT/14 conference: Strategies for Education in a New Context. Editorial Universitat Politècnica de València, Valencia, pp 142–157

Woolnough BE (2001) Of 'knowing science'and of 'doing science': a reaffirmation of the tacit and the affective in science and science education. Can J Math Sci Technol Educ 1(3):255–270

Zollman D, Fuller R (1994) Teaching and learning physics with interactive video. Phys Today 47:41–47

Chapter 13
Designing Teaching Learning Sequences Based on Design-Based Research

Jenaro Guisasola, Kristina Zuza, and Ane Leniz

Abstract In this chapter, we put forward a proposal for the design of teaching and learning sequences in high school and university. We will connect our proposal with relevant contributions on the design of teaching sequences, ground it on the design-based research methodology, and discuss how teaching and learning sequences designed according to our proposal supposes an research on physics education. An iterative methodology for designing the teaching and learning sequence (TLS) is presented in the teaching of fundamentals of dc circuits.

13.1 Introduction

Since the 1980s of the last century, research in physics education has repeatedly shown that students do not achieve the expected learning. Secondary and university students, despite years of schooling, still have alternative conceptions to scientists (Duit 2009). From these works, research in physics education has developed design-based research approaches. The Design-Based Research is based on the fact that we can learn important knowledge about the nature and conditions of teaching and learning by trying to design and develop educational innovation in classroom settings. Design-Based Research (hereinafter DBR) necessarily includes the design, implementation, and evaluation of Teaching/Learning Sequences (TLS) as interventional research that generates new pedagogical knowledge (Kortland and Klaassen 2010).

TLS can be broadly defined as "an interventionist research activity and product, similar to a traditional curriculum unit, that includes well-researched teaching-learning activities empirically adapted to student reasoning" (Méheut and Psillos 2004). Therefore, TLS design reflects the interrelationship between the development of tools and learning environments and the development of theory. This

J. Guisasola (✉) · K. Zuza · A. Leniz
Department of Applied Physics, Escuela de Ingeniería de Gipuzkoa, DOPER Research Group, University of the Basque Country, Donostia-San Sebastian, Spain
e-mail: jenaro.guisasola@ehu.es; kristina.zuza@ehu.es

B. G. Sidharth et al. (eds.), *Fundamental Physics and Physics Education Research*,
https://doi.org/10.1007/978-3-030-52923-9_13

interconnection is a complex and cyclical process in which the principles of general education are applied to the teaching of specific subjects in specific contexts (Lijnse 2004). At each stage of development there are opportunities to test conjectures about student learning and to refine those conjectures based on experience, as well as to redesign the SEA proposal. As a consequence, researchers have defined development frameworks, to be used by designers, as interfaces between major theories and the needs associated with developing the SEA on specific topics. The improvement obtained in research based on the use of TLSs has been shown in some cases to be significant, even for teachers with little experience in using TLSs (Ametller et al. 2007; Savall-Alemany et al. 2019; Savinainen et al. 2017).

A broad coherence between theoretical or methodological efforts in experimental design approaches could be expected, but as we have seen in the literature this is not the case. Partly due to the great breadth of research covering a wide variety of school interventions with different specific aspects (cognition, cognitive development, teaching strategies, classroom interactions). DBR approaches have involved improving existing material by designing research-based teaching activities. Therefore, the teaching materials are designed based on the results of the research and can be evaluated accordingly. Most of these approaches share some common characteristics. They are generally informed by empirical evidence of students' previous conceptions, an epistemological analysis of objectives, theoretical perspectives on learning that are aligned with the cognitive constructivist approach and the specific educational context. Articles, such as those cited above, generally present evidence of student learning. However, less frequently it is discussed how the above considerations informed the design, implementation, and evaluation of TLS. Therefore, there are still significant gaps in this field of study. In particular, many articles on TLS design lack: (a) a detailed explanation of the implicit and explicit decisions made regarding design and implementation; (b) a detailed specification of teaching strategies, which are often implicitly addressed under the label of "active teaching" or "active learning"; (c) a comprehensive assessment procedure (that is, one that goes beyond the learning achieved); and (d) a detailed description of the iterative refinement process. The lack of such explicit descriptions makes it difficult for the science education community to interpret the results presented, propose a systematic design improvement, and build on the findings.

Although research based on DBR design has the potential to offer a useful set of methodological tools to deal with the above mentioned problems, there are issues that must be addressed in order for this methodology to be credible for the community. Questions such as: what are the key elements that define design-based research and what differentiates it from other types of research, or what evidence is offered to support the achievements of this type of research should be addressed. In this chapter, we intend to present the key elements of the DBR methodology that have been agreed by the community and that characterize this methodology. Second, we will present an example of the application of the DBR methodology focusing on the topic of "Fundamentals circuits" for introductory physics courses in the area of Mechanics. In this contribution we will show the part of the design of the TLS. In particular, examples will be given on three aspects: (a) to justify the contents of TLS;

(b) To evaluate students 'difficulties; (c) To define the learning demands. We will finish with a discussion on the advantages of the DBR methodology in order to make the TLS design criteria explicit. Making these criteria explicit helps teachers to understand the advantages and usefulness of the TLS and, furthermore, contributes to developing their pedagogical content knowledge.

13.2 DBR Phases and the TLS Design

DBR is a cyclical process that typically comprises three phases: design, teaching experiments (implementation), and retrospective analysis (assessment) (Anderson and Shattuck 2012; Easterday et al. 2014; Barab and Squire 2004).

The design phase leads to an initial product (the TLS) that includes a hypothetical learning path consistent with the theory. This phase starts with an evaluation of the learning goals for the target audience and the epistemological analysis of the topic. In the design phase, the clear and explicit definition of the learning objectives is key if we want to evaluate the achievements obtained. The epistemological analysis and findings about students 'difficulties guide the formulation of provisional learning objectives, which in turn shape the learning path of the TLS. Next, we use the "learning demand design tool" (Ametller et al. 2007) to analyze the ontological and epistemic differences between the students' ideas and the defined learning objectives. These gaps will guide the TLS learning path by highlighting both the type and degree of difficulty that we can expect the students to encounter.

In summary, the phase of design includes learning goals, information on students' relevant difficulties, structured activities to implement in classroom and guidance on how the teacher can support the teaching processes (see Table 13.1).

Once the objectives and sequence of the program have been established and justified, a first version of the activities that made up the sequence to be implemented is designed. The documents that are needed for the implementation of the SEA are generated, usually include work materials, based on epistemological analysis and learning demands, assessment guidelines related to defined learning objectives, and material for teachers with information on the use of work materials. Not all aspects of these materials are directly derived from the different elements of the design and, therefore, it is important to inform the teachers who are going to use the TLS, which

Table 13.1 Design phase

Analysis of the instructional context and of the epistemological aspects of the topic	Analysis of students' prior ideas, and conceptual and reasoning difficulties	Need for interactive environments that reflect the skills and attitudes of scientific research
Learning objectives	*Learning difficulties* and *Learning demands*	*Teaching strategies*
Building specific tasks that lead to a proposed learning trajectory *TLS activities* *Teacher's guide for implementing the TLS*		

aspects are referents of theory and research, and which are contributions personal designers based on their Pedagogical Knowledge of Content.

In the DBR methodology, implementation is the part where TLS is applied in the classroom. The evaluation stage of the DBR methodology involves an iterative process that carries out tests throughout the design and redesign process. In our proposal, the design of the TLSs is empirically validated, through two dimensions (Nieveen 2009). In the first dimension, the quality of the sequence is analyzed, that is, the ability of the teaching material to involve students in the task. In particular, implementation difficulties are analyzed in relation to the students' understanding due to drafting problems of the activities. Implementation time problems and difficulties in designing new content activities are evaluated. In the second dimension, the learning achieved by the students in relation to the defined objectives is evaluated. This evaluation includes understanding concepts, laws and theoretical models and, using the procedure of scientific activity to solve problems.

In this chapter, we will show two examples. First example of the design of the DBR, that is, about the first phase of the DBR (design TLS) in the topic of DC circuits for introductory physics courses. Second example is about testing the quality of design by the redesigning of the activity.

13.3 Theoretical Informed Learning Goals of the Sequence

This section provides a general description of the design process, which shows how the epistemological and cognitive analysis underpins the election of the design learning aims. On the first place, we will show how general educational theories underpin the design tools, which we use to carry out the "fine grain" analysis of the specific contents which will be taught in the TLS during the first phased of the DBR. The first step is focused in teaching context. In our case the context is the topic of the "Fundamentals of DC circuits" addressed to Physics introductory courses at University. The aims of the syllabus are defined according to the educational level of application.

Firstly, the contents of the curriculum set out in the standards for the chosen level of education are taken into account. Secondly, we consider the need to justify the sequence of contents of the program. In these first two steps, the epistemological analysis of the content will help to justify the contents and laws included, as well as their sequence. The epistemological analysis includes the historical development of the topic, the difficulties that the scientific community had to overcome and the arguments that were used to build new explanatory concepts and models. Epistemological analysis allows us to select the key aspects of the topic that contribute to defining learning objectives. The epistemological analysis of the content is used as a didactic tool to avoid definitions of the program based on the idiosyncrasies of the designers or traditional curricular choices. Furthermore, the defined objectives can be used to sequence the main problems that must be solved to learn the contents of the program.

Next, we proceed to explicitly justify the chosen content and its sequence. As we have indicated previously, many of the design-based research proposals do not explicitly justify the decisions made regarding the content program and its sequence. The design of a TLS based on the DBR methodology must show the most outstanding elements of epistemological analysis—paradigm shifts, justification of new theories, events of the emergence of new interpretive models—in order to show the relationships between the theoretical framework of the discipline and the objectives chosen to teach the subject. In the example we are considering different discussions in the Physics Education literature about teaching the concepts involved in the explicative model of how DC circuits work (Closset 1983; Liegeois and Mullet 2002; Smith and van Kampen 2011).

The mentioned studies show empirical evidences on the difficulties that students have in learning a scientific model of fundamental of DC circuits with resistors. A summary of the literature is indicated below:

- The fact that most of the students who participated in the aforementioned studies avoid using the concepts of potential to analyze the current in a circuit, indicates confusion in the meaning of this concept.
- A significant percentage of students indicate as a cause of the current the difference in the amount of charge or sign of charges, between the ends of the battery.
- Students often use reasoning based on the "formula" in their description of how a circuit works.
- Many students consider Ohm's law as the general law of the circuit that explains its current and energy balance.
- Most of the students do not relate the macroscopic phenomena (current, potential measurements, resistances ...) and the microscopic phenomena (movement of electrons, action of the electric field on the electrons ...).

The beginnings of electrical theory in the eighteenth and nineteenth centuries are linked to electrostatic phenomena. In this way, one of the first theories on the flow of current in a closed circuit refers to the "electrical conflict" based on the model of the two fluids that leave the battery and are neutralized along the wire. During the eighteenth century, studies were carried out in transient current discharges, in accordance with the development of experimental (Leyden jar) (Bensheguir and Closset 1996). After Volta's discovery, the experimental base expands and new explanatory theories of electric current appear. The concepts of "degree of electrical tension" of a charged conductor and "electromotive force" were introduced by Volta to explain the voltage in electrical conductors that caused the movement of current. During the 1920s, Ohm investigated the conductivity of materials and their resistance to current flow. In this context, Ohm explained the different roles of the current and the potential at a time when they were both quite confused. Ohm used the analogy of the temperature gradient that drives heat transfer to explain the flow of electricity in a circuit through lead wires (Schagrin 1963; Taton 1988). Kirchhoff synthesized Ohm's work and went further by proposing a theory for current flow in electrical circuits. This theory was based on the development of the concept of potential and the analysis of the complete circuit. According to the theory in electrostatics, there could be no volume charge gradient inside the conductor, and Kirchhoff solved the problem by putting a charge gradient on the surface (Whittaker 1987). Furthermore, Kirchhoff experimentally demonstrated that Volta's "electrical voltage" and Poisson's potential function were numerically identical in a conductor and therefore could be reduced to a single concept. Since advances in Kirchhoff's electrical theory, electrostatics and electrokinetics have converged into a single electrical theory (Heilbron 1979). Taking Coulomb's law into account, the theory explained that direct current is due to the flow of electrons under the influence of electrical forces and under the influence of a potential difference across the poles of a battery. Later, with works on the field model of Faraday and, Maxwell (1865), the electrical theory of current circuits received a new impulse. In this interpretation, the electrical currents in cables, resistors, etc., are driven by electric fields. The electric field has its source only in the surface charge distributions on the cable (Jefimenko 1966; Härtel 1993).

The epistemological analysis identifies three key problems: (a) relations between electrostatics and current; (b) relations between macroscopic phenomena and microscopic level models; (c) relations between operative definitions of charge, potential, and electric capacity and their meaning in electrostatics and current.

We consider explanations at macroscopic level those that are focused on the physical quantities, such as resistance, current, and potential difference from the overall perspective of the circuit. Quantities such as conventional current and those defined at the level of operational definitions (for example, Ohms' law or the formula that relates potential and field). Instead, we consider explanations at the microscopic level those that are focused at the local level in terms of electrons, electron current, or charge density (Griffith 2014). After covering electrostatics, students are often confronted with a new situation (e.g., electrokinetic phenomena) that not only leads to new phenomena (charges of movement in a wire) but also challenges their view that an electric field is required for a current to flow in a wire. However, literature show that students do not use to seeing circuits as being driven by an electric field that acts on electrons at the microscopic level (whether or not this results in a conventional current at the macroscopic level) (Hirvonen 2007).

Taking into account the epistemological key problems and students learning difficulties, we defined the learning objectives for the TLS on "Fundamental of DC circuits". Reasoning in qualitative and quantitative terms about the phenomena that occur in electrical circuits involves to develop robust models of the microscopic processes underlying them, which lead to the observed phenomena and their macroscopic explanations (Leniz et al. 2017). Consequently, first of all, it is a question of learning a microscopic model of electrical current in a simple DC circuit with the following characteristics:

LO.1: Recognize that the magnitudes of electric field and electric potential defined in electrostatics are the same as those used in electrodynamics. Knowing how to apply these magnitudes in both contexts.

LO.2: Relate the microscopic electron current model, which is related to the electric field magnitude in the conductor ($i = n\ A\ v_d = n\ A\ u\ E = n\ A\ u\ \Delta V/L$), with the conventional current ($I = \Delta Q/\Delta t$). Knowing how to use the magnitudes of each model in the appropriate context and field of validity ($i = n\ A\ u\ E$; $I = \Delta Q/\Delta t = n\ q\ A\ v_d = n\ q\ A\ u\ E$).

Secondly, here is often a need to find an alternative to a purely macroscopic description. We are often left at the macroscopic scale, with explanations such as "this law, or this other law, tells us that things have to be this way." These types of explanations are insufficient to satisfy students, especially when alternative conceptions appear on the way of learning. So, it will be necessary to define the causal mechanism that generates the electric field inside the wire:

LO.3: Knowing and applying the model of Gradient of Surface Charges (GSC) in relation to the role played by the battery in the production of the distribution of surface charges in the wire. This distribution of surface charges (GCS) produces an electric field inside the wire and consequently, produces a potential difference in the different parts of the circuit.

Thirdly, it is necessary to know how to relate the electrical current microscopic model with the measurements of the ammeters and voltmeters in the different parts of the circuit, for an understanding of the context in which the different laws in a DC circuit are applied. In consequence:

LO.4: Know how to explain the conservation of charge and energy (Kirchhoff's laws) in a simple DC circuit, both at the microscopic level (the electric field that drives the electron sea) and macroscopic (the potential difference between the parts of the circuit). This implies knowing how to relate the conventional current, the electric field inside the wire and the potential difference in different parts of the circuit ($\Delta V = E\ L$; $I = \Delta V/R$; $i = n\ A\ u\ \Delta V/L$; $\varepsilon = \Delta V + rI$).

Once the learning indicators have been defined, it is necessary to identify the learning demands. That is, it is necessary to know the magnitude of the gap between the defined learning objectives and the learning difficulties that students usually have in those objectives. This allows the teachers to get an idea of how big or small the cognitive demand is. This will lead to justify the greater or lesser number of activities to work the learning indicator and the process of construction of the concept, procedure or theory. At the beginning of this section, a summary of previous studies showing the learning difficulties on the subject has been made. However, some of the defined indicators have not been specifically investigated and, therefore, it was necessary to carry out a study of learning difficulties.

In a previous study (Goikoetxea 2017), we designed a questionnaire with an emphasis on explanations. We gave 120 students at the University of the Basque Country (Spain) a questionnaire after they had studied the topic in class. We describe here, as an example, one of the questions completed by the students and summarize the results. In the second question Q2 (see Fig. 13.1), the difficulty of students' understanding on the concept of potential difference in electrostatic to electrokinetic

Q2. Student S1 states that the potential difference between points A and B is calculated in the same way by the equation $\Delta V = \int_A^B \vec{E}\, dl$ in the two situations.

Student E2 disagrees and states that the equation $\Delta V = \int_A^B \vec{E}\, dl$, only is used to calculate the potential difference in electrostatics (situation 2), but for the situation 1 the potential difference is calculated by the Ohm's law: ΔV = I R.

Situation 1

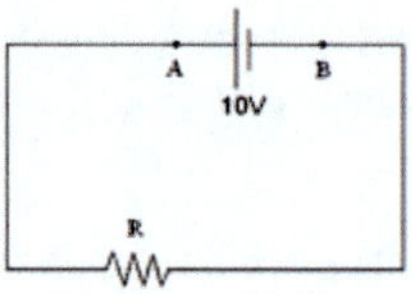

Situation 2

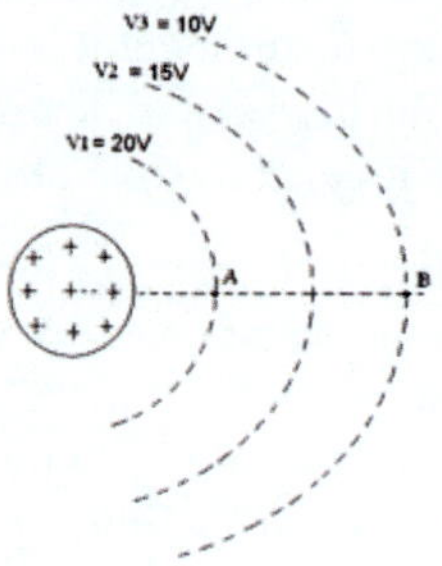

With what student do you agree? Explain your answer

Fig. 13.1 Question Q2 related to the content of learning indicator i1. *Q2*. Student S1 states that the potential difference between points *A* and *B* is calculated in the same way by the equation $\Delta V = \int_A^B \vec{E}\, dl$ in the two situations. Student *E*2 disagrees and states that the equation $\Delta V = \int_A^B \vec{E}\, dl$, only is used to calculate the potential difference in electrostatics (situation 2), but for the situation 1 the potential difference is calculated by the Ohm's law: $\Delta V = I\,R$

Table 13.2 Percentages of categories of answers in question Q2

Categories of answers		Percentage (%)
A	The magnitude is the same in both contexts	16.5
B	Two different definitions of the magnitude according to context	75.0
D	Does not answer the question	8.5

is investigated. The question Q2 required students to recognize that the magnitude of potential difference is the same in both electrostatic and electrokinetic and so, the definition of the magnitude is the same (LO.1).

The results of the students' answers to the Q2 question are given as percentages in Table 13.2. The detailed description categories are provided below.

In category A, a minority of students' answers (16.5%) recognize the relations between the equations and students argument about the definition of potential difference. An example:

> *I agree with the student E1. The general definition of potential difference is expressed by the equation,* $\Delta V = \int_A^B \vec{E}\, dl$*, and we know that* $V = E{\cdot}d$*. So, we calculate the* ΔV *using one way or another depending on each situation.*

In the category B, a distinction is made between the definition of the difference of potential in the context of electrostatic and electrical circuits. Three quarter of the students' answers do explain the calculation of the potential difference in each context but with no relation between the two equations. It seems that they consider two different definitions of the same magnitude. A standard example:

I agree with the student E2 because the nature of the field is not the same. In an ohmic circuit we use macroscopic laws ($\Delta V = I\,R$) while in electrostatic they are microscopic".

The results of this question show that students have difficulty in recognizing that the magnitude difference of potential has the same meaning, although different ways of calculating, in the context of electrostatic and electrokinetic. It should be borne in mind that learning objective LO.1 requires a high level of comprehension effort for students and that therefore the demand for learning in this indicator is high. Of course, only one example is shown here, in other works the size of the learning demands for each defined indicator will be shown with a greater number of data.

In the next section we will present an example of the evaluation of the implementation of the TLS, according to the phase test of DBR.

13.4 Testing the Quality of the TLS

As a product-oriented project, one of the essential characteristics of the TLS design and evaluation projects is the re-elaboration of the teaching sequence based on empirical data obtained during its implementation in an educational intervention. The TLS design must be confronted empirically in the evaluation of the proposal itself and the achieved learning results. In this chapter, we only will show how we evaluate the TLS itself, which is related to the quality of the design (Nieveen 2009). On the basis of collected data during the implementation, we will infer problematic aspects of the activities. Following this analysis, we will define types of students' difficulties (metacognitive difficulties, learning difficulties, related to interpretation and comprehension of information, etc.) and we will proceed to introduce modifications to the activities and their sequencing. The instruments for the iterative development of the TLS are standards instruments from the educational research such as Teacher's diary, Student's workbook and External observers' reports. In Table 13.3 shows an example redesign of an activity due to students' metacognitive difficulties. The analysis of implementation difficulties led to consider the difficulty of students to understand the objective of the activity. This type of metacognitive difficulty causes students to not adequately approach the activity, even when its objective has been explained to them (Treagust et al. 2002).

In the activity of the first version, students' standard answers when establishing the relation between current of electrons was to substitute the data in the equation $I = e\,i_e = n_e\,e\,v_d\,A$, but they do not give explanations of the meaning of each magnitude and they do not comment the result obtained, whether correct or

Table 13.3 Students' metacognitive difficulties encountered when implementing the TLS and its re-elaboration (rewriting and rethinking of the approach of the activities)

Activity 3.1 (version 1)	Activity 3.2 (version 2)	
In a copper wire, the current of electrons is about $i = 3.4 \times 10^{18}$ electrons/s. This is an incredible number of electrons to pass through a section of the wire every second. But, How long it takes for an electron traveling a meter in a copper wire? Make hypothesis: A: 10^6 s B: 1 s C: 10^{-6} s D: Another one Suppose that 100 mA is drifting through a copper wire. The diameter of the wire is 1 mm ($A = 8 \times 10^{-7}$ m^2) and the density of mobile electrons in a copper is 8.4×10^{28} m^{-3} (a) What is the drift speed of electrons? (b) How long is a single electron in the electron sea to drift 1 m of the wire? (c) If the speed of the electrons in the wire is 9.41×10^{-6} m/s and then, electron takes 29.5 h to move 1 m. Why does the light in a football stadium come on almost instantly when you flip a switch 300 m away?	Microscopic point of view 1. The velocity of the electrons in the wire in a circuit is: (a) Slow (b) Fast (c) Almost light speed Explain: 2. The force that acts on electrons is produced by: (a) E (b) ΔV (c) Other Explain:	Macroscopic point of view 1. The effects of the current in a circuit are: (a) Slow (b) Fast (c) Almost light speed Explain: 2. In a circuit with I current, the multimeter measures: (a) E (b) ΔV (c) Other Explain:

incorrect. However, in the activity of the second version, students have to argument about the option that they choose. Students define the meaning of the current of electrons and hypothetize about the cause of the current. The new version is more according with the learning objective LO1. The change made in the rewriting of the activity leads to students focusing on that aim when answering the questions.

13.5 Discussion

Our design-based research proposal includes common elements agreed by the community. These elements of design, implementation and evaluation allow the definition of didactic tools that help to make explicit the decisions made at the level of definition of objectives and in the design of activities to be implemented in the classroom. The TLS as a final product requires making explicit both the school context and the use of design tools, such as epistemological analysis or learning demands.

We are aware that we have only presented two examples of our design. An example of analyzing difficulties and taking them into account in the design of activities. Another example is of redesign as a consequence of implementation analysis. The objective of this chapter is not to show all the evidence obtained in the design research process, but rather to show the consensus reached by the community in the research that follows a DBR methodology and define its main characteristics. The examples try to illustrate the research theory for a specific topic.

Acknowledgment The authors would like to thank the economical support provided by the Project MINECO EDU2015-65359-P and by the Project of the University of the Basque Country GIU19/071.

References

Ametller J, Leach J, Scott P (2007) Using perspectives on subject learning to inform the design of subject teaching: an example from science education. Curric J 18(4):479–492

Anderson T, Shattuck J (2012) Design-based research: a decade of progress in education research? Educ Res 41(1):16–25

Barab S, Squire K (2004) Design-based research: putting a stake in the ground. J Learn Sci 13 (1):1–14

Bensheguir A, Closset JL (1996) The electrostatics-electrokinetics transition: historical and educational difficulties. Int J Sci Educ 18(2):179–191

Closset JL (1983) Sequential reasoning in electricity. In: Delacote G, Tiberghien A, Schwartz J (eds) Research on physics education: proceedings of the first international workshop. Editions du CNRS, Paris, pp 313–319

Duit R (2009) Bibliography–STCSE. Students' and teachers' conceptions and science education. http://www.ipn.uni-kiel.de/aktuell/stcse/download_stcse.html

Easterday MW, Lewis DR, Gerber EM (2014) Design-based research process: problems, phases, and applications. International Society of the Learning Sciences, Boulder, CO

Goikoetxea A (2017) Dificultades de aprendizaje de los estudiantes de universidad en la transición entre electrostática y circuitos simples de corriente continua. Thesis of master. Programa de Psicodidáctica, San Sebastián

Griffith DJ (2014) Introduction to electrodynamics, 4th edn. Pearson Education Ltd., Harlow

Härtel H (1993) New approach to introduce basic concepts in electricity. In: Caillot TM (ed) Learning electricity and electronic with advance educational technology. Springer, Berlin, pp 5–21

Heilbron JL (1979) Electricity in the 17th and 18th centuries. A study of early modern physics. University of California Press, Berkeley, CA

Hirvonen PE (2007) Surface-charge-based micro-models a solid foundation for learning about direct current circuits. Eur J Phys 28:581–592

Jefimenko OD (1966) Electricity and magnetism. Appleton-Century Crofts, New York, NY

Kortland K, Klaassen K (2010) Designing theory-based teaching-learning sequences for science education. FSME, Utrecht

Leniz A, Zuza K, Guisasola J (2017) Students' reasoning when tackling electric field and potential in explanation of dc resistive circuits. Phys Rev Phys Educ Res 13:010128

Liegeois L, Mullet E (2002) High school students' understanding of resistance in simple series electric circuits. Int J Sci Educ 24(6):551–564

Lijnse P (2004) Didactical structures as an outcome of research on teaching–learning sequences? Int J Sci Educ 26(5):537–554

Maxwell JC (1865) A dynamical theory of the electromagnetic field. Philos Trans R Soc Lond 155:459–512

Méheut M, Psillos D (2004) Teaching–learning sequences: aims and tools for science education research. Int J Sci Educ 26(5):515–535
Méheut M, Psillos D (2004) Teaching–learning sequences: aims and tools for science education research. Int J Sci Educ 26(5):515–535
Nieveen N (2009) Formative evaluation in educational design research. In: Plomp T, Nieveen N (eds) An introduction to educational design research. SLO, Enschede, pp 89–101
Savall-Alemany F, Guisasola J, Cintas SR, Martínez-Torregrosa J (2019) Problem-based structure for a teaching-learning sequence to overcome students' difficulties when learning about atomic spectra. Phys Rev Phys Educ Res 15(2):020138
Savinainen A, Mäkynen A, Nieminen P, Viiri J (2017) The effect of using a visual representation tool in a teaching-learning sequence for teaching Newton's third law. Res Sci Educ 47 (1):119–135
Schagrin ML (1963) Resistance to Ohm's law. Am J Phys 31(7):536–547
Smith DP, van Kampen P (2011) Teaching electric circuits with multiple batteries: a qualitative approach. Phys Rev ST Phys Educ Res 7:020115
Taton R (1988) Histoire generale des sciences. Presses Universitaires de France, Paris
Treagust DF, Chittleborough G, Mamiala TL (2002) Students' understanding of the role of scientific models in learning science. Int J Sci Educ 24(4):357–368
Whittaker E (1987) A history of the theories of aether and electricity. American Institute of Physics, College Park, MD

Chapter 14
IDIFO6 MQ_P: A Course for In-Service Secondary School Teachers Education on Modern Physics

Daniele Buongiorno, Marisa Michelini, Lorenzo Santi, and Alberto Stefanel

Abstract The recent demanding for innovation with respect to modern physics topics in secondary school curricula and the possibility for a new final exam in the school year 2016/2017 in which modern physics could be a predominant part drove Italian secondary school teachers in requesting a specific training. Project IDIFO6, coordinated by Physics Education Research Unit (PERU) from University of Udine (IT) allowed to respond positively to the request of 25 physics and mathematics secondary school teachers from a network of school from Veneto, proposing them a specific course named MQ_P on modern physics topics. The structure and contents of the course have been designed in a tight cooperation between PERU and the involved teachers, sharing with them topics, teaching sequences, methodologies, and research questions. A setting integrating experiential and situated models in teacher formation allowed participants to experience actively scientific-cultural problems that the introduction of modern physics sets on conceptual plan in designing didactical proposals integrating operative aspects in the form of experimental, interpretative, and problem solving activities. At the end of the course, every participant revised individually the contents addressed and two rubrics guided the planning of the teaching intervention. Here we analyze teachers' works in which they put into practice in their classes the outcomes of their formation.

14.1 Introduction

In-service secondary school teacher formation in Italy is mainly due to self-training during school activities (Luzzato 1999; Pugliese et al. 1999; Dutto 2001; Bonetta et al. 2002; Dutto et al. 2003a, b) and, despite a law dating back to year 1990 (Law

D. Buongiorno (✉) · M. Michelini · A. Stefanel
Physics Education Research Unit, DMIF, University of Udine, Udine, Italy
e-mail: buongiorno.daniele@spes.uniud.it; marisa.michelini@uniud.it; alberto.stefanel@uniud.it

L. Santi
DMIF-URDF, Università degli Studi di Udine, Udine, Italy
e-mail: lorenzo.santi@uniud.it

B. G. Sidharth et al. (eds.), *Fundamental Physics and Physics Education Research*,
https://doi.org/10.1007/978-3-030-52923-9_14

341/90) expects initial university teacher formation, only in 1999 preservice teacher education effectively started.

The problem of the specific disciplinary formation of secondary school teachers arises from the fact that they have a degree obtained after four years at university, resulting in good disciplinary formation, but without professional formation. Younger teachers can rely on preservice educational programs which is a bi-annual (S.S.I. S., 1999–2009) or annual (T.F.A., 2011–2015) post-degree master consisting in 30% antro/psico/pedagogical areas, 30% educational labs for analysis of didactical planning and review of class intervention and 40% apprenticeship with final thesis.[1] The employment process took usually place through procedures ascertaining mainly cultural preparation and transmission capacities (Dutto et al. 2003a, b). Sometimes teachers attended fragmentary actions in their in-service training, but for the main part they formed their professionalism through direct experience in the class (Luzzato 1999; Dutto 2001). Despite in 2016 a law makes mandatory professional development of in-service secondary school teachers, very often the total responsibility of their training is continued to be attributed to schools, resulting in a poor organic unity of the activities. Most teachers thus continue to be self-trained on the basis of their willingness.

Researches concerning pre- and in-service secondary school teacher formation on scientific topics evidenced formative needs on methodological and disciplinary plans (Michelini 1997; Pugliese et al. 1999). In-service formation has to be able to face specific educative and formative problems: the nature and role of interaction between contents and the organization of the educational activity, the managing of curricula, of the learning processes, of the methods of innovation and of the overcoming of the conceptual knots (Dutto et al. 2003a, b; Michelini and Schiavi 2001).

The intuitive dimension drives teachers in developing a useful sensitivity for the choice of strategies and methods to be adopted, according to their experiences, and very often it remains their only reference for educational choices. Proposals coming from didactical tradition, scholastic publishing, dispersed and differentiated forms of in-service training are mostly disorienting because of their fragmented setting regarding contents, methods, duration and offers (Dutto et al. 2003a, b) and lead teachers to think that it is necessary to adapt their approach to a consolidate praxis relying on textbooks or on the experience of older teachers (Eraut 1994).

The re-qualification of teachers of the profession of the teacher cannot be satisfied with theoretical notions (Anderson 1995), but it has to be contextualized. The situated dimension of contextualized analysis, of experimentation and of implementation in context of educational proposals allows the development of the reflection in professional practice, which is an indispensable condition to master innovation (Woolnough 2001).

As concern modern physics topics, recent reforms (Law 107/2015) re-designed secondary school curriculum introducing modern physics, with the possibility for a

[1]A new model for preservice teacher formation (F.I.T.) has been now adopted according to law 107/2015 consisting in a three-year formation and apprenticeship in tight cooperation with university reality.

new final exam in the school year 2016/17 in which, for the first time, physics could be a predominant part. Secondary school physics teachers have a degree in math, phys, engineering, resulting in a not-homogeneous good disciplinary formation, accompanying the aforementioned problems regarding the specific professional competences.

New teaching profession is composed by an articulate set of disciplinary, technical, pedagogical, social, and organizational competences (Michelini 2001; Michelini et al. 2002) and Physics Education Research (PER) has to contribute to in-service teacher formation, since research represents the most effective instrument for teachers involvement in their formation, integrated with didactical commitment. Experiences of collaboration with professionals in educative and didactical research is therefore an important condition for the joining of research and professional practice (Dutto et al. 2003a, b) that turns to be a mixture of pure, applied and action research.

14.2 The IDIFO6 Project and the MQ_P Course Peculiarities

Since 1996, Physics Education Research Unit (PERU) from University of Udine (Italy) coordinates IDIFO (Didactical Innovation In Physics and Guidance) project[2] with 20 Italian universities involved. A master offering 198 CTS on modern physics topics for in-service teacher education and 2, 12, and 60-CTS educational modular paths, both in frontal lessons and in e-learning making use of a specific web platform, allow the a choice for a personal profile formation. IDIFO6 project offer allowed to respond positively to the request of 25 physics secondary school teachers asking for specific training on modern physics from a network of three schools from Veneto, one of the biggest Italian region, proposing them a specific course named MQ_P. Teachers requested this specific training, moved by their responsible engagement, in order to best accomplish their mission in the formation of the new generations of students.

Teacher formation model in MQ_P course employed integrates phases of study, critical reflections and actions, sharing with the researchers' discussions of problematic issues and proposals of intervention in a school setting. Contents and methods of in-service teacher education involved the educational reconstruction of fundamental disciplinary contents, analysis of alternative teaching paths, formal deepening of content structures and problem solving activities together with the planning of teaching interventions.

Three phases were planned corresponding to three specific formative models for teachers (Michelini et al. 2013, 2015):

[2] http://www.fisica.uniud.it/URDF/laurea/idifo6.htm.

- *(A) Metacultural*: an organic and coherent research-based path on a specific topic is proposed to teachers. During this phase, time is dedicated to discussion of disciplinary contents, educational choices, methods, and instruments with respect to the rationale of the path.
- *(B) Experiential*: teachers play the role of students, testing materials, exercises, lab activities.
- *(C) Situated*: teachers choose and design a customized FIM (4–12 h) to be experimented in their classes (addressing specific crucial aspects and/or the whole path).

At the end of the course, every participant revised individually the contents addressed and two rubrics guided the planning of the teaching intervention, demanding them to point out the founding cores and the conceptual knots together with a conceptual map, the proposed activities and the sequence of stimulus question for at last one topic among the ones addressed. The outcomes of the various experimentations have been shared with our research group, allowing the monitoring both of the work done by the teachers in designing a didactical proposal on modern physics, and of the learning outcomes of the students involved.

14.3 Contents and Setting

The structure of MQ_P course aimed at addressing the following topics:

- Analysis of guidelines for innovation in curriculum on modern physics.
- Educational proposals on photoelectric effect, wave optics, optical spectroscopy and Bohr's interpretation.
- Focus on crucial experiments interpretation[3] (Compton and photoelectric effect, Bertozzi's experiment on relativistic energy of electrons).
- Laboratorial activities on measurement of the ratio e/m (electric charge and mass of the electron), Franck and Hertz experiment, optical polarization and Malus' law, measure of the speed of light, measurements of atomic spectra wavelengths and energies.
- From structured to open problem solving (Maloney 2001) on Compton effect, photoelectric effect, de Broglie wavelength, Heisenberg's uncertainty principle and spectroscopy (derivation of energy level structure from discrete spectra and prevision of spectral line formation in different physical conditions).

Twenty-five Italian physics secondary school teachers from three different scientific high schools attended MQ_P course. 12/25 teachers attended only (A) and/or (B) phases in order to grasp the main issue concerning educational approach; 13/25

[3]Crucial experiment interpretation consists in discussion of those experiments which interpretation founds the physics of quanta interpretational proposal in the passage from classic physics to quantum mechanics.

teachers attended all the phases, and among them, eight teachers conducted an experimentation in their classes on the designed education proposals.

14.4 Educational Materials

In order to monitor teachers' project, two rubrics were used. Those rubrics, widely used in teacher education, allow a synthetic and reliable analysis of the projects in order to compare the different types of rationale used:

- S1 (Sheet 1): fundamental/pivotal concepts of the proposals and expected learning difficulties:
- S2 (Sheet 2): key questions according to a direction in developing the addressed topics and relative conceptual map, rationale, and structure of the proposal.

Educational material provided to teachers was a collection of research-based problem solving exercises on optical spectroscopy, in particular concerning the way in which is possible to infer the energy level model from an experimental spectra and vice versa.

14.5 Data Analysis

Teachers' production consisted in two kind of educational projects: 8/13 specific on optical spectroscopy and 5/13 general on modern physics.

The structure of the proposals on optical spectroscopy was conducted first qualitatively, identifying the contents, organized subsequently according to Table 14.1 and then qualitatively, and coding each project according to the structure of the addressed contents. A graph allows to see the whole structure of each proposal (Fig. 14.1). Other graphs are thus constructed in a way in which it is possible to identify, for each content, the number of teachers that included that topic in the proposal and the position of that topic in the proposal (Fig. 14.2). Representation regards the number and the position of the addressed contents, but does not provide any information concerning duration and importance given to each content.

Five teachers who did not specifically address optical spectroscopy designed a more general proposal to modern physics topics, addressing contents categorized in Table 14.2 divided in the categories "crucial experiments," "key aspects of quantum physics," "interpretative hypothesis," "properties of a massless particle," "electromagnetic waves and spectra," and "case studies." Since no organic path emerged from the educational proposals on modern physics, the analysis was conducted qualitatively accounting only for the frequencies of the included contents. The more frequently included crucial experiments (Fig. 14.3a) are the photoelectric effect, specifically addressed in the MQ_P course, the Compton effect and the black body spectrum. The Frank-Hertz experiment, despite its importance, is

Table 14.1 Rubric for the analysis of contents included by teachers in the designed educational projects on optical spectroscopy

A. Light sources A1: technological/structural aspects A2: emission mechanism A3: primary/secondary sources A4: light sources as systems transforming energy A5: incandescent bodies A6: Flame tests	*D. Light–matter interaction* D1: photoelectric effect D2: transmission D3: reflection/diffusion D4: absorption	*G. Interpretation of discrete spectra* G1: energy level model G2: spectra as tracers of energetic exchanges G3: from levels to line G4: from lines to level
B. Characteristics of the emitted light B1: color B2: intensity B3: nature of white light B4: light as an EM radiation B5: light as an entity carrying energy B6: color as energy B7: quantization of radiation	*E. Formal constructs, laws, principles* E1: Balmer's formula E2: Rydberg's formula E3: Bohr's model E4: negative binding energy in H atom E5: classical atomic models E6: Kirchhoff's laws E7: Planck's hypothesis E8: Wien's law E9: Stefan-Boltzmann law E10v Rayleigh-Jeans formula E11: Ritz' combination principle	*H. Role of diffraction grating and slit* H1: shape of the slit H2: diffraction grating

C. Light dispersion phenomena C1: refraction C2: diffraction	*F. Kind and role of spectra* F1: continuous F2: discrete F3: spectra role in characterizing elements	A Light sources B Characteristics of the emitted light C Light dispersion phenomena D Light-matter interaction E Formal constructs, laws, principles F Kind and role of spectra G Interpretation of discrete spectra H Role of diffraction grating and slit

In the very last cell, the code color for each thematic area is shown in order to interpret the graph shown in Fig. 14.1

Fig. 14.1 Structure of the eight designed educational proposals on optical spectroscopy. Each line represents a teacher's proposals, divided into included topics, as reported in Table 14.1

included in the proposal only by one teacher. Every teacher quotes at least two crucial experiments and only one quotes them all. The majority of the involved teachers includes crucial theoretical aspects of quantum physics (Fig. 14.3b) quoting at least two aspects, the most addressed one is de Broglie's contribution to the interpretation of the wave nature of matter. Only one teacher does not consider important to integrate any of those aspects in the proposal. A bigger importance is given to the interpretative hypothesis or models accounting for the different phenomenology as Bohr, Einstein, or Plank's hypothesis (Fig. 14.3c): every teacher quote at least one of those aspects and one of them quotes them all. The analysis of relativistic dynamic is taken into account only from three teachers, addressing concepts like energy, velocity and momentum for a photon, while the other two do not believe to be an important part of the proposal. Two teachers begin the educational proposal on modern physics addressing the electromagnetic spectrum and the wave nature of light using the Maxwell's equations. Three teachers include specific case studies and applications as the tunneling effect and the potential well phenomenology for a massive particle in the middle or at the end of the proposal.

14.6 Discussion

The emission mechanism is one of the most quoted arguments in the proposal on optical spectroscopy. In the proposal structure, it is placed at the beginning, usually at the first, second or third place; this is quite surprisingly since, in our educational path, it is presented as an arrival point of a sequence of reasoning and not presented directly to students. Technological aspects are included by five teachers: this is not a natural setting, since generally applicative aspects are addressed at the end of standard lessons. Probably this peculiar characteristic is due to our setting, in which technological aspects of different light sources are used as an approach. A general tendency of focusing on light sources and emitted light characteristics at the beginning is clearly recognizable. Only in one case the characteristics of the emitted light come before the sources, and in this peculiar case, because characterization of the emitted light is given a big importance. Analysis of incandescent emission occupy an equally relevant position in teachers' proposals and it is placed immediately after approaches to phenomena, sometimes alone, sometimes accompanied

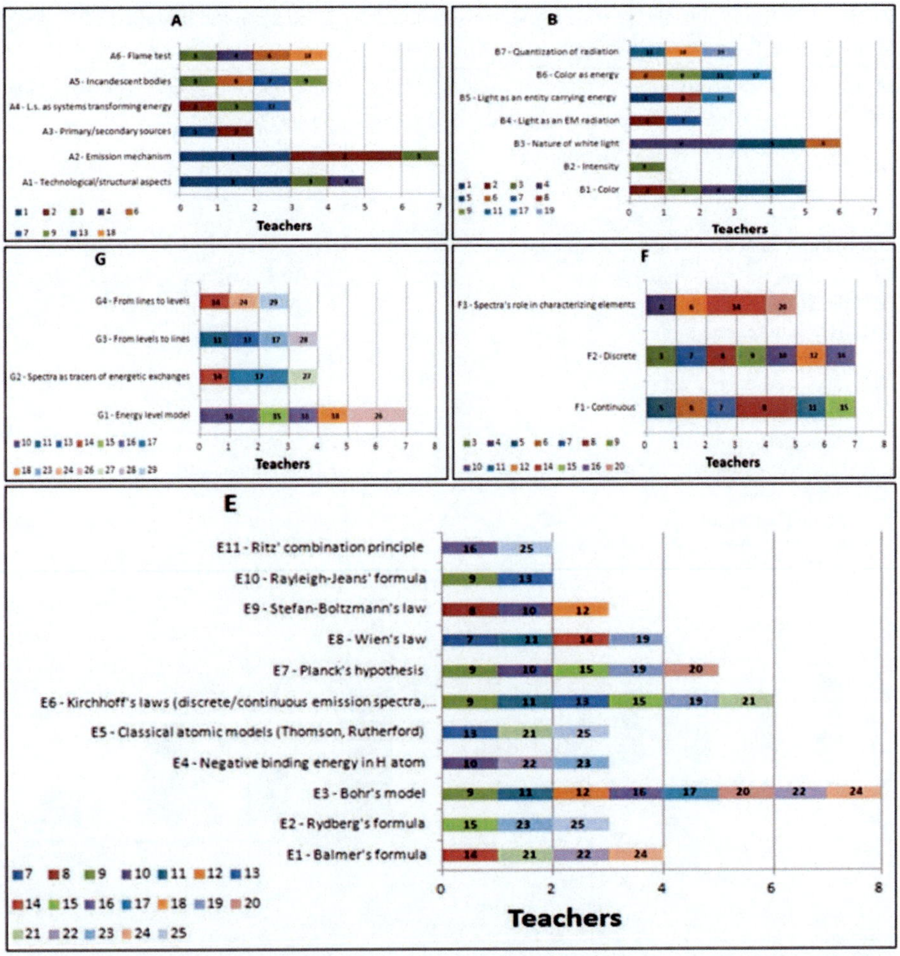

Fig. 14.2 For every addressed topic (vertical axes) it is possible to observe the number of teachers who included it in the project (horizontal axes) and its position in the sequence of the educational proposals, according to the color code (or the equivalent numbers). For example, topic A2 is included by seven teachers: three of them placed it in first position, three placed it in second position and only one teacher placed it in the third position. Here we report the most significant contents: (**a**) Light sources. (**b**) Characteristics of the emitted light. (**g**) Interpretation of discrete spectra. (**f**) Spectra. (**e**) Formal constructs, laws, principles

from an insight of the characterization of a light source as a system transforming energy. Flame test is a content that teachers believe important to be addressed, even if in different moment during the proposal. Composite nature of white light is another of the topic that teachers consider important to be addressed, not immediately, but after clarifying that light is an entity carrying energy rather than an electromagnetic radiation and/or the discussion on the energetic nature of colors, which is another topic mainly included in the initial phase, as the characterization of

Table 14.2 Rubric for the analysis of contents included by teachers in the designed educational projects on modern physics

A. Crucial experiments A1: photoelectric effect A2: Compton effect A3: Franck and Hertz experiment A4: black body spectrum A5: Devisson and Germer experiment A6: Atomic spectra	*D. Properties of a massless particle* D1: energy D2: speed D3: momentum
B. Key aspects of quantum physics B1: de Broglie wavelength B2: Schrodinger equation B3: Heisenberg's uncertainty principle B4: quantized energy exchanges	*E. EM waves + spectra* E1: Maxwell equation E2: EM spectra
C. Interpretative hypothesis C1: Planck's hypothesis C2: Einstein's photoelectric effect interpretation C3: Bohr's model C4: Balmer's formula C5: energy level model	*F. Case studies* F1: tunneling F2: potential well

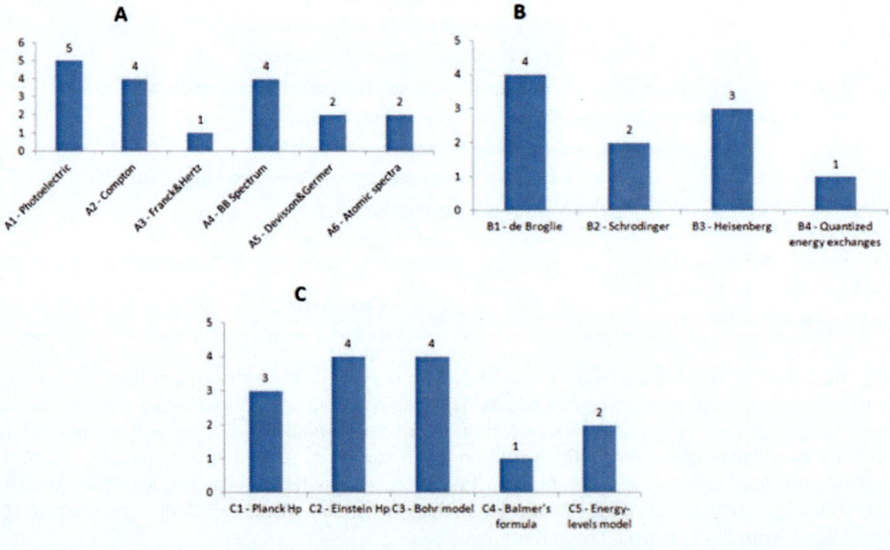

Fig. 14.3 Some addressed contents in modern physics proposals. On the vertical axes the number of teacher is reported. (**a**) Crucial experiments. (**b**) Key aspects of quantum physics. (**c**) Interpretative hypothesis

light sources by color and intensity of the emitted light. No one of the formal constructs is addressed before entering into the discussion on the nature of colors, white light peculiar characteristics or the nature of light and its description in energetic terms. In particular, among the main addressed constructs, Bohr's model,

Kirchhoff's laws, Planck's hypothesis, Wien's law and Balmer's formula emerge, ordered with respect to their frequencies. Only in one case, formal aspects are addressed using the classical atomic model, at the beginning of the proposal. In structuring the proposals, the majority of the teachers deserve to Balmer's formula and Planck's hypothesis the final positions. Very few teachers address Ritz' combination principle, usually at the end, despite its utility in interpreting spectra. Concerning the position of spectra in the rationale, their continuous or discrete nature or their peculiarity in characterizing elements occupy the first positions only in few proposals: teachers address the phenomenological exploration together with the analysis of the formal aspects. The majority of the proposals addresses the role of spectra in characterizing elements in the middle or at the end of their proposals. As expected, the majority of teachers address the energy level model in the final part of their proposals, as well as the interpretation of spectra as tracers of energetic exchanges. The link between levels and lines is addressed by the majority of the teachers, but the reverse and more difficult task to derive levels from a discrete spectrum is addressed by few of them. Optical spectroscopy proposed as a bridge between classical and modern physics deserves attention and insight, generally in the final part of the proposals. Dispersion phenomena is immediately subsequent to the analysis of the characteristics of the emitted light, followed by an exploration with spectroscopes. Formal constructs are always addressed in the following, except when an insight into the role of diffraction grating and slit is addressed, which is not a common aspect.

Five teachers' main approach to modern physics is based upon the analysis of the crucial experiments leading to the revising of classical physics principles, in particular in the light of Planck and Einstein's hypothesis and only in one case with the analysis of the quantization of the energetic exchanges coming before these themes. Bohr's model follows a review of the different aspects of quantum physics, and only in one case optical spectroscopy is used as a bridge for the analysis of the quantized nature of light and the emission processes. Wave nature of matter and Heisenberg's principles are generally addressed subsequently. To be noticed that in one case the rationale of the optical spectroscopy proposal is quoted: the coherence of a path thus influences also teachers who wants to address more general issues. Elements are embedded in the proposals without any justification in a more organic rationale and the design appears to be an effort of treating, via story-telling approach, every characterization of key aspects, crucial elements and typical cases of modern physics, with a coherence given mainly from the perspective of addressing every topic rather than of an organic interpretation of the phenomena: rather than build an organic frame for quantum theory, teachers adopt approaches in which fragmented peculiar elements of the theory are shown to students. A need of supporting teachers in insert the interpretative perspective on disciplinary plan emerges, in order to overcome the narrative one for the identification of the relevant elements.

14.7 Conclusions

A considerable number of secondary school teachers in Italy aims at improving modalities to teach modern physics in the school. Our Physics Education Research Unit (PERU) from University of Udine was therefore asked to contribute to in-service teachers' professional development on this topic.

In this paper both modalities according to which a specific course, MQ_P, was addressed and some formative outcomes are described in detail. MQ_P was requested by a group of serious and motivated secondary school teachers asking for guidelines for the innovation in the curriculum, including crucial aspects of physics of quanta and relative experiments, to be addressed. Particular attention was paid to single aspects' critical analysis and to educational setting of problem solving and laboratorial activities. In implementing the request we inserted a more organic part addressing a specific educational proposal on optical spectroscopy, whose organic unity was re-elaborated to teachers according to designing rubrics requesting the detection of both educational goals and key questions linked to activities in an organic educational proposal.

The accurate writing of the educational projects disappointed the expectations regarding the organic unity in favor of a careful detection of the educational goals that, for teachers, represent the reference pattern of the main innovative aspects of the transition between classical and modern physics. Interpretative, modeling and formal aspects are addressed in conceptual rather than in disciplinary terms for a multiplicity of nonorganic interpretations. Applicative contexts, as the case of the potential well and the tunneling effect, are rarely addressed. Addressing and organization of proposals on optical spectroscopy turned out to be quite complete and significant, since they result to be highly fertilized from the presentation and discussion of our educational proposal, which was re-elaborated in different ways maintaining, at the same time, a coherent structure.

Needs to deepen conceptual knots that physics education research (PER) is pointing out concerning modern physics and optical spectroscopy remain evident, in particular the conceptual role of every part of an apparatus performing spectroscopic measures and the relative physics is underestimated, the need to separate macroscopic and microscopic plans, the emission mechanisms and the nature of the radiation needs to be specifically addressed.

Class experimentation performed by five teachers, designing the optical spectroscopy proposal, ended with the administration of exercises, evaluated only in terms of correct/incorrect answers without analyzing learning trajectories of the students.

References

Anderson LW (ed) (1995) International encyclopaedia of teaching and teacher education. Elsevier Science Ltd., Oxford

Bonetta G, Luzzato G, Michelini M, Pieri MT (eds) (2002) Università e formazione degli insegnanti: non si parte da zero. Forum, Udine

Dutto MG (2001) La professionalità nel sistema dell'autonomia. Q6 La formazione dei docenti/1 Treccani Iter 9(Suppl):11

Dutto MG, Michelini M, Schiavi S (2003a) Research grant for in-service teacher formation:pilot experience in University of Udine. In: Michelini M (ed) Quality development in teacher education and training, Second international GIREP seminar, selected contributions. Forum, Udine, pp 205–210

Dutto MG, Michelini M, Schiavi S (2003b) Reinventing in-service teacher education and training: research grants for teachers. In: L'educazione plurilingue. Dalla ricerca di base alla pratica didattica, selected papers. Forum, Udine, pp 213–266

Eraut M (1994) Developing professional knowledge and competence. Falmer Press, London

Luzzato G (1999) Insegnare a insegnare. Carocci, Roma

Maloney D.P. (2001). An overview of physics education research on problem solving, Getting started in physics education research. 2(2)

Michelini M (1997) Didattica disciplinare e bisogni formativi: risultati di un'indagine tra i frequentanti di un corso di perfezionamento per la formazione degli insegnanti. La Fisica Nella Scuola 30(2):68

Michelini M (2001) Supporting scientific knowledge by structures and curricula which integrates research into teaching. In: Pinto R, Surinach S (eds) GIREP book – selected contributions of the physics teacher education beyond 2000 international conference. Elsevier, Paris, p 77

Michelini M, Schiavi S (2001) La ricerca degli insegnanti: le prime esperienze di borse di ricerca per insegnanti. Q6 La Formazione dei docenti/1 Treccani Iter 9(Suppl):106–125

Michelini M, Rossi PG, Stefanel A (2002) Integrazione fra formazione iniziale degli insegnanti e ricerca didattica: un modello per la fisica. In: Università e formazione degli insegnanti: non si parte da zero. Forum, Udine, pp 124–140

Michelini M, Santi L, Stefanel A (2013) La formación docente: un reto para la investigación. Rev Eureka Sobre Enseñanza Divulg Cien 10(Núm. Extraordinario):846–870

Michelini M, Santi L, Stefanel A (2015) La formazione degli insegnanti in fisica come sfida di ricerca: problematiche, modelli, pratiche. Giornale Ital Della Ricerca Educ 14:91–207

Pugliese JS, Michelini M, Mancini AM (1999) Physics teachers in secondary school in Italy. In: Ferdinande H, Pugliese Jona S, Latal H (eds) The training needs of physics teachers in five European countries: an inquiry, vol 4. European Consortium, European Physical Society Technology Research and Development, Mulhouse

Woolnough BE (2001) Physics teachers as self-evaluating professionals. In: Physics teacher education beyond 2000. Barcelona, GIREP Book

Chapter 15
Freshman Engineering' Reasoning Strategies When Answering FCI Questions: A Case Study

Onofrio R. Battaglia and Claudio Fazio

Abstract Force Concept Inventory (FCI) is a questionnaire commonly used to assess students' conceptual understanding of Newtonian Mechanics. We show that Cluster Analysis methods can be used to study student answers to FCI by finding their reasoning strategies on Newtonian Mechanics. Our analysis is performed to data obtained by a sample of freshman engineering students just at the beginning of their first General Physics course. The analysis takes into account the decomposition of the force concept into the conceptual dimensions suggested by test authors and successive researches. We identified groups of students with similar answering strategies, characterised by correct answers, as well as by non-correct answers showing student misconceptions/nonnormative conceptions. Such answering strategies give insights into the relationships between the student force concepts and their ability to describe and/or explain motions.

15.1 Introduction

Force Concept Inventory (FCI) is a multiple-choice test useful to evaluate the efficacy of teaching approaches (Hake 1998) and to validate proposed learning progressions on force and motion (Fulmer 2015). Some studies focused on students' coherence when they answer FCI questions in different contexts (Bao and Redish 2006; Savinainen and Viiri 2008). Some authors (Bao et al. 2002; Bao and Redish 2006) used Model Analysis to analyse FCI students' answers. A recent paper (Brewe et al. 2016) has described a new methodology for carrying out network analysis on answers to multiple-choice conceptual inventories. Moreover, studies using Factor Analysis (Scott et al. 2012; Scott and Schumayer 2017; Semak et al. 2017) analysed the possible conceptual associations made by students among FCI questions.

O. R. Battaglia (✉) · C. Fazio
Dipartimento di Fisica e Chimica—Emilio Segrè, Università degli Studi di Palermo, Palermo, Italy
e-mail: onofriorosario.battaglia@unipa.it; claudio.fazio@unipa.it

B. G. Sidharth et al. (eds.), *Fundamental Physics and Physics Education Research*,
https://doi.org/10.1007/978-3-030-52923-9_15

Commonly, instructors and researchers are interested to the correctness of the students' answers on the FCI. Here we want to use FCI as a "tool" to find relationships among student answers to questions related to different aspects of Newtonian mechanics in order to give more complete information about their understanding.

In this study, we propose a quantitative analysis of FCI student answers and our results can help teachers and educators in designing appropriate teaching approaches. In it we apply a Cluster Analysis (*ClA*) method, based on the *k-means* algorithm (Everitt et al. 2011) to study students' answers to FCI questions, with the aim to point out reasoning strategies based on Newtonian conceptions as well as alternative conceptions (misconceptions/nonnormative conceptions). We show that our analysis can provide new insights into the students' conceptions of the different dimensions of the force concept, as defined by FCI's authors (Hestenes and Halloun 1995).

15.2 Theoretical Framework

Several pieces of research studied in detail the FCI test, some of them have shown its use as a diagnostic instrument. Among these latter ones, we would like to note the following that are, in our opinion, the most significant.

The first one, based on Factor Analysis is discussed in a recent paper (Semak et al. 2017). It has shown that this method is able to gauge the changes in conceptual associations made by students when the evolution of their answer patterns is known.

It was already pointed out (Bao and Redish 2006) that a way to use the FCI test as diagnostic instruments is given by methods that allow the researchers to analyse wrong answers. Several studies have shown that different alternative knowledge frameworks can coexist in college students and that the development and use of such knowledge are context-dependent. Moreover, the reasons why a student may deploy the correct knowledge in some situations and revert to use alternative kinds of knowledge in other ones can be hidden if students' alternative knowledge is not assessed. Such behaviour is often treated in the literature as random noise.

The second method we want to cite was introduced by Bao and Redish in 2006. It is called "Model Analysis", and provides a quantitative representation framework. In this framework, this method is able to quantitatively assess students' alternative knowledge and the probabilities for students to use such knowledge in a range of equivalent contexts. They take into account five FCI questions to study what kind of student's models can be associated with the force-motion ideas. Model Analysis allowed them to analyse students' answers in terms of predetermined mental models and graphically represented the probability that students can use each of these models.

The third method was developed by Brewe et al. (2016) to study relationships among nonnormative answers to the 30 FCI questions. By using network analysis and techniques of community detection (Grunspan et al. 2014) they were able to discover structures into patterns of answers. The result of this analysis allows the researchers to identify Modules of nonnormative answers which can highlight important underlying structures of the whole nonnormative answer network. Alternative reasoning procedures that involve student alternative conceptions are represented in Modules including high correlations among nonnormative answers to different questions, and some interesting interpretations are supplied. However, it is worth to note that the analysis provides information about patterns of nonnormative answers, but "the drawback is that we cannot investigate how these patterns relate to normative responses", as the authors state (Brewe et al. 2016).

In the literature some studies using Cluster Analysis (*ClA*) methods and concerning research in education are found. *ClA* methods can separate a sample of students into subgroups so that students belonging to the same subgroup are more similar to each other than those are not belonging in the same ones. These subgroups can be studied to characterise students' answers of open-ended questionnaires (Springuel et al. 2007; Fazio et al. 2013; Battaglia and Di Paola 2015; Di Paola et al. 2016; Battaglia et al. 2017a, b; Battaglia et al. 2019) or multiple-choice tests (Stewart et al. 2012). All these papers show that the use of *ClA* leads to individuate groups of students whose characterisation makes sense to researchers. In a recent paper, Stewart et al. (2012) analyse the student answers to seven questions by using Model Analysis. They study the state of student's knowledge and *ClA* methods to characterise the distribution of students' answers. They show that *ClA* is an effective method to inquiry the student understanding and to discover subgroups of a data set mathematically well-defined and meaningful for the researcher.

15.3 The Research Question

Hestenes et al. (1992) as well as successive researchers have divided the FCI test in different conceptual dimensions (Hestenes and Halloun 1995). We want to investigate the student understanding of two of such dimensions, and making diagnostic inferences about student knowledge. The research question that guided our study is:

To what extent can a *ClA* method reveal students' reasoning profiles of Newtonian mechanics understanding when they answer FCI questions about the first and second laws and the concept of force?

15.4 Methodology and Sample

15.4.1 The Sample

We administered the FCI test just at the beginning of an activity that the authors proposed as an optional course about Newtonian Mechanics. The sample was composed of 148 freshman engineering students. We analysed all the students who answered to more than 80% of the questions. For this reason, we analysed a subsample composed of 116 students (73.3% male and 26.7% female).

15.4.2 FCI Questionnaire

FCI is a multiple-choice questionnaire and it was presented for the first time in a paper published in 1995 (Hestenes and Halloun 1995). It is made of 30 multiple-choice questions (each question has five possible answers). The 1995 version included a classification of the Newtonian force concept investigated in the questionnaire in different dimensions. The FC authors reported a decomposition of the force concept into six conceptual dimensions by highlighting that all six dimensions are fundamental for the Newtonian force concept. In another paper (Hestenes and Jackson 2007) another table with a taxonomy of common-sense misconceptions related to the Newtonian mechanics, and the corresponding Inventory questions is proposed.

Here, we define a different categorisation of the 30 FCI questions in four categories or subtests as listed below.

- SubA includes 15 questions. Each question requires to apply Newton's first and/or second laws. Students have to choose among the different five answers that contains description of trajectories, kinematics quantities and/or explanations.[1]
- SubB includes questions that allow the researcher to evaluate student ability in identifying interactions between bodies and related forces in different dynamical or static contexts.
- SubC includes questions involving the Newton's third law. They require to identify action and reaction forces acting on bodies in dynamical or static conditions.
- SubD includes questions requiring student to describe motions by using the kinematics quantities and laws.

Table 15.1 reports our classification of FCI questions according to the criteria above described.

15.4.3 Data Coding and k-Means Algorithm

We quantitatively analyse the answers students give to the questionnaire by using Cluster Analysis (ClA). *ClA* methods allow us to generate groups of students by partitioning it and producing a set of non-overlapping clusters. We decided to apply the *k-means* algorithm (MacQueen 1967), because of its efficiency and simplicity.

This algorithm requires a coding of the answers. We code the FCI students' answers by using a binary code and generate a binary matrix, called "matrix of answering choices". It is made of *N* rows and *M* columns. In this matrix, each row

[1]We included questions involving Newton's first or second law in the same subtest. Such a view can be also pointed out from the analysis of other interview studies (Brokes and Etkina 2009).

Table 15.1 FCI questions divided into the four subtests

Subtest	Question														
SubA	Q3	Q6	Q7	Q8	Q9	Q10	Q12	Q17	Q21	Q22	Q23	Q24	Q25	Q26	Q27
SubB	Q5	Q11	Q13	Q18	Q29	Q30									
SubC	Q4	Q15	Q16	Q28											
SubD	Q1	Q2	Q14	Q19	Q20										

represents a student and has M components where only one is 1 while the other ones are zero. If in a generic row, a component is 1 in the column j means that the student associated with that row chose the j answering choice. For each question, each student can choose among five answering options (A–E), or to not answer at all. So, the total number of answering options is equal to 6. Therefore, for instance, in the case of SubA M is equal to 90 (6×15) and N to the number of students, is equal to 116.

In order to apply k-means algorithm we have to choose a distance index through which calculate the "similarity" between a couple of students. According to literature (Everitt et al. 2011; Mantegna 1999; Di Paola et al. 2016; Battaglia et al. 2017b) we chose a Euclidean metric (Gower 1966) to calculate this distance index and we obtain an $N \times N$ symmetric matrix that contains all the mutual similarity between our students.[2]

The results of k-means algorithm, several group or clusters, can be represented in a Cartesian plane by using a well-known procedure called Multidimensional Scaling (Borg and Groenen 1997).

Each cluster is made of students that are represented as points according to the mutual distances between them.

Once an appropriate partition of students has been found, each cluster has to be characterised in terms of the student behaviours. To do this we find for each cluster and each question the most frequent answers given by the students in that cluster. Those answers characterise that cluster, and according to Springuel et al. (2007) we call them "prominent" answers.

15.5 The Results

We separately analysed only the first two subsets to discover common student reasoning strategies into the different conceptual dimensions of the force concept. In the case of the 15 questions classified as SubA, the number, q, of clusters that best partitions our student sample was obtained through the maximisation of the mean value of *S-function* (Rouseeuw 1987), for different numbers of clusters ($2 \leq q \leq 4$). Their values and their 95% confidence intervals (C.I.) are the following: $\langle S(2)\rangle = 0.76$ (C.I. = 0.73–0.78), $\langle S(3)\rangle = 0.72$ (C.I. = 0.68–0.75), $\langle S(4)\rangle = 0.63$ (C.I. = 0.58–0.68).

The values reported above show that there are two statistically equivalent best clustering solutions (into two or three clusters). In order to choose one, we apply the Variation Ratio Criterion (*VRC*) (Calinski and Harabasz 1974). We calculate the Calinski index for the partition into three and two clusters and obtained the following

[2]In this case the distance between two students, is: $d_{ij} = \sqrt{2 \cdot (1 - R_{\text{bin}})}$, where R_{bin} is the correlation coefficient. A distance d_{ij} between two students equal to zero means that they are completely similar ($R_{\text{bin}} = 1$), while a distance $d_{ij} = 2$ shows that the students are completely dissimilar ($R_{\text{bin}} = -1$). When the correlation between two students is 0 their distance is $\sqrt{2}$.

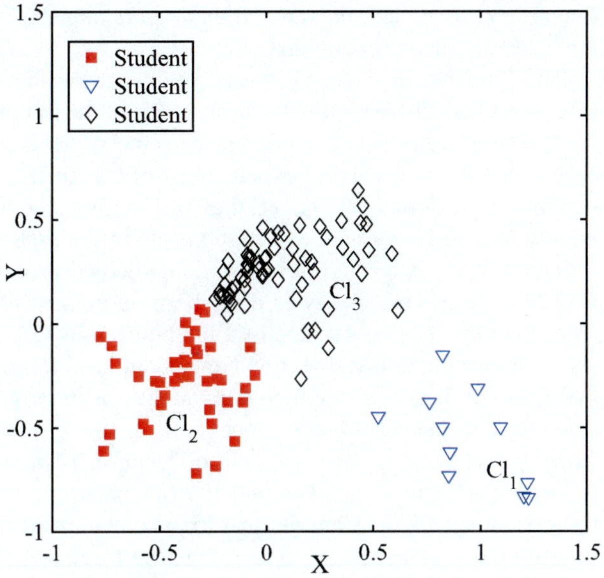

Fig. 15.1 *K-means* solution for SubA answers where three clusters are identified. Each point in this Cartesian plane represents a student

Table 15.2 Results for the SubA

Cluster	Cl_1subA	Cl_2subA	Cl_3subA
Number of students	66	39	11
Most frequent answers	**3C**, **6B**, **7B**, **12B**, 17A, **21E**, 25D, **27C**	**6B**, **8B**, **12B**, **21E**, 22A, 25F, 26F, 27F	**3C**, **6B**, **7B**, 8E, **9E**, **12B**, 17A, **22B**, **24A**, 25D, 26B, **27C**

Number of students belonging to each cluster and most frequent answers shared by more than 60% of students belonging to each cluster. The correct answers are reported as bold characters

values. (VRC(3) = 219 and VRC(2) = 149). So, we chose the three-clusters solution as the best one.

Figure 15.1 reports this solution in a Cartesian plane. Table 15.2 shows the most frequent answers supplied by students in the different clusters, with correct answers in bold characters.

The arrays in Table 15.2 allow us to describe the dominant behaviours of students in each given cluster.

The Students in Cl_1subA supplied correct answers for questions that require the qualitative description of the motion of dynamic systems (question Q27) and the prediction of trajectories (as in questions Q6, Q7, Q12 and Q21). Only question Q3 requires also a Newtonian explanation of the phenomenon. Moreover, students in Cl_1subA incorrectly answer questions Q17 and Q25 involving more forces applied to the same body, yet when the body has a constant velocity. In such cases, we think that misconceptions as "*a force due to the motion*" or "*a force in the direction of motion*" or "*an impetus force which is acting on the object after the object is no longer in contact with the agent applying the impulse*" may have been the cause of the students' incorrect answers.

The Students in *Cl_2subA* cluster supply correct answers to questions Q6, Q8, Q12, and Q21. These require the description of the motion of dynamic systems or the prediction of trajectories. However, the same students give incorrect answers in the case of questions involving explanations of the kinematic variable and the individuation of trajectories. Moreover, they fail to give a correct answer to questions that mainly require Newtonian explanations about the motions.

The students in by *Cl_3subA cluster* supplied correct answers to questions Q6, Q7, Q12 that require the ability to describe motions and predicting trajectories. Moreover, the same students correctly solved questions Q3, Q9, Q22, Q24 requiring the ability to use the first and second Newtonian laws to find an explanation. Questions Q8, Q17, Q25, Q26 show high percentages of incorrect answers indicating in our students a persistent of naïve conceptions as that "*persistence of original impetus*", "*largest force determines motion*", or "*motion is possible when forces overcome resistance*" (Hestenes and Halloun 1995). Coherently with the previous description, the students in *Cl_3subA* answer that for an elevator lifting up by a rope at a constant velocity the rope tension is greater than the force of gravity (17A). The idea that a constant force makes a body moving with a constant velocity is made explicit by the students in answers 25D and 26B. Differently, such students use a different model in answering question Q3 (3C). In this case, they clearly see an increase in the velocity for a falling object since the force of gravity is constant. Many nonnormative conceptions above described are strictly connected with the "Impetus Module" pointed out by Brewe et al. (2016).

As a second step, we performed the same analysis described above to obtain a partition of our student sample according to their answers to the six questions classified as SubB. The number, q, of clusters that best partitions our student sample was obtained through the maximisation of the mean value of S-function, $\langle S(q) \rangle$, calculated for q values from 2 to 4 and their 95% C.I.

We obtained $\langle S(2) \rangle = 0.68$ (C.I. = 0.65–0.70), $\langle S(3) \rangle = 0.69$ (C.I. = 0.64–0.73), $\langle S(4) \rangle = 0.62$ (C.I. = 0.59–0.64).

As in the analysis of SubA, in this case, we found that the $\langle S(q) \rangle$ values for $q = 2$ and $q = 3$ are comparable. Again, we perform a further analysis by using the VRC and obtained for the results into three clusters the highest value (VRC(3) = 219 and VRC(2) = 149).

Figure 15.2 reports the clustering solution into three different groups called *Cl_1subB, Cl_2subB, Cl_3subB*. Table 15.3 shows the most frequent answers supplied by students in different clusters, with correct answers in bold characters

Students in cluster *Cl_1sub* correctly interpret the forces as quantities that describe the interaction between bodies. However, they also cite a kind of force that seems to be directly connected to the body velocity or to some hit supplied to it as a "*force due to the motion*" or "*supplied by a hit*"). We can conclude that students comprised in this cluster are characterised by a "*hybrid*" (Ding and Beichner 2009*) or "synthetic"* (Gilbert and Boulter 1998) conception of force. This idea of force unifies different features of the naïve conceptions *("obstacles exert no force*" or "*motion implies*

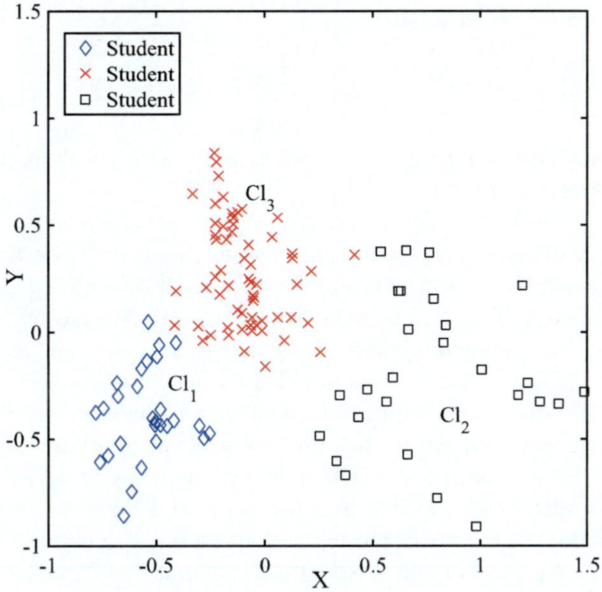

Fig. 15.2 *K-means* graph for SubB answers where three clusters are identified. Each point in this Cartesian plane represents a student

Table 15.3 Results for the SubB

Cluster	Cl_1subB	Cl_2subB	Cl_3subB
Number of students	28	27	61
Most frequent answers	5E, 11C, 13A, **29B**, 30E	5C, 11B, 18F, 29F, 30F	11B, 13C, 18C, 30E

Number of students belonging to each cluster and most frequent answers shared by more than 60% of students belonging to each cluster. The correct answers are reported as bold characters

active force" or "*velocity proportional to active force*" (Hestenes and Jackson 2007) and the scientifically accepted one that considers force as a quantity to model interactions between bodies. Finally, they are able to give correct answers for question 29 in which a static situation is considered.

Cl_3subB contains students that almost completely forget the interaction forces. They consider as forces acting on a moving body only the gravitational force and the one due to the motion. We think that a naive conception of force, widely discussed in the literature (see Brookes and Etkina 2009 and included references) is on the basis of the answers given by these students.

Students in Cl_2subB do not answer half of the subtest questions and this is the main difference by those belonging to Cl_3subB.

15.6 Discussion

The results above described allow us to answer our research question: "To what extent can a *ClA* method reveal students' reasoning profiles of Newtonian mechanics understanding when they answer FCI questions about the first and second laws and the concept of force?"

The results above described show that the *k-means* algorithm can be a useful tool to discover latent structures within the answers on the FCI test. We highlight a cluster of students with similar answers and such clusters are characterised by correct answers as well as by non- correct answers. Moreover, these latter have allowed us to identify student's misconceptions/nonnormative conceptions.

The method we used can identify in detail the status of knowledge of the students. We discover dominant/prevalent behaviours, and these can allow us to infer the characteristics of models used by students in different contexts.

The main results, obtained by our analysis of FCI answers supplied by our students characterised by a low level of understanding of the Newtonian dynamics, allow us to answer our research question, as reported below.

By analysing the student understanding of dynamical characteristics of motions we note that the majority of our students can't supply a Newtonian explanation of such motions (by applying the first and/or second Newtonian laws). They show in some contexts a correct ability to describe trajectories, but not to explain them in the theoretical framework of the Newtonian Mechanics. This ability can be attributed mostly to their common life experience. Only a small number of students are able to supply correct answers to questions requiring explicative skills, but only in given contexts.

The majority of students in our sample when is required to identify forces acting on a moving body only think on the gravitational force and/or "*motion force*" or "*impetus force*". They almost completely omit the reaction forces. Such students are characterised by a naïve idea of force, widely discussed in the literature (see Brookes and Etkina 2009 and included references) and connected with several other misconceptions ("*obstacles exert no force*" or "*motion always implies active force*" or "*velocity proportional to active force*") (Hestenes and Jackson 2007). Together with this naïve idea of force, only a small group of students has a Newtonian concept of force as a quantity to describe the interaction between objects. In their minds, these two absolutely different ideas seem to coexist, as several studies also pointed out (Brookes and Etkina 2009). Finally, it is worth to highlight that, despite such a mixed idea of force, these students are able to use the first and second Newton laws and to identify correct relationships between force and motion only in some contexts.

15.7 Conclusions

We would like to outline that the results obtained here for FCI could not be generalised to all Concept Inventories (CIs). At the same time, we think that *ClA* applied to CIs can be useful at the level of the classroom. For instance, we see the use of CIs as diagnostic instruments that could be useful to connect instruction with student ideas on a more fine-grained level and to help them in developing pedagogical approaches that target the ideas in classes. Specifically, we found that a fundamental role in the correct analysis of mechanics problems is given by the concept of force, investigated using student ability to individuate forces acting in statics and dynamical contexts.

Educational implications from our results can be also drawn. These mostly point to the relevance for teachers and instructors to be aware that students' difficulties with the ideas of force and motion mainly arise from their difficulty in understanding the ontological status of the force concept in physics (Jammer 1957). The contexts supplied by FCI items can provide teachers with useful examples, as well as the acknowledgement of reasoning strategies used by physicists in the historical development of force and motion concepts (Jammer 1957). Teaching approaches to Newtonian mechanics can be implemented by thinking that student alternative conceptions and mainly their reasoning strategies applied in different mechanical contexts can make the students able to speed up their mastering of such an important theoretical framework.

References

Bao L, Redish EF (2006) Model analysis: representing and assessing the dynamics of student learning. Phys Rev ST Phys Educ Res 2:010103

Bao L, Hogg K, Zollman D (2002) Model analysis of fine structures of students models: an example with Newton's third law. Am J Phys 70:766

Battaglia OR, Di Paola B (2015) A quantitative method to analyse an open-ended questionnaire: a case study about the Boltzmann Factor. Nuovo Cimento Soc Ital Fis C 38(3)

Battaglia OR, Di Paola B, Fazio C (2017a) K-means clustering to study how student reasoning lines can be modified by a learning activity based on feynman's unifying approach. EURASIA J Math Sci Technol Educ 13(6)

Battaglia OR, Di Paola B, Fazio C (2017b) A quantitative analysis of educational data through the comparison between hierarchical and not-hierarchical clustering. EURASIA J Math Sci Technol Educ 13(8)

Battaglia OR, Di Paola B, Adorno DP, Pizzolato N, Fazio C (2019) Evaluating the effectiveness of modelling-oriented workshops for engineering undergraduates in the field of thermally activated phenomena. Res Sci Educ 49(5):1395–1413

Borg I, Groenen P (1997) Modern multidimensional scaling. Springer, New York, NY

Brewe E, Bruun J, Bearden IG (2016) Using module analysis for multiple choice responses: a new method applied to Force Concept Inventory data. Phys Rev Phys Educ Res 12:020131

Brookes DT, Etkina E (2009) "Force," ontology, and language. Phys Rev Spec Top Phys Educ Res 5(1)

Calinski T, Harabasz J (1974) A dendrite method for cluster analysis. Commun Stat 3(1):1–27

Di Paola B, Battaglia OR, Fazio C (2016) Non-Hierarchical Clustering as a method to analyse an open-ended questionnaire on algebraic thinking. S Afr J Educ 36(1):1–13

Ding L, Beichner R (2009) Approaches to data analysis of multiple-choice questions. Phys Rev ST Phys Educ Res 5:020103

Everitt BS, Landau S, Leese M, Stahl D (2011) Cluster analysis. John Wiley & Sons, Ltd, Chichester

Fazio C, Battaglia OR, Di Paola B (2013) Investigating the quality of mental models deployed by undergraduate engineering students in creating explanations: the case of thermally activated phenomena. Phys Rev Spec Top Phys Educ Res 9(2)

Fulmer GW (2015) Validating proposed learning progressions on force and motion using the Force Concept Inventory: finding from Singapore secondary schools. Int J Sci Math Educ 13 (6):1235–1254. https://doi.org/10.1007/s10763-014-9553-x

Gilbert JK, Boulter CJ (1998) Learning science through models and modelling. In: Frazer B, Tobin K (eds) The international handbook of science education. Kluwer, Dordrecht, pp 53–66

Gower JC (1966) Some distance properties of latent root and vector methods used in multivariate analysis. Biometrika Trust 53:3–4

Grunspan DZ, Wiggins BL, Goodreau SM (2014) Understanding classrooms through social network analysis: a primer for social network analysis in education research. Cell Biol Educ 13:167

Hake RR (1998) Interactive engagement versus traditional methods: a six-thousand-student survey of mechanics test data for introductory physics courses. Am J Phys 66:64

Hestenes D, Halloun I (1995) Interpreting the force concept inventory: a response to March 1995 critique by Huffman and Heller. Phys Teach 33:502–506

Hestenes D, Jackson J (2007) Revised table II for the Force Concept Inventory (Unpublished). http://modeling.asu.edu/R&E/Research.html

Hestenes D, Wells M, Swackhammer G (1992) Force concept inventory. Phys Teach 30:141–151

Jammer M (1957) Concepts of force. Harvard University Press, Cambridge, MA

MacQueen J (1967) Some methods for classification and analysis of multivariate observations. In: LeCam LM, Neyman J (eds) Proceedings of the 5th Berkely symposium on mathematical statistics and probability 1965/66, vol I. University of California Press, Berkeley, CA, pp 281–297

Mantegna RN (1999) Hierarchical structure in financial markets. Eur Phys J B 11:193197

Rouseeuw PJ (1987) Silhouttes: a graphical aid to the interpretation and validation of cluster analysis. J Comput Appl Math 20:53–65

Savinainen A, Viiri J (2008) The force concept inventory as a measure of students conceptual coherence. Int J Sci Math Educ 6(4):719–740

Scott TF, Schumayer D (2017) Conceptual coherence of non-Newtonian worldviews in Force Concept Inventory data. Phys Rev Phys Educ Res 13:010126. https://doi.org/10.1103/PhysRevPhysEducRes.13.010126

Scott TF, Schumayer D, Gray AR (2012) Exploratory factor analysis of a Force Concept Inventory data set. Phys Rev ST Phys Educ Res 2012:020105

Semak MR, Dietz RD, Pearson RH, Willis CW (2017) Examining evolving performance on the Force Concept Inventory using factor analysis. Phys Rev Phys Educ Res 13:019903. https://doi.org/10.1103/PhysRevPhysEducRes.13.010103

Springuel RP, Wittmann MC, Thompson JR (2007) Applying clustering to statistical analysis of student reasoning about two-dimensional kinematics. Phys Rev ST Phys Educ Res 3:2

Stewart J, Miller M, Audo C, Stewart G (2012) Using cluster analysis to identify patterns in students' responses to contextually different conceptual problems. Phys Rev Spec Top Phys Educ Res 8(2)

Chapter 16
Inquiry-Based Approach and Numerical Simulations: A Powerful Integration in Condensed Matter Physics Education

Dominique Persano-Adorno

Abstract In this chapter, we present and discuss two inquiry-based learning paths on condensed matter physics topics in which numerical simulations play a relevant role. The first one addresses the study of the electron transport dynamics via simulative explorations in 3D semiconductors. His emphasis is not on student modeling skills, but rather on a chain of reasoned investigations performed within a learning environment aimed at supporting a valuable understanding of the physics concepts underlying the complex world of semiconductor electronics. The second learning path is a 5E-cycle-based workshop of advanced physics targeted to strengthen student's understanding of the various aspects of the Hall Effect. In this latter, the instructors stimulate a discussion about the classical, integer and fractional quantum Hall effects. Both learning paths represent a powerful instrument for educators introducing young undergraduates to the effectiveness of numerical simulations to investigate a physical system where the theoretical processes are well known, but analytical methods of examination still provide only approximate results. Our findings show that the stimulated activation of the inquiry process, also by means of numerical simulations, can represent an effective teaching/learning method. This approach successfully engages students into active learning and, at the same time, supports the clarifying of important experimental and technological aspects of material science, representing a feasible example of combination of a traditional lecture-based teaching method with efficacious teaching/learning strategies.

D. Persano-Adorno (✉)
Dipartimento di Fisica e Chimica "E. Segrè", Università di Palermo, Palermo, Italy
e-mail: dominique.persanoadorno@unipa.it

B. G. Sidharth et al. (eds.), *Fundamental Physics and Physics Education Research*,
https://doi.org/10.1007/978-3-030-52923-9_16

16.1 State of Art in Condensed Matter Physics Instruction at Undergraduate and Graduate Levels

A traditional lecture-based teaching on Condensed Matter Physics provides Physics/Engineering students with a theoretical background regarding relevant concepts as the effective mass, the band structure, the dispersion relation, the phonon-induced scattering mechanisms, the correlated particles behavior, etc. However, a valid and efficacious instruction should train the students toward a full comprehension of the fundamental notions of materials science and, at the same time, strengthen their reasoning abilities and transversal skills. Graduate Scientists should demonstrate specialist-discipline knowledge, capability to solve everyday problems, and modeling skills based on innovative thinking (Borrego and Bernhard 2011). At the University, Condensed Matter Physics courses offer learners a theoretical application of mathematical methods, sometimes forgetting to focus on hands-on projects to improve the students' learning curve. Besides, a mere theoretical approach is hardly successful in teaching science, because any mental construction is based on experience and learners seldom fully understand a theory if it is left far from an experimental investigation (Greca and Moreira 2000). Recent research studies ask for university curricula including "*integrative laboratory experiences that promote inquiry, relevance, and hands-on activities*" and recommend the learning experiences replace the lecture, embracing active learning, i.e., laboratories, internships, and all forms of cooperative learning (from NSF Report 1996). A new concept of active construction of significant knowledge and inspiration of high levels of critical thinking skills has been suggested, switching from a passive lecture-style teaching to a more active and student-centered teaching approach (Altbach et al. 2009). The analysis of "how people learn" should guide design, assessment, and estimation of teaching and learning transformations. The most suitable scaffolding of a learning environment should foster learning by inquiry and transform the experience at University from a "*culture of receivers into a culture of inquirers, in which faculty, graduate and undergraduate students share in an adventure of discovery*" (from Boyer Commission 1998).

16.1.1 Inquiry-Based Science Education: IBSE

In this framework, inquiry-based education represents the natural context to generate possibilities of learning science notions in terms of an active building of effective understanding and promotion of high levels of critical reasoning abilities. Inquiry is a method of active learning in which students are involved in numerous integrated activities of identifying queries, gathering experimental data in a laboratory or real life setting, constructing descriptions and explicative models, communicating and sharing their results (Llewellyn 2002; Pizzolato et al. 2014; Wei et al. 2014; Persano-Adorno et al. 2018a). Rather than a distributor of information, the educator is a mentor, facilitator, and co-discoverer who fosters students to inquiry, contest, and formulate their own theories, models, and conclusions. Deep knowledge is obtained through active engagement rather than through imitation or repetition.

16.1.2 Active Learning and Numerical Simulations: A Powerful Integration

Condensed Matter Physics instruction requires the knowledge of many physics' concepts, techniques and phenomena in a limited time. Unfortunately, the experimental setup of inquiry-based experiences on Condensed Matter Physics topics is not easily exploitable in most university laboratories for great numbers of students. At this regard, numerical simulation, *being considered a practice in between theory and experiment, could represent a valid alternative* (Li et al. 2012). The main advantage of using computer simulations, video clips and movies is the super visualization feature, especially helpful in the description of Condensed Matter Physics evidences, because certain complex phenomena due to the action of abstract fields could be difficult to imagine. In these "augmented" learning settings, students can explore theoretical and experimental features through numerical simulation and real experiments (Silsbee and Draeger 1997). Furthermore, at present a variety of technologies is accessible to create computer simulations actively promoting student interest.

In this contribution, two inquiry-based learning paths on Solid State Physics are presented and discussed. The first one, focused on the analysis of the electron transport dynamics via Monte Carlo explorations in 3D semiconductors, has been experimented by a sample of students in the MsC in Electronic Engineering (Persano-Adorno et al. 2016b). In this learning experience, the inquiry approach stimulated the undergraduates to follow a question-driven path of investigation, starting from the validation of the model used for simulating the electron transport within the semiconductor bulk, up to performing reasoned questions about the observed features of carrier transport (Persano-Adorno et al. 2015a). The second learning path is a 5E-cycle-based laboratory of advanced physics designed with the purpose of boosting Physics/Engineering student knowledge of the different features of the Hall Effect (Persano-Adorno et al. 2019). The aim of this learning path is to stimulate a discussion about the classical, integer and fractional quantum Hall effects and to introduce a unified visualization based on the notion of composite fermions and interacting quasiparticles exhibiting a quantum behavior (Persano-Adorno et al. 2019).

16.2 Workshop 1: An Experience of Elicited Inquiry Elucidating the Electron Transport in Semiconductor Crystals

The understanding of semiconductor transport properties is a necessary tool for any physicist or engineer involved in semiconductor technology. In last years, there has been a significant attention toward indium phosphide (InP) because of its usage in several optoelectronic and photonic devices (Persano-Adorno et al. 2015b, 2016a). Thus, a greater knowledge of the characteristics of the electron transport dynamics in InP devices is crucial in undergraduate instruction in electronic engineering as well as in semiconductor science.

16.2.1 Method

A sample of ten undergraduates in electronic engineering at the Laboratory of Condensed Matter Physics of the Department of Physics and Chemistry, University of Palermo, Italy, took part in this learning experience. Two tutors, having more than 15 years of expertise in scientific research and on teaching physics at both University and high-school, supported student scientist-like activities. The undergraduates, selected among those who attended more than 80% of the traditional lectures on Physics of Materials for Electronics, were engaged in an inquiry-based learning environment about the investigation of the carrier dynamics in InP semiconductor bulk via Monte Carlo (MC) simulations. This method is one of the most powerful simulative techniques allowing the numerical simulation of the carrier dynamics in semiconductors, away from the quasi-equilibrium approximation (Moglestue 1993). The MC technique represents a space-time continuous solution of the Boltzmann transport equation and provides an exhaustive description of particle motion in a semiconductor. Therefore, it is appropriate for investigating both the steady state and the dynamic features of a device (Persano-Adorno 2010). The MC method allows to account for the main aspects of band structure, scattering processes and heating effects, device design and material parameters.

16.2.2 Activity Description

The task requested to the students was to investigate the carrier dynamics in an InP crystal by using MC simulations, with a focus on the role of the effective mass, intervalley and intravalley scattering, crystal impurities, and lattice temperature on carrier dynamics (Persano-Adorno et al. 2015a). The final challenge guiding learners' inquiry was the exploration of real possibilities of improving the transport dynamics, in terms of an increase of the signal transmission speed, i.e., the charge velocity, with respect to the lower attainable cost of preserving (the driving electric power).

The students, working in groups, first tried to design their protocol of investigation, as required in a traditional *guided inquiry* (Banchi and Bell 2008). Although our students had already received a traditional lecture-based instruction on condensed matter physics and attended a session on the use of MC procedures, when they were engaged in the learning environment, experienced numerous problems on planning and executing a fruitful sequence of numerical experiments, frequently coming to a standstill. At this point, the two instructors decided to actively participate to the students' discussion on the physics explaining the observed phenomena, not once providing comprehensive explications to the learners, but giving comments and suggestions, occasionally definitely incorrect, but "*effective to stimulate students' reasoning and activating a proficient scientific inquiry*" (Persano-Adorno and Pizzolato 2015). The active involvement of the instructors to the debate (as peers) triggered student scientific questioning through the start of an efficacious inquiry: after the initial validation of the adopted model, the stimulated inquiry learning path articulated in three successive levels. Each one initiated from a reasoned query and consisted of a sequence of simulative experiments whose findings were explanatory at some level of understanding and, at the same time, stimulating the students' thinking with additional questions to address by a deeper scientific inquiry.

16.2.3 Preliminary Phase: Model Validation

Prior to start exploiting a model developed by others, our students assessed its validity, by comparing the computational results with experimental findings reported in literature. In this initial phase, the learners thoroughly examined the conditions under which real experiments have been achieved (lattice temperature, electron density, impurity concentration, etc.). With the aim to set up the right parameters, first concentrated their attention on the capability of the simulated data to carefully reproduce the corresponding experimental data, by leaving the real understanding of the physics beneath their numerical results to a successive explicative phase. In this stage, the instructors pushed students' inquiry toward the examination of those model parameters, that could be suitably adjusted to achieve their objective.

16.2.4 Stage 1: Inquiry About the Physical Quantities Affecting the Velocity-Field Characteristic

The students observed a nonlinear velocity-field characteristic (see panel (a) of Fig. 16.1): an initial increasing trend of the electron drift velocity, followed by a maximum (at ~10 kV/cm) and a successive region—for higher intensities of the

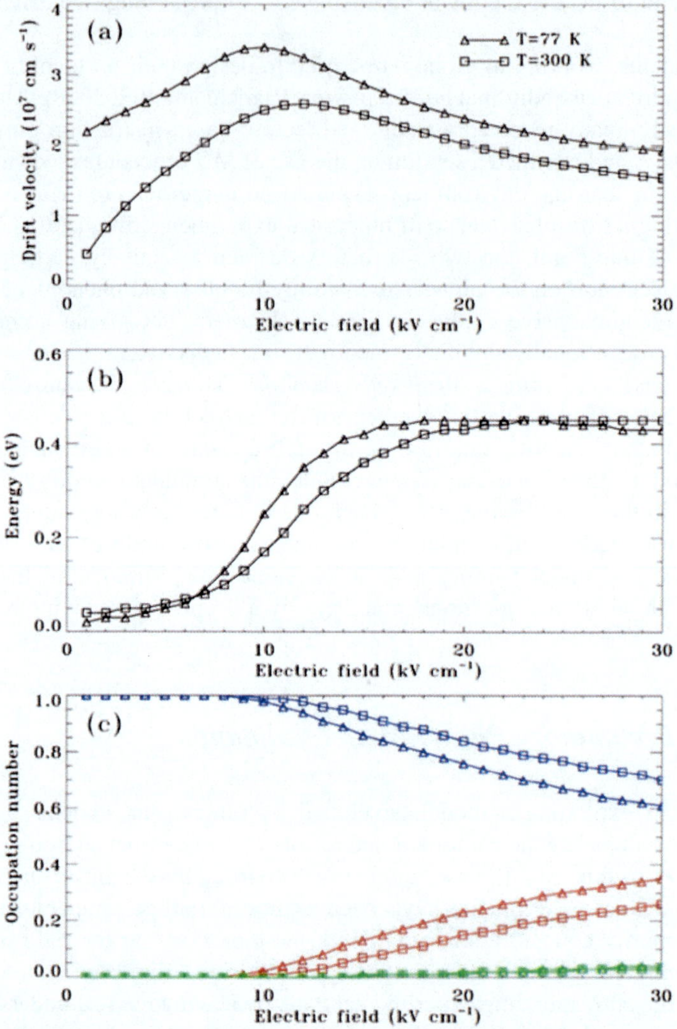

Fig. 16.1 Electron mean drift velocity (**a**), mean kinetic energy (**b**), and occupation number (**c**) as a function of the driving electric field amplitude, obtained by MC simulations of electron transport in InP with impurity density $n = 10^{13}$ cm^{-3}, at lattice temperatures $T = 77$ K (triangles) and 300 K (squares), respectively. Panel (**c**): Blue—Γ valley; Red—L valleys; Green—X valleys. (Figure from Persano-Adorno et al. 2015a)

electric field—characterized by a diminishing velocity (Persano-Adorno et al. 2015a). This result astonished students, who almost certainly expected to find the well-known ohmic behavior. The professors stimulated the learners to questioning about this phenomenon, with the purpose to identify the physical explanation of the observed reduction of the electron drift velocity.

As shown in panel (b) of Fig. 16.1, in the range 1–8 kV/cm, the mean energy slowly rises to ~0.1 eV. Higher values of driving fields produce a rapid increase of the electron energy up to a saturation regime at about 0.4 eV. This outcome was twice surprising for learners who first expected that the mean energy trend follows that of the electron velocity and accordingly mean energy decreases as the mean velocity does, and then they did not expect a saturation of the energy levels, but at most a growth for higher values of the driving fields. The instructors encouraged a debate on how this phenomenon could be physically justified.

From the theory the students learned that charge carriers traveling within a semiconductor, depending on their energy, may reside in different valleys, in which are characterized by different effective masses. The application of an electric voltage causes the electrons are not more in equilibrium with the crystal lattice and increase their energy; this happens until they have the opportunity to migrate from the Γ valley to the higher energy valleys (L- and X-valleys), where the effective mass is greater (heavy electrons). This carrier transfer was corroborated by the numerical investigation of the electron occupancy in each valley as a function of the electric field, reported in panel (c) of Fig. 16.1. The students observed that electrons begin to occupy the higher energy valleys when the amplitude of the electric field achieves values greater than ≈10 kV/cm, the same intensity characterizing the maximum in the velocity-field curve. This evidence pointed up the significance of considering the effective mass of charge carriers and the important role played by scattering events, accountable for intervalley transitions.

16.2.5 Stage 2: Investigation of the Role Played by the Effective Mass

A cogent query drove the students through an extensive examination of the importance of the part played by the effective mass of drifting electrons. The students conducted various simulations and made a comparison between the numerical outcomes achieved by employing a three-valley model and those coming from the one-valley (Γ) model, in which the electron shifts to higher energy valleys are interdicted. Furthermore, they explored the effects of considering all electrons having a same mass equal to the mean value among the effective masses of the different valleys (Persano-Adorno et al. 2016b).

16.2.6 Stage 3: Study of the Effects Due to a Change of the Doping Density

With the aim to point out the consequence of the interactions among free electrons and ionized impurities, randomly disseminated inside the InP bulk, the instructors encouraged the undergraduates to explore the electron transport for different values of the doping density. As ionized impurities scattering appear to be important principally at low field amplitudes and/or at low temperatures, the learners concluded that the impurity scattering rate reduces when the electron energy growths, becoming neglectable in the high-field range (Persano-Adorno et al. 2015a).

The inquiry-based learning cycle, generated by unforeseen outcomes from numerical simulations, developing through a sequence of explorations and promoting working hypothesis, was enriched by an explicative stage where learners were stimulated to share their opinions and debate about the conclusive evidence. Students' reasoned inquiries, successful in overcoming the impasses, together with the procedure for the numerical simulations and the evaluation of the outcomes from the simulated experiments, strengthened the concepts on the physics underlying the charge transport in semiconductors.

16.3 Workshop 2: The Different Features of the Hall Effect

The different features of the Hall Effect and the related quantum transport phenomena in two-dimensional structures received considerable attention in last decades. Nevertheless, the quantum Hall effect explication is frequently away from the program of the typical undergraduate courses and beyond the knowledge of most nonspecialist physicists. The Hall Effect, characteristic of systems with free or nearly free charged particles, whose motion is affected by the concurrent action of a magnetic field and an electric one, represents the most amazing demonstration of the Landau quantization (Stormer 1999; Tsui 1999). Furthermore, it represents a rare example of microscopic effects detectable on a macroscopic scale; it permits to determine very accurate values of microscopic quantities, such as the electron charge and the Planck constant (Persano-Adorno et al. 2019).

The 5E-learning cycle is a pedagogical approach aimed to develop student critical thinking and to help them to explore and assess their own understanding (Banchi and Bell 2008). In particular, the 5E cycle is a constructivist student-centered instructional model where the learner fulfills five "steps" of instruction. These stages are *Engagement* (learners switch on and evaluate previous knowledge linking "*the new to the known*"), *Exploration* (learners analyze a real-world problem), *Explanation* (learners enlighten their thinking), *Elaboration/Extention* (learners process their reasoning and reinforce their knowledge), and *Evaluation* (learners evaluate their own learning).

16.3.1 Engagement Phase

In the ENGAGE phase, the classical Hall Effect is reminded. This effect, discovered by Edwin Hall in 1879, describes the onset of a voltage difference across a thin

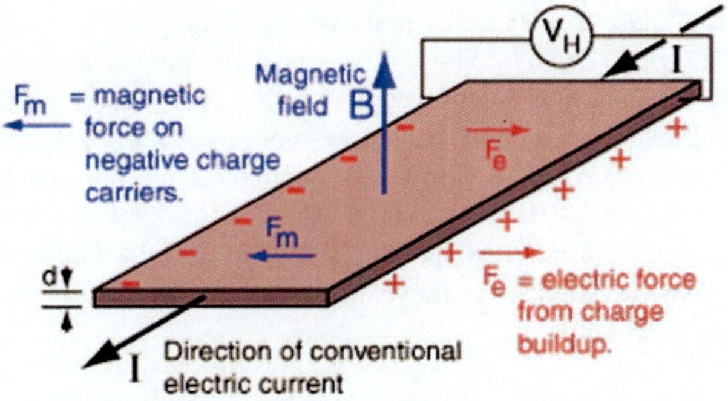

Fig. 16.2 The classical Hall Effect. (From the HyperPhysics site, Department of Physics and Astronomy—Georgia State University; http://hyperphysics.phy-astr.gsu.edu)

metallic bar, transverse to an electric current and to an external magnetic field perpendicular to the current (Persano-Adorno et al. 2018b). It occurs also at room temperature. The negative charges begin to pile up on one side of the bar, leaving the opposite side depleted of electrons. Accordingly, an electric field rises in the transverse direction; at equilibrium, this field gets to an amount such as to inhibit additional charges to accumulate (see Fig. 16.2). The *Hall coefficient*, $R_H = E_y/j_xB_z = -1/\text{nec}$, is proportional to the Hall field and depends on the sign and value of the surface charge concentration. The *material resistivity* or *magnetoresistance* ρ_{xx} is a constant, that depends on carrier density and mobility; the *Hall resistivity* $\rho_{xy} = V_H/I$ (where V_H is the *Hall voltage* along the y-side) is proportional to B_z (Kittel 2005).

By using numerical simulations, learners could investigate the classical Hall Effect and its usefulness in obtaining sign, density and mobility of the charge carriers.

The classical Hall Effect can be considered the precursor of a set of quantum Hall effects, typical of 2D systems discovered successively thanks to the developments in quantum mechanics and in nanotechnology.

In particular, we will discuss (1) the Integer Quantum Hall Effect (IQHE), which occurs at low temperature in presence of strong transverse magnetic fields, and (2) the Fractional Quantum Hall Effect (FQHE), which is a particular case of the previous one, occurring in a small number of 2D materials where electrons collectively behave, showing unexpected features.

16.3.2 Exploration Phase

In a 2D electron system at low temperatures, in the presence of a large perpendicular magnetic field, one can detect the integer quantum Hall Effect, where the Hall resistivity ρ_{xy} experiences quantum Hall transitions, quantized in integer units of h/e^2 (Stormer 1999). This effect has been revealed about a century after the discovery of the classical Hall Effect. In the EXPLORE phase, by means of the examination and interpretation of the experimental findings achieved in different heterostructures, the learners will have the chance to get directly engaged with phenomena and new materials. The experimental evidences show that ρ_{xy} varies by steps regulated by the law $\rho_{xy} = h/(se^2)$, where s is a positive integer dependent on B_z (Tsui 1999). The coefficient h/e^2 is a universal constant independent on the used sample. Experimental results acquired in a low-mobility two-dimensional electron gas in GaAs/AlxGa1-xAs at a lattice temperature $T = 66$ mK and a carrier concentration $n = 1.93 \times 10^{11}/\text{cm}^3$, shows the concomitant quantization of the Hall resistivity ρ_{xy} characterized by the step behavior and the fall of the magnetoresistance ρ_{xx} to very low values (for more details, see Fig. 7 and its description in Persano-Adorno et al. 2018b).

16.3.3 Explanation Phase

In the EXPLAIN phase, the learners stimulated by the educator should understand that the IQHE is due to the circumstance that the closed circular orbits covered by the electrons in a gas of free (independent) electrons only assume quantized energies, called *Landau Levels*, whose values are solutions of the Schrödinger equation: $E_n = (n + 1/2)\hbar\omega_c$, with $n = 1,2,3 \ldots$. The mechanism originating the quantized Hall resistivity depends on how many of these energy levels are occupied by electrons (Kittel 2005). The number of permitted orbitals on each orbit is constant for a given amplitude of the applied magnetic field. The electrons cannot dwell in the energy gaps in between or in the quasi-continuous k_z (unaffected by B_z) states that do not occur in a 2D system. Figure 16.3 shows how the transition from the quasi-continuous Fermi levels to the highly degenerate Landau levels arises. Since the degeneration rises with the B_z amplitude, the population of the filled Landau levels will depend on B_z and on the size of the conductor. The electrons fill these levels up to the last one that may be entirely filled or not.

We point out that these properties pertain to an ideal crystal structure.

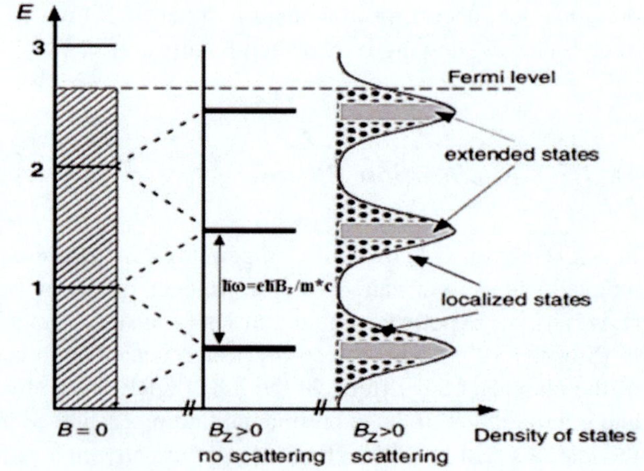

Fig. 16.3 Left: quasi-continuous states of a Fermi electron gas at $T = 0$ in absence of perturbations; center: rearrangement of the electron distribution in presence of a magnetic field without impurity/phonon scattering (Landau levels); right: rearrangement of the electron states when impurities cause scattering and electron trapping (localized states). (Adapted with permission from Longo 2011)

In a real crystal, in which defects and impurities are inevitable, things may change. Let suppose that only the lowest Landau level is entirely occupied and start to reduce B_z. Accordingly, the degeneration decreases, and a growing number of electrons is forced to move to the adjacent level that may be too high. From the energetic point of view, for them it becomes more convenient to be entrapped into localized impurity sites. Therefore, such electrons are not movable and, consequently, they cannot participate to the conductivity. A further decrease of B_z reduces the Landau energy separation and for trapped electrons becomes convenient to fill an available lower vacant level. It comes out that the Landau levels in a real crystal are entirely occupied or entirely unfilled over a broad range of B_z. The plateau in every step of ρ_{xy} is described in the framework that there are always occupied Landau levels with the exclusion of those B_z amplitudes, falling in the narrow regions in which a transition from the last filled Landau level to the next unoccupied one occurs. Therefore, the magnetoresistance ρ_{xx} is equal to zero because when some Landau levels are entirely full and others completely void, there is no chance of electron scattering. In the narrow transition range in which s takes the contiguous value there will be non-completely filled levels, scattering becomes possible, consequently ρ_{xx} is not more null (Persano-Adorno et al. 2018b).

Students should observe that in an ideal structure, without defects, the IQHE would not be possible.

16.3.4 Extension Phase to the Fractional Quantum Hall Effect

The fractional quantum Hall effect is a particular case of quantum Hall effect which arises in a small number of low-dimensionality systems, (for example in graphene) where electrons assume a collective behavior, exhibiting unexpected features. The FQHE is observable in these 2D materials at very high magnetic fields ($>$150 T) and very low temperatures ($<$1 K). The principal difference between the IQHE and the FQHE lies on the basic electron feature: they behave as free (noninteracting) particles in IQHE and correlated interacting electrons in the FQHE. In FQHE the equal spacing of the Landau levels is substituted by a collective behavior of composite quasiparticles whose charge is smaller than e and the level separation is proportional to $B_z^{1/2}$. The unexpected circumstance that the charge of these composite quasiparticles is smaller than e has been ascribed to bits of magnetic field, that attached to each electron, give rise to a new object whose properties are different from those of a free or a bound electron. These quasiparticles are affected by an effective magnetic field B_z*, different from the external applied field B_z. Their movement appears do not depend strictly on the magnetic field amplitude and they may behave as bosons or fermions depending on the magnetic field (Laughlin 1983).

For instance, in graphene the transversal conductance σ_{xy} is discretized in steps with half integer multiples of $4e^2/h$ and the longitudinal resistivity ρ_{xx} falls to zero in each plateau (Novoselov et al. 2005)

16.3.5 Evaluation Phase

The quantum Hall Effects result from an intriguing interplay among disorder and interactions, but the basic physics is that describing an electron subjected to a magnetic field. Additionally, in a specific case (FQHE), there is indication of the existence of interaction processes that manifest as a fractional charge and could facilitate the understanding of physics governing interacting particles. At the end of the learning experience learners should have consolidated their knowledge about the different features of the Hall Effect and accomplished a more meaningful knowledge of the important concepts concerning the quantum behavior of correlated particles (Persano-Adorno et al. 2019). In particular, they should be capable to make a comparison between the Cooper pairs in superconductors and the composite electrons in graphene, and to recognize analogies or differences among the concepts of effective mass of electrons in semiconductors and effective charge of Dirac fermions in graphene (Persano-Adorno et al. 2018b).

In this perspective, the integration between active learning and numerical simulations could be very fruitful.

16.4 Conclusion

A real understanding of scientific concepts is accomplished when learners can utilize those concepts to solve problems never encountered before. An appropriate instruction on problem solving should be based on the development and reinforcing of student reasoning abilities. Higher levels of thinking skills can be established by "forcing" the students to directly experiment the world and strive for finding answers to real problems. This can be achieved by training students to pose scientifically appropriate inquiries, perform scientific explorations, obtain meaningful measurements and analyze experimental data, build explicative models, check findings with further investigations, share and discuss results with peers. Teacher's responsibility on supplying an appropriate scaffolding is central because Inquiry-based learning settings with a lower teacher supervision may stimulate higher reasoning competencies, but occasionally may produce negative feelings due, for example, to run into mistakes or achieve unexpected findings. Our results show that the integration between the use of numerical simulations and a specific activation of the inquiry process can represent an effective teaching approach to successfully engage undergraduates into an active learning. This may represent a possible example of merging between a traditional lecture-based teaching approach and effective discovery-based learning strategies.

Acknowledgment We gratefully acknowledge Prof. Leonardo Bellomonte and Dr. Nicola Pizzolato for their invaluable support and encouragement.

References

Altbach PG, Reisberg L, Rumbley L (2009) Trends in global higher education: tracking an academic revolution. In: Report for the UNESCO 2009 World Conference on Higher Education, pp 113–121

Banchi H, Bell R (2008) The many levels of inquiry. Sci Child 46(2):26–29

Borrego M, Bernhard J (2011) The emergence of engineering education research as an international connected field of inquiry. J Eng Educ 100(1):14–47

Boyer Commission (1998) Reinventing undergraduate education, a Blueprint for America's research universities. State University of New York, Stony Brook, NY

Greca IM, Moreira MA (2000) Mental models, conceptual models, and modelling. Int J Sci Educ 22 (1):1–11

Kittel C (2005) Introduction to solid state physics, 8th edn. Wiley, New York

Laughlin RB (1983) Anomalous quantum Hall effect: an incompressible quantum fluid with fractionally charged excitations. Phys Rev Lett 50(18):1395

Li Y, Ma L, Shi Y (2012) Exploration on computer simulation method in physics education. In: Wang Y (ed) Education and educational technology, vol 108. Springer, Berlin, pp 517–522

Llewellyn D (2002) Inquiry within: implementing inquiry-based science standards. Corwin Press Inc., Thousand Oaks, CA

Longo M (2011) Effetto Hall su graphene. Tesi di laurea in Fisica, Università degli Studi della Calabria

Moglestue C (1993) Monte Carlo simulation of semiconductor devices. Chapman & Hall, London

National Science Foundation (NSF) Report. (1996) Shaping the future: new expectations for undergraduate radication in science, mathematics, engineering, and technology, Education and Human Resources Directorate, National Science Foundation, Washington, DC (pp 96–139)

Novoselov KS et al (2005) Two-dimensional gas of massless Dirac fermions in graphene. Nature 438(10):197–200

Persano-Adorno D (2010) Polarization of the radiation emitted in GaAs semiconductors driven by far-infrared fields. Laser Phys 20:1061–1067

Persano-Adorno D, Pizzolato N (2015) An inquiry-based approach to the Franck-Hertz experiment. Il Nuovo Cimento 38C:109

Persano-Adorno D, Pizzolato N, Fazio C (2015a) Elucidating the electron transport in semiconductors via Monte Carlo simulations: an inquiry-driven learning path for engineering undergraduates. Eur J Phys 36(5):1–19

Persano-Adorno D, Alaimo P, Pizzolato N, Spagnolo B (2015b) Electron dynamical response in InP semiconductors driven by fluctuating electric fields. Chaos, Solitons Fractals 81:425–431

Persano-Adorno D, Pizzolato N, Spagnolo B (2016a) Noise-induced resonance-like phenomena in InP crystals embedded in fluctuating electric fields. J Stat Mech Theor Exp 2016:P054021

Persano-Adorno D, Pizzolato N, Fazio C, Battaglia OR (2016b) An experience of elicited inquiry elucidating the electron transport in semiconductor crystals, Chapter 63. In: Lavonen J, Juuti K, Lampiselkä J, Uitto A, Hahl K (eds) ESERA Conference Proceedings series Science education research: Engaging learners for a sustainable future, Part 3, Strand 3: Science teaching processes. University of Helsinki, Helsinki, Finland, pp 510–520

Persano-Adorno D, Pizzolato N, Fazio C (2018a) Long term stability of learning outcomes in undergraduates after an open-inquiry instruction on thermal science. Phys Rev Phys Educ Res 14:010108. (11pp)

Persano-Adorno D, Bellomonte L, Pizzolato N (2018b) The amazing graphene: an educational bridge connecting different physics concepts. Eur J Phys 39:013001. (25pp)

Persano-Adorno D, Bellomonte L, Pizzolato N (2019) A 5E-based learning workshop on various aspects of the Hall effect. In: McLoughlin E, van Kampen P (eds) Concepts, strategies and models to enhance physics teaching and learning. Springer, Cham, pp 61–71

Pizzolato N, Fazio C, Sperandeo-Mineo RM, Persano-Adorno D (2014) Open-inquiry driven overcoming of epistemological difficulties in engineering undergraduates: a case study in the context of thermal science. Phys Rev Spec Top Phys Educ Res 10(1):010107

Silsbee RH, Draeger J (1997) Simulations for solid state physics: an interactive resource for students and teachers. Cambridge University Press, Cambridge

Stormer HL (1999) Nobel lecture: The fractional quantum Hall effect. Rev Mod Phys 71:875–889

Tsui DC (1999) Nobel lecture: Interplay of disorder and interaction in two-dimensional electron gas in intense magnetic fields. Rev Mod Phys 71:891–895

Wei Y, Zhai Z, Gunnarsson K, Svedlindh P (2014) A guided enquiry approach to introduce basic concepts concerning magnetic hysteresis to minimize student misconceptions. Eur J Phys 35 (6):065015

Chapter 17
A View on High School Students' Knowledge About Nanotechnology

Octavian Vasile Caroaie and Ovidiu Florin Caltun

Abstract Nowadays, almost each member of this society, not just high school students, uses cutting edge technology, whether those devices are used for fun or work. Technological evolution registered in different fields of knowledge imposes challenging technological situations for all of them, proposing learning situations that belong to informal education, which sometimes do not cover the real scientific facts.

17.1 Introduction

Interacting with freshly gained scientific knowledge could become a big problem, these interactions suggesting various learning situations, each developing a specific set of cognitive links based as well on previous interactions with technology. This informal learning leads to the development of cognitive structures which sometimes do not cover the real scientific facts, while through formal education students would be able to form cognitive connections that reflect scientific reality. The main problem in this situation is that school curriculum does not contain updated information regarding scientific progress, information that could become the foundation for a coherent understanding (Stabback 2016). The adaptation of education can be achieved by changing the structure of contents and skills taught in the classrooms or transferred in extracurricular activities so that it reflects current society requirements (Darling-Hammond et al. 2019).

O. V. Caroaie · O. F. Caltun (✉)
Faculty of Physics, Alexandru Ioan Cuza University, Iasi, Romania
e-mail: caltun@uaic.ro

B. G. Sidharth et al. (eds.), *Fundamental Physics and Physics Education Research*,
https://doi.org/10.1007/978-3-030-52923-9_17

17.2 Experimental Framework

Following this hypothesis we conducted a study whose purpose was to determine the level of knowledge that high school students have about nanotechnologies and their applications (Srinivas 2014). The instrument used in this study was a questionnaire containing 12 items referring to various aspects regarding nanotechnologies and some peculiar addressing the knowledge about magnetic materials applications.

17.2.1 Developing the Analysis Tool

The questionnaires used were focused on three main issues: magnetism, fluid state, and nanotechnologies. Some questions wanted to determine the quantity and quality of knowledge obtained in formal education in the classrooms, and others were built to probe students' interest on nanotechnologies and knowledge obtained during informal education sequences. Through this study, we managed to determine the level of knowledge and interest that high school students manifest toward the field of nanotechnology (Sebastian and Gimenez 2016). Addressing a representative sample of schools/pupils proved to be a difficult task for our educational research. To ensure that the sample resembles as much as the population from which it was extracted, this survey included both pupils of schools with high ranked results and those with difficulties in improving scholars' performances, both urban and rural areas and representatives of both sexes. The questionnaires were applied between February and March 2016 on a sample of 650 pupils from 7 high schools in the southern part of Vaslui county, 4 from urban areas and 3 from rural areas.

17.2.2 The Questionnaire

Including too many questions could create a certain level of boredom among pupils, while too few wouldn't have gathered enough information. This being said, we chose 12 questions that most of the respondents could answer. The sentences were built using carefully chosen words, in order to keep them as short and as pointed as one can be. Each question referred to a piece of the puzzle, some of the questions were open format so that pupils could respond as they felt at that moment.

17.2.3 Centralizing Responses

The first two questions allow the readers (respondents) to familiarize with survey, referring to simple aspects such as the usefulness of studying physics and identifying

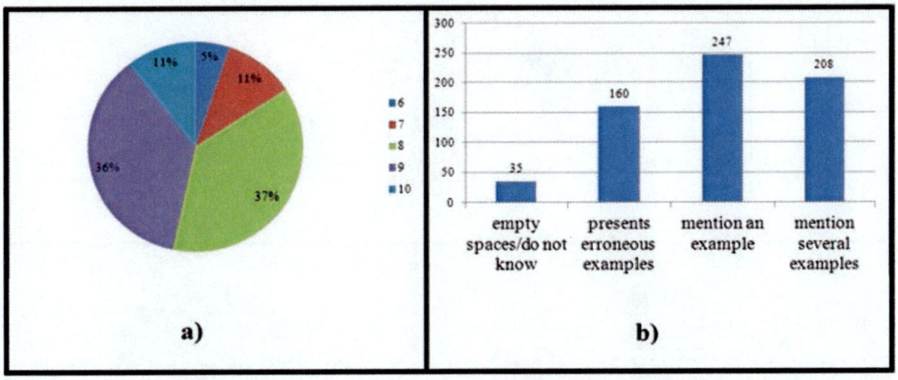

Fig. 17.1 (**a**) Grades given by the students to the how important is Physics in their life. (**b**) Number of the respondents answers regarding the contribution of physics to the revolutioning of society

areas revolutionized by it. The answer options for the first question were grades 1 to 10, 1 marking the uselessness of Physics, and 10 its importance. A small number of students awarded notes near the middle of the scale presented in the questionnaire, while 478 of those questioned giving grades 8 and 9. To the second question, 70% of the questioned students are able to present a correct example of a fields revolutionized by the knowledge gained within the borders of progress of science. The answers provided by students include medicine, robotics, astronomy, computing, acoustics, various types of engineering, optics, or telecommunication (Fig. 17.1).

Questions 3, 4 and 7 allowed a quantitative determination of the knowledge that students have about basics of magnetism, fluids, and nano-sized materials (Sederberg and Bryan 2006; Guisasol et al. 2004; Li and Singh 2016). The knowledge scrutinized by these three questions could be found in school curricula. The third question conceived as an open-ended item asks the students to described some properties of the magnets. The students had had a variety of ways to formulate the answer. The most often given answer of 86.2% of the students expressed the property of the magnet to have two poles. The second answer with a frequency of 73.5% of the respondents was about the interaction in between magnets repelling or attracting. The majority of the answers given by students was incomplete presenting only part of the aspects taught in the classroom and provided by the curriculum. The fourth question, also an open-ended item moved the focus on fluids and asked the students to provide some properties of the fluids (Besson 2004), with the aim to predict if students can explain why fluid can behave different in magnetic field. Only 27.5% of the students defined correctly the properties of fluids while 45.8% left the space blank or answered "do not know." At least one correct example of a fluid was provided by 56.9% of the students.

The presentation of correct examples correlated with the inability to present the properties of the fluids indicates the existence of loopholes that lead to the impossibility of a coherent expression of the concept. Surprisingly, as can be seen from

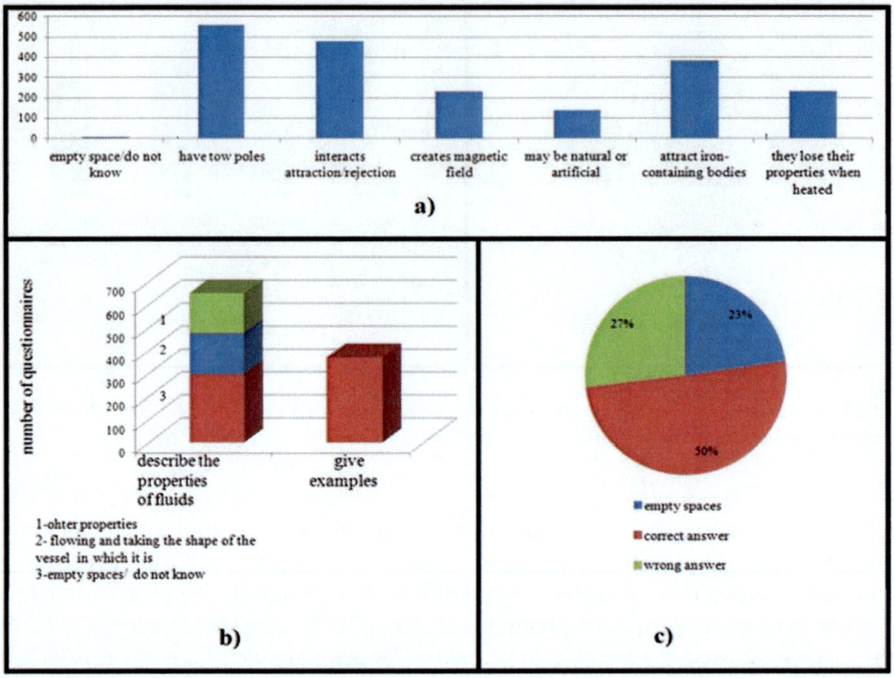

Fig. 17.2 (**a**) Distribution of the answers of students to the open-ended item on magnets. (**b**) Distribution of the answers on fluids' properties. (**c**) Distribution of the open-ended item on ferrofluid or magnetic fluid

Fig. 17.2c, half of the respondents managed to describe a ferrofluid and some properties even in the school curricula ferrofluids are not mentioned. The survey targeted both rural and urban schools. The majority of the correct answers in describing magnetic liquids came from the students from urban area. The justification for such high percentage is the exposure of these students to the informal activities.

The questions 5, 6, and 8 concern information that is not part of the curriculum implemented in Romania nowadays. In order to answer these questions correctly, it would have been necessary to have information about this field from other sources as informal experiences outreach activities, science fairs, science night, science center, etc. The fifth item addressed the complex concept of Ferrofluids that is not taught in Romanian high school. This explains why 48% of the respondent left blank spaces. Only 4% of the respondents had correctly defined the concept and only six students imagined application for ferrofluid. The number of those who considering appreciate that these "substances" interact with magnets is significantly higher than those who provide a correct definition, possibly the "name" of the substance suggested the "behavior." To the sixth question, 76.5% of the respondents failed to provide an answer, 20% presented the definition of the concept, and only about half of them

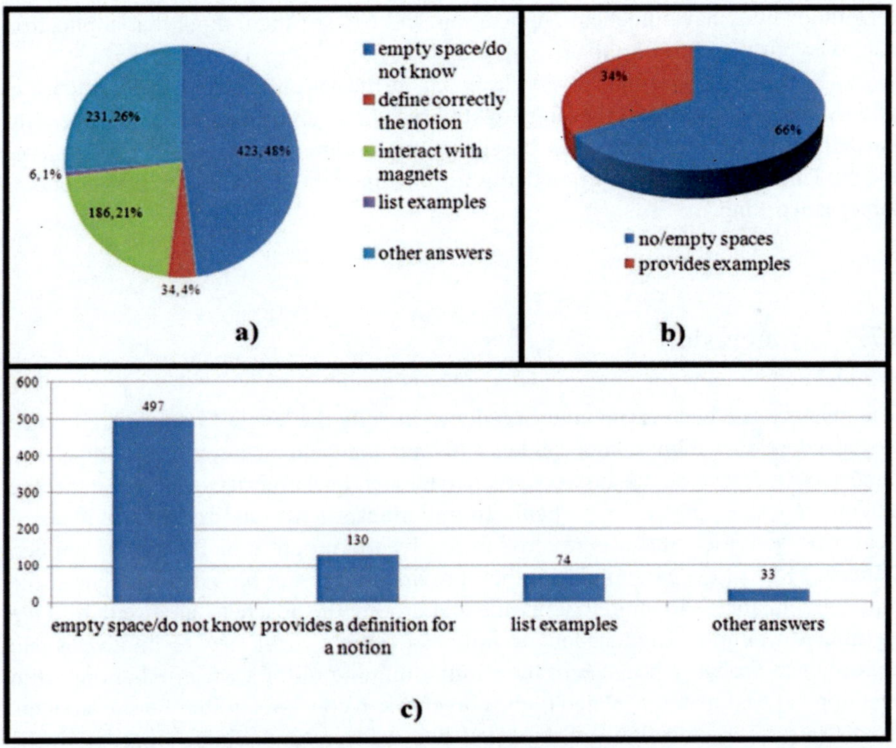

Fig. 17.3 (**a**) Distribution of the answers providing a definition of ferrofluid; (**b**) distribution of the answers describing the interaction in between magnets and ferrofluid; (**c**) students providing examples for ferrofluids' application

gave at least an example. Students who provided other answers mentioned that they had heard about this notion on TV/film, but did not look for more information. Many of the students surveyed fail to present practical applications of nanotechnologies, although it can be said that each of them uses a mobile phone at least. The motivation to study a certain amount of information is the most difficult to achieve, making a correlation between this information and everyday life could be a bridge that could underpin this motivation (Fig. 17.3).

Questions 9 and 10 relate to the pupils' interest in "new technologies" a term that is in the day by day life used to describe the last progresses in science and industry. The students can have a vague idea of the concept and only few of them have read systematic on this topic. The item nine intended to qualitatively mirroring the students' interest to observe/understand/practice something with and for new technologies. For 67% of the students, participating in a laboratory activity of hands on ferrofluids, implicit in new technologies, as suggested by question 9 would be interesting. The majority of students following the questionnaire realized that ferrofluids must have practical applications and 61% of them think that application can exist but they have no idea where and how.

La last item of the survey was designed to measure the interest of students in understanding how a device or gadget is working and about their readiness to participate in informal activities based on nanotechnology topics. About 70% of the students are interested in practicing for understanding how their smart phone or tablet is working.

17.3 Conclusions

The research has been carried out in order to identify the level of knowledge as well as the interest of high school students toward the field of new technologies and nanoscience. Based on the answers given to the first questions it can be stated that the students consider physics a particularly important science and have clear ideas of fields and activities that was revolutionized by the progress of Physics as applied science. Most of the examples that they provided represent big advancement in the humanity history showing that in the classroom the teachers are used to give significant example. The teachers do not use examples from new technologies and nanoscience probably because in their initial training did not became familiar with such topics. All the items related to the contents and concept familiar for the students from the formal education got expected god answers underlining once more the importance of formal education in achieving desired scientific key competence the fundament for social and professional insertion. The items related to the new technologies and students' opinion demonstrate that students are not familiarized with basic concepts in the field but they manifest high interest in hands on activities using devices or gadgets incorporating these new technologies. The students' answers emphasize the students need in knowing more about science progresses and the necessity of introducing more and more formal and informal activities for and with new technologies.

References

Besson U (2004) Students' conceptions of fluids. Int J Sci Educ 26(14):1683–1714. https://doi.org/10.1080/0950069042000243745

Darling-Hammond L, Flook L, Cook-Harvey C, Barron B, Osher D (2019) Implications for educational practice of the science of learning and development. Appl Dev Sci. https://doi.org/10.1080/10888691.2018.1537791

Guisasol J, Almudí JM, Zubimendi JL (2004) Difficulties in learning the introductory magnetic field theory in the first years of university. Sci Educ 88(3):443–464. https://doi.org/10.1002/sce.10119

Li J, Singh C (2016) Developing and validating a conceptual survey to assess introductory physics students' understanding of magnetism. Eur J Phys 38(2):025702. https://doi.org/10.1088/1361-6404/38/2/025702

Sebastian V, Gimenez M (2016) Teaching nanoscience and thinking nano at the macroscale nanocapsules of wisdom. Procedia Soc Behav Sci 228:489–495. https://doi.org/10.1016/j.sbspro.2016.07.075

Sederberg D, Bryan LA (2006) Tracing a prospective learning progression for magnetism with implications at the nanoscale. In: Proceedings of learning progressions in science (LeaPS) conference, June 2009, Iowa, pp 1–34. http://education.msu.edu/projects/leaps/proceedings/Sederberg.pdf / accessed 11/14/2019

Srinivas K (2014) Need of nanotechnology in education. Sci J Educ 2(2):58–64

Stabback P (2016) What makes a quality curriculum? Series current and critical issues in curriculum and learning, March 2016, No. 2IBE/2016/WP/CD/02 UNESCO Digital Library, pp 19–28

Chapter 18
Student Learning Paths from Exploration of Optical Diffraction with Online Sensors to Formal Interpretative Models

Alberto Stefanel

Abstract Physical optics is a part of many national curricula, usually treated in a reductive way, not effective for students learning. A wide research literature points out students learning difficulties on the features of the phenomena, the concepts involved and propose different research-based approaches. The PERG of the University of Udine studied how to exploit the potentialities of new technologies, to enable students to acquire a coherent vision of optical diffraction. The conceptions of students were studied involving them in educational labs on light diffraction and monitoring their learning progressions with inquiry-based learning tutorials. The qualitative data analysis showed that many students developed initial models based on a rectilinear path light propagation, progressively adapted to include the main features of the phenomenology and creating the condition for the construction of a coherent model.

18.1 Introduction

Physical optics is part of many national curricula (TIMMS 2015), treated usually according to a minimal approach focused on the relation between wavelength and maxima/minima positions. This simplified approach provides students of a tool only apparently effective to tackle the phenomenology. Researches point out many students' difficulties concerning for instance the features of the phenomena considered, the maxima/minima interference conditions, the wave models. Inquiry-based learning (IBL) approaches attack the related learning knots (Wosilait et al. 1999; Ambrose et al. 1999; Romdhane and Maurines 2007; Colin and Viennot 2000; McDermott et al. 2012). The new technology offers to that goal many opportunities for both experiment and modelling (Hirata 1998; Corni et al. 1993; Santi et al. 1993; COMPADRE 2019).

A. Stefanel (✉)
Physics Education Research Unit, DMIF, University of Udine, Udine, Italy
e-mail: alberto.stefanel@uniud.it

B. G. Sidharth et al. (eds.), *Fundamental Physics and Physics Education Research*,
https://doi.org/10.1007/978-3-030-52923-9_18

The PERG of the University of Udine studied how to exploit the potentiality of ICT to build effective educational proposals enabling students to acquire a coherent vision of optical diffraction and a functional understanding of the wave model of light (Corni et al. 1993; Santi et al. 1993; Michelini et al. 2004, 2014). The present work regards a study on how high school students face light diffraction phenomenon produced by a single slit, the model they activate exploring the phenomenology, passing from geometrical optics, to a wave model of light. After a brief review on researches on wave optics, the research instruments and context will be presented, discussing then the students learning outcomes and concluding remarks.

18.2 Theoretical Background and Research Questions

Literature on how students face physical optics evidences a generalized and persistent tendency to use a geometrical model to interpret a diffraction phenomenon, or hybrid models according to the different situations analysed (Wosilait et al. 1999; Ambrose et al. 1999; Romdhane and Maurines 2007; Colin and Viennot 2000; Horn et al. 2002). Students attribute a particle-like behaviour to a wave rather than the behaviour of a perturbation propagating in space-time is evidently (Wittmann 2002; Barnioli and Zavala 2016). Most students' evidence difficulties in manage coherently wave model, for instance considering diffraction produced only by the edge waves, often because of educational rituals: "...every slit is a point like wave source..." (Ambrose et al. 1999). Many students do not have clear features of the phenomena considered, confusing for instance refraction and diffraction, identifying diffraction as an angular spread (Ambrose et al. 1999; Michelini et al. 2014; Kryjevskaia et al. 2013).

In the specific of wave optics, some researches evidence that many students have difficulties in applying the maxima/minima condition to solve also the simple problem of interference (Ambrose et al. 1999; Romdhane and Maurines 2007). This problem seems related to the difficulties of students to measure a distance in terms of wavelength (Kryjevskaia et al. 2013). Other not well-studied conceptual knots are the following: concepts of phase, optical path, wave front, optical ray in a wave model; interpretative role of Huygens-Fresnel principle in explaining the wave behaviour; distinction between amplitude and intensity of the light; constant uniform light intensity observed/measured and oscillating nature of light perturbation.

To overcome these learning difficulties, different groups developed and tested educational approaches based on research. The tutorials of the Washington University suggest an operative analysis of the interference starting from waves on a water surface and simulating the interference of waves produced by two point sources superposing the circles drawn on two transparent foils (McDermott et al. 2012). The Maurines proposal builds the tools for optical interference studying different phenomenological contexts (Romdhane and Maurines 2007). Viennot stress the role of the superposition principle and suggest analysing situations merging geometric and waving optics image formation (Colin and Viennot 2000). Several proposals suggest finally the opportunities offered by computer online sensors and modelling (Hirata 1998; Mayes and Melton 1994; Hinrichsen 2001; Chauvat et al. 2003; Grove 2003).

The approach followed in the present research adopts an inquiry-based learning strategy, similar to that suggested by McDermott group (McDermott 1991; McDermott et al. 2012), emphasizing the opportunity offered by online sensors and computer modelling to grasp the learning difficulties discussed, to construct the phenomenological laws of diffraction as bridge toward an interpretation based on a wave model (Corni et al. 1993; Michelini et al. 2004, 2014).

The research questions here focused are the following:

RQ1: What are the students' models activated by the exploration of diffraction phenomena?

RQ2a: How they represent the diffraction distribution before and after performing a quantitative experiment?

RQ2b: Which kind of model underlay these representations?

RQ3a: Which conceptions they have of order of interference, superposition, sum of waves?

RQ3b: How affects these conceptions the construction of a wave model on diffraction and on the wave behaviour of light?

18.3 Instruments and Methods

To answer these research questions, a study was conducted on high school students involved in educational laboratories of operative exploration (Michelini 2006), following with students a research-based educational approach developed in previous works (Corni et al. 1993; Michelini et al. 2004, 2014), aimed to get students familiar with physical optics and to gain in modelling construction. The path was adapted to the specific contexts according to a project shared between university researchers and schoolteachers. This approach was implemented in educational laboratories providing two main stages, described in the following.

18.3.1 Conceptual Lab of Operative Exploration (CLOE) on Physical Optics

The first stage occurred at the University of Udine and involved the students in a Conceptual Laboratory of Operative Exploration (CLOE) (Michelini 2006) on light diffraction. They explored the diffraction pattern produced by a laser light passing through a single slit, following the stimuli of an Inquiry-Based Learning (McDermott 1991; McDermott et al. 2012) open tutorial (Michelini et al. 2014), promoting a PEC Strategy (Theodorakakos and Psillos 2010). Students in little groups explored first (15 min) the phenomenology of diffraction, to recognize extension (contexts) and intension (main features). Then (30 min), they explored qualitatively the diffraction figure collected on a white screen and produced by a single slit, to recognize the general features of the phenomenon: the angular nature and symmetry of the distribution, the regular alternation of maxima and minima, the very intense central maximum and the progressively decreasing intensity of other maxima, parameters affecting the pattern (D—distance screen-slit, λ—laser colour or wavelength, a—width of the slit). The partial conclusions obtained by each group were shared in a large group discussion. Students were requested to design alone (15–30 min) and then perform experiments in little groups to characterize quantitatively the phenomenology of diffraction (1.5–2 h). They used in lab the online sensor system Lucegrafo (Gervasio and Michelini 2009), to perform the real-time acquisition of the diffraction distributions, analysed at school under the supervision of the schoolteachers after.

The second stage, conducted by a researcher at school, included three parts. In the first part (1–1.5 h), the data analysis were resumed discussing the data collected by the students in the lab, posing the question to model the phenomenon with an assumption on the wave nature of light. In the second part (0.5 h), students simulated the interference of waves produced by two point sources superposing and moving the circles drawn on a transparent foil over these drawn on a white paper and exploring the condition for maxima/minima nodal lines (McDermott et al. 2012). The third part (0.5 h) involved students in the discussion of the meaning of superposition of wave and its mathematical implementation to acquire the minima instruments to construct a formal model accounting and interpreting the single slit diffraction distribution analysed in the lab. In the fourth part (0.5 h), the layout of the model based on Huygens principle was discussed with students, considering the following steps: (A) the slit can be considered as n secondary light sources each of them located in a point x_i inside the slit and producing a secondary wave with equal wave length λ and period T; (B) the ith source produces a secondary waves of amplitude given by $A_{ij}(t, X_j) = (A_0/R_{ij})$ sen $[2\pi(R_{ij}/\lambda - t/T)]$, in the point X_j of detection and at the time t, where $R_{ij} = [(X_j - x_i)^2 + D^2]^{0.5}$ is the distance between the ith point source and the jth point on the screen, D the distance slit-plane of detection; (C) the total amplitude $A_j(t, X_j)$ in the point X_j is given by $A_j\ (t, X_j) = \sum_{i = 1...n} A_{ij}$; (D) The intensity is given by the time average of the amplitude square: $I_j = \langle (A_j(t, X_j)^2 \rangle$. This sequence can be transformed in a flux flow, implemented in a code, or in a modelling environment, or in an electronic sheet to fit the experimental data (Santi et al. 1993). The final evaluation carried out under the responsibility of the schoolteachers.

18.3.2 Context for Research

This chapter documents the results of the analysis of the tutorial filled by 168 K11 students aged 16–17 of eight groups of Scientific Lyceum classes of three schools of North Italy towns, in the stages 1 and 3 of the CLOE labs on light diffraction described in the previous section. The students were of middle level in physics according to the evaluation of their schoolteachers. They possessed just basic knowledge on geometric and wave optics (Young experiment), on wave's phenomenology (quantities characterizing a wave, equation of a plan wave, concepts of superposition of waves and interferences) and no experience in lab.

18.3.3 Monitoring Tutorials/Tools and Methodology of Analysis

The students learning paths in the stages 1 and 3 were monitored using two tutorials.

The first tutorial suggests to students the following explorative challenges: (T1A) Preview the light pattern on the screen when light passes a single slit width 1 cm and another slit width less than 1 mm, (T1B) after observing the light diffraction pattern produced by red light diffracted by a 0.12 mm slit, students were requested to draw the pattern, to describe it, to sketch the corresponding light intensity vs position graph expected, describing the main characteristics of that graph; (T1C) exploration of how the diffraction pattern is affected by the change of parameters of the systems (D, λ, a); (T1D) design a quantitative experiment aimed to extract the phenomenological laws of diffraction; (T1E) design a data analysis.

The second tutorial is divided into two parts. The first follows the suggestions of McDermott group (2012), asking: the minimum distance between two sources to obtain (T2A1) only a single (minima) nodal line or (T2A2) only one line of maxima (besides the axis); (T2A3) assuming $|F1 - F2| = 6\lambda$, which relation among PF1, PF2 and λ provides the set of the maximum interference points of order 2? Condition to have second order maximum; (T2A4–5) Constructive interference points for arbitrary distance between sources. The second part proposed to students the following questions: (T2B1) How do you write the equation of a wave? (T2B2) What does it mean to superpose two waves? (T2B3) What does it mean to "sum up" two waves? (T2B4/T2B5) Role of time and spatial parts of the phase wave to obtain a stable interference pattern.

The analysis of students' answers to the tutorial questions followed the qualitative research criteria (Erickson 1998), distinguishing between interpretative and descriptive explanations, models underlay interpretation; conceptual references adopted. The categories were defined a priori, according the previous research outcomes (Michelini et al. 2014), and then redefined a posteriori, assuming typical students' answers as operative definition of categories.

18.4 Data from Tutorials

When students were requested to preview the pattern collected on a screen by light passing through two slits, respectively of width 0.5–1 cm and less than 1 mm (Tutorial 1—question T1A), they answered according to the four categories summarized in Table 18.1.

The majority of the 168 students of the sample expected some features typical of a diffraction phenomenon, where the "expansion" of light after the slit is the aspect more quoted (65%). This is a consequence of the fact that these students faced the basis of diffraction and interference in school before the CLOE lab. Some students (21%) expected an "enlargement" of the light pattern in a direction parallel to that of the slit, underlying a particle-like conception of wave propagation of light. A great group (49%) previewed that the narrower the slit is, the narrower the illuminated area on the screen will be, according to a geometric rectilinear path model.

Concerning the light intensity versus position distributions expected, the Table 18.2 resumes the main categories. In the DA category was putted all the representation catching the main features of the experimental graph. The second category (DB—equal peaks graph) differs from first because the central maximum is of the almost the same intensity of the first order maxima (changes from DB to DA occurred frequently in the representation of the observed graphs). The presence of maxima and minima characterize also the subcategory DB′ exemplified in Fig. 18.1 and showing discontinuity in the intensity distribution. The other three categories of Table 18.1 (DC, DD and DE) differ for the mathematical function used (bell or parabolic shape, inverse power shape, linear shape) to represent in any cases the envelope of the maxima intensity.

After observing the diffraction figure produced by a 0.12 mm vertical slit, the majority of students (72%) stressed the presence of an enlargement of the pattern, and added in half of cases that the distribution is "perpendicular" to the slit direction. Just 17% of the sample evidenced the presence of points/lines. The "enlargement" of the light pattern is quoted more frequently, because it can be included in a straight-line model of light. Highlighting maxima/minima requires almost activating a change in the standard point of view of geometrical optics. The model that students had in mind affects strongly the recognition of the main features of the phenomenon observed (Karmiloff-Smith 1988; Chinn and Brewer 2001).

Figure 18.2 shows the distributions of the representation categories drawn by students before performing the experimental acquisition with sensor (prevision) and after observing the experimental graph on the screen (observed). The prevision

Table 18.1 Categories of previsions on the behaviour of light passing through a slit

Cat	Large slit (0.5–1 cm)	Tin slit (less than 1 mm)	%
A	Light more large	Light less large	49
B	Light remains the same	Light is expanded	44
C	Light remains the same	Light is expanded vertically (same direction of the slit)	21
D	Light remains the same	More lines inside	7

Table 18.2 Categories of representation of the light diffraction pattern and graphs intensity vs position

Cat	Example	Description
DA	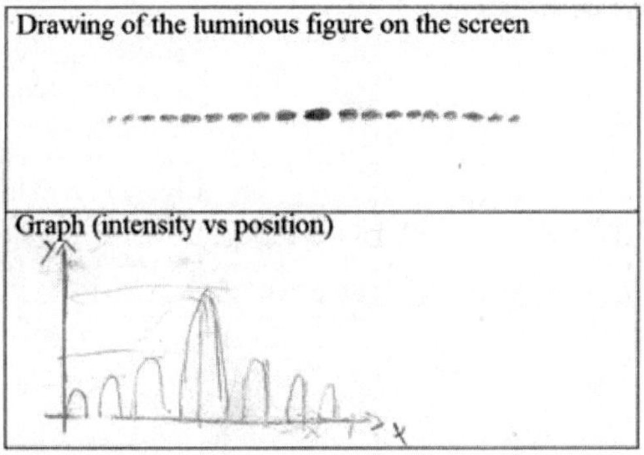	Experimental graph. Characterized by a central peak with $I > 2I_1$
DB	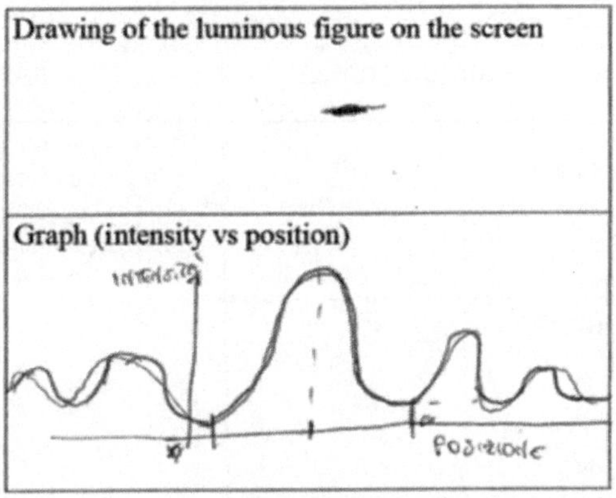	Graph characterized by peaks of equal/ almost equal intensity (the central maximum $I < 2xI_1$)
DC	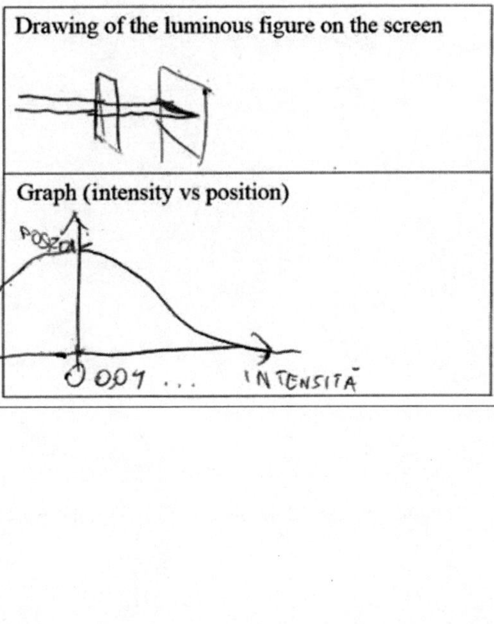	Bell shape graph, characterized by a single peak covering all peaks. Within this category also parabolic graphs where the intensity of light suddenly becomes 0 beyond a certain distance from the central maximum

(continued)

Table 18.2 (continued)

Cat	Example	Description
DD	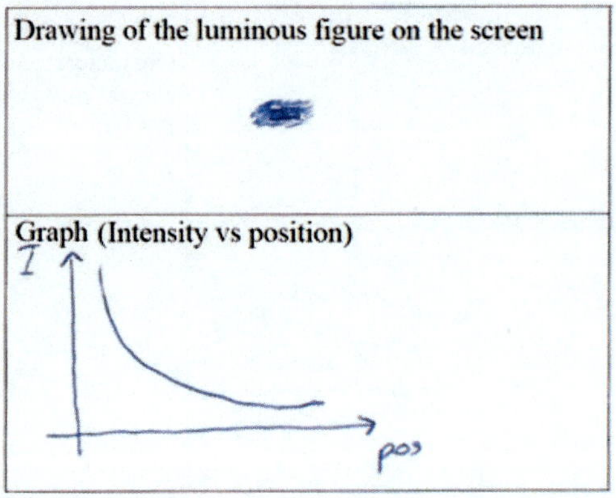	Sharp/cusp/asymptote graph, where the graph is always concave (reverse power type). Maximum: cusp. Within this category also right arm, graphs with the only branch of the right
DE	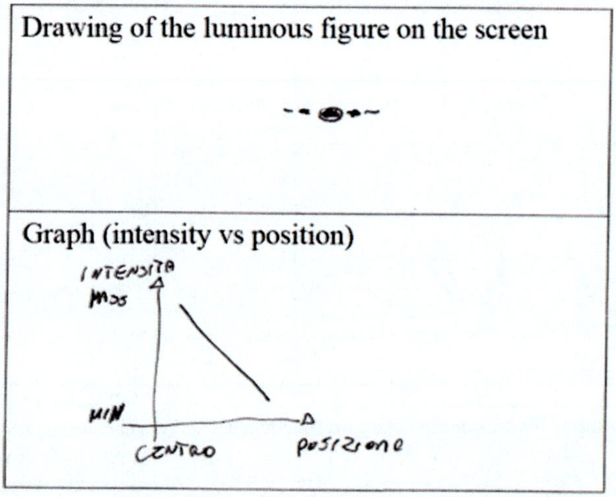	Straight-line graph, descending line that intersects the axes in both positive values. Within this category also graphs with only the right branch

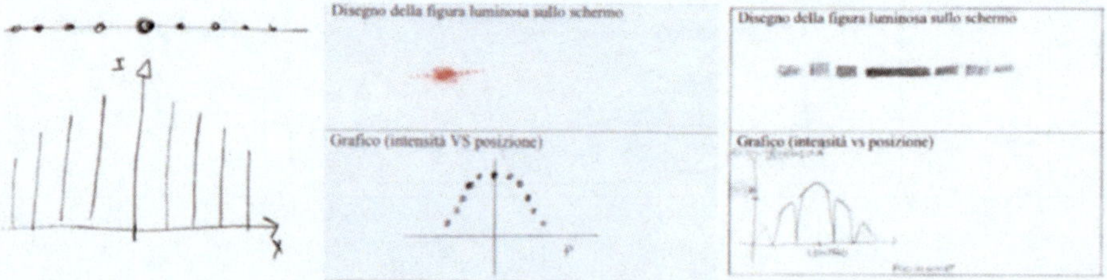

Fig. 18.1 *I* vs *x* distribution of a DB′ cat, including discontinuity in light intensity distribution

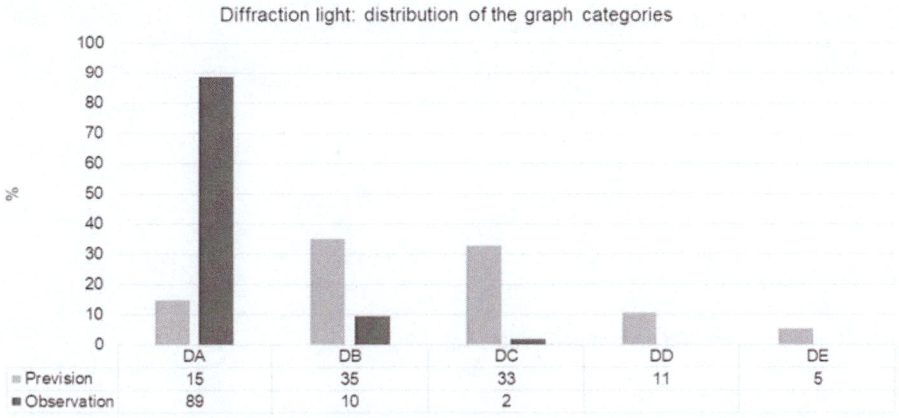

Fig. 18.2 Distribution of the categories of representation of the light intensity versus position graphs expected by students before perform the experimental acquisition with sensor (prevision) and after observing the experimental graph on the screen (observation)

graphs were in the majority of cases with maxima of the same size order (cat. DB) or of bell shape type (cat. DC). Half of the students included maxima and minima in their representations (cat. DA and DB), the other half represented only the envelope (cat. DC–DD–DE). After performing the experiment, the large majority of students (89%) represented graphs of type DA. A minority represented the observed graph or not emphasizing the central maxima (cat. DB) or without minima (cat. DC). The presence of these categories is an indicator of the difficulties of some students to include also the experimental evidence in a conceptual framework able to give meaning to that evidence.

To understand which conceptual paths students activated during the experimental exploration of diffraction patterns it is useful to summarize the models that a small group of students have used to account the observed phenomenology, and documented in the tutorials.

The first two models of Fig. 18.3 are representative examples of the majority group (14/25). The first shows an enlargement of the light beam similar to that of a jet of water coming out of a tube. A student described in words a similar model "In my opinion the light passes through the slit and bounces on the wall of it, since it is not possible that they are perfectly smooth, light is not reflected in the same direction interrupting reflection with a constant period."

The second model based on the reflection of light on the inside edge of the slit. A further model expressed in words by a student provides that "the signal is blocked every affixed interval of time, because the intensity is not continuous, that is, it has time spaces determined". In this model, the student speaks of an unspecified signal in which time plays a central role in determining the observed phenomenon. The third image of Fig. 18.3 shows a wave model, where two travelling wave fronts superpose in a defined point and time. The general (trivial) condition for maxima or minima of

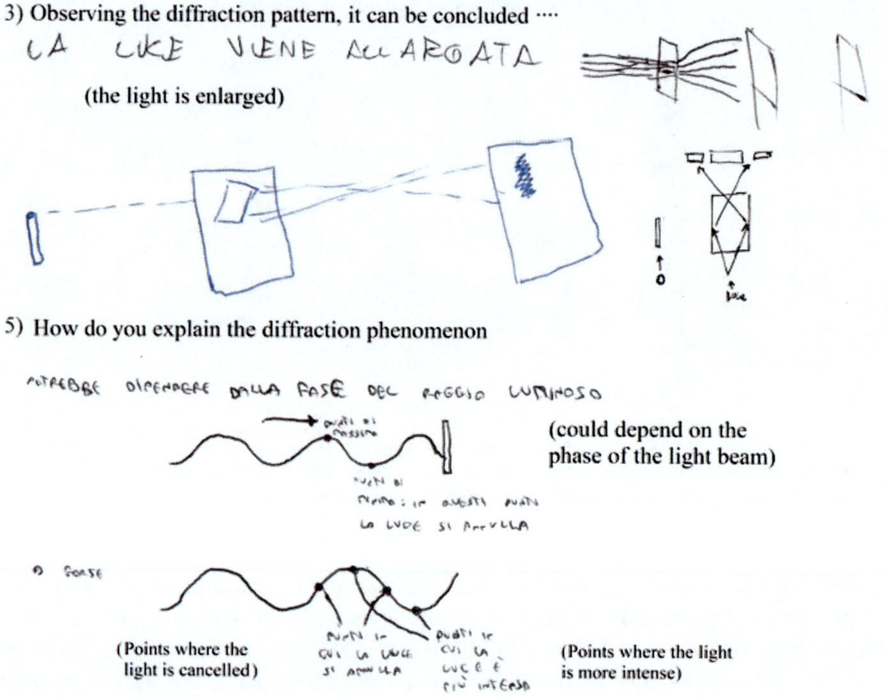

Fig. 18.3 Models activated by students to explain the observed diffraction pattern

course is at the base also of the optical diffraction pattern, but it does not explain alone the specific diffraction pattern. The pictures of Fig. 18.3 evidence also the problematic aspect concerning the different roles of the spatial and time parts of the wave equation. In the last model, the diffraction is related to the waves hitting the edges (Ambrose et al. 1999): "The diffraction causes that the part of the light wave that hits the edges of the slit will be curved and so the wave is wider than the slit".

As described above, the students explored in the third stage of the CLOE lab the conditions for maximum/minimum, according to the tutorial 2 (adapted from McDermott et al. 2012). Table 18.3 resumes categories and percentages related to the questions T2A1, T2A2, T2A3 (minimum distance between sources to have a (single) node line of minima; only a line of maximum, the second order maximum). The percentages of correct answers is high regarding the first two points, showing that the majority of students apply correctly the basic conditions of maximum and minimum. The renunciation of addressing the third point (T2A3) highlights the difficulty of many students on the concept of order of interference and with the condition producing it.

Regarding how to write the wave equation (Question T2B1), 84% of the sample provided a response of the type $Y = A_o \cos(\omega t - kR)$, while the remaining 16% either evaded the request or wrote expressions in which only the spatial part or only the

Table 18.3 Frequency of answers to the question (T2A1) Minimum distance between the sources to have a (single) node line of minima; (T2A2) Minimum distance between the sources to have one (only) line of maximum (besides the axis); (T2A3) Condition to have the second order maximum ($N = 168$)

T2A1 Cat	%	T2A2 Cat	%	T2A3 Cat	%
0	6	λ	78	2λ	14
λ	8	$\lambda/2$	2		
$\lambda/2$	71	2λ	2		
$0 < d < \lambda$	9	$0 < d < \lambda$	4		
NA	6	NA	14	NA	86

Table 18.4 Frequencies of answers to the question T2B2. What does it mean to superpose two waves? T2B3. What does it mean to "sum up" two or more waves? ($N = 168$)

Category	T2B2 (%)	T2B3 (%)
The two wave interfere	31	24
Total perturbation = sum of the single perturbation/amplitude/wave front/ equation	38	72
Wave with the same phase/amplitude	5	
Wave with some points in common/their graphs overlap	16	2
Sources in the same point	8	2
Same λ	2	

temporal part appears. Finally, Table 18.4 summarizes the outcomes concerning the two questions T2B2 and T2B3, highlighting the distinction between superposition and sum of waves.

A quarter of students (25%) respond only to one of the questions, with a slight prevalence for the first. For 37% the meaning of the two expressions is substantially the same (almost all to add up and only in two cases to interfere). Among the students who distinguished the meaning attributed to superpose and to add (38%), only a part (18%) distinguished overlapping, as a physical process of interference, and sum, as a sum of the amplitudes. A non-negligible part (12%) has literally inverted the meaning or attributed distinct meanings.

It should be added that 22% of the sample also used the concepts of interference to indicate only the maximum/minimum interference situations. These criticalities highlighted in the formal translation of the concepts of interference and superposition constitute an indicator of difficulty in understanding the wave model of light, that is an obstacle to the construction of a functional understanding of this model.

The data obtained with a small sample group ($N' = 26$ students of two classes) lead in this direction. To these students, the model based on the Huygens principle discussed above which accounts for the diffraction distribution were presented and it was requested to summarize the elements of this model. The students, rather than describing the structure of the model by elements, responded by list the model elements they had grasped or that had remained more impressed. More than half (16/26) recalled some formal elements of the wave model (the equation of the amplitude of a sinusoidal wave, the sum of amplitudes as an implementation of the overlap, the average of the square of this amplitude to obtain the "intensity") without defining the symbols or explaining the meaning of the written equations, but explaining most often the conditions of maximum/minimum interference (9/16) and recalling other elements of the model such as the fact that the slit is much greater than the wavelength of the light used.

18.5 Outcomes and Conclusion

A study carried out on how a sample of 186 high school students' analysed the phenomenology of light diffraction and how they passed from a geometrical, to a wave model of light. Students in little groups explored light diffraction, in educational labs of operative exploration (Michelini 2006). Their conceptions on light diffraction and learning paths were monitored using tutorials implementing IBL strategy, analysed according to the qualitative research criteria.

Concerning, the students' models activated by qualitative exploration of phenomena (RQ1), students looks initially at diffraction more often as a "light enlargement" (76%). The presence of maxima and minima, crucial for that phenomenology, is not always recognized or is considered of second order by students, because a lack in their physical or mathematical models of the phenomenon. This shows that the students' model affects strongly what they look at/observe. In accord with the theoretical perspective of Karmiloff-Smith (1988) and other researchers (Chinn and Brewer 2001), students "seen" the aspects that find meaning in their conceptual model. A new model is needed to include unexpected features/data. In the CLOE Lab, the overcoming of the first partial models activated was promoted thanks to the discussion in large group, which allowed to share a complete picture of the observed phenomenon.

Concerning the light intensity distributions expected by students before performing the experiment, in half of cases they have drawn a bell/parabolic shape graph or other form of envelope of the distribution (linear or inverse power). Two profiles can be distinguished: The first profile includes students representing envelope distributions in accord to their representation of the diffraction pattern as a unique large spot, and the second profile includes students representing an envelope of the distribution, but having drawn the diffraction pattern as a discontinuous sequence of points/lines. In this case, students were not able to activate a formal model accounting both the periodic variation and the decreasing of the intensity. The use of discontinuous distributions is another aspect of the same problem. The lack of a physical interpretative model of the phenomenon added to the poverty of the formal models possessed by the majority of students. Few students in fact, less than 10%, were able to mastery function as the sync square that is the product of two elementary functions (RQ2).

These outcomes highlight that the representation are influenced by at least three elements, not always harmonized and coherent: what a student has seen of the pattern observed on the screen (a luminous spot of gradually decreasing intensity; an alternation of maxima and minima); what is the student mental model of the observed physical phenomenon; what are the mathematical/formal tools (analytical, graphical) that the student is able to put in place (RQ2a).

The students of our sample activated at least five different models to account the phenomenology. The first two remaster the geometric perspective (a spread of the light as that of water flowing outside a pipe; reflection of light on the edge). A third hybrid model attributes role for diffraction to the time part of a wave. Two other models evoke elements of wave optics, without an effective modelling: the first related diffraction to the waves hitting the slit edges; the second generically to the constructive and destructive interference condition (RQ2b).

Regarding the basics of modelling, students demonstrated competence in knowing how to "measure" distances in wavelengths thanks to the tool used in the CLOE lab and know how to set the condition for a maximum/minimum (71–78%). The concept of order of interference and the condition producing it are clear only for a minority of students. More than half of the samples (60%) identified the concepts of superposition and sum of waves that is a spy of the problematic relation between a concept and its formal representation (RQ3a).

The results discussed here indicate the activation of a process of revision of the conceptions of the students based on geometric optics, but also highlight some criticalities for instance related to overcome the phenomenological plane and to build an effective wave model that we can presume can be faced by more active role of students in lab (RQ3b).

The approach followed has shown to activate significant competences in the students on the phenomenology of the optical diffraction and on the construction of the wave model. Further work is required to make the students masters both physical and math aspects involved. At the same time the formal aspects involved for instance both in building of the phenomenological laws from the experimental diffraction pattern and in the construction of the model, applying first principles seems quite important to promote students' acquisition of a functional understanding of high level competencies both in math and in physics. Future researches will focus on the interlacing of math and physics in the context of optical diffraction, in particular concerning the phase of a wave and the role of time and spatial part for interference.

Acknowledgement I thank for this work the discussion with Marisa Michelini and the PERG of Uniud, and school, teachers and students participating in the research.

References

Ambrose S, Shaffer P, Steinberg N, Mcdermott L (1999) An investigation of student understanding of single-slit diffraction. Am J Phys 67:146–155

Barnioli P, Zavala G (2016) Mechanical waves conceptual survey. Phys Rev Phys Educ Res 12:010107

Chauvat D, Emile O, Brunel M, Le Floch A (2003) Huygens' principle and Young's experiment in the propagation of light beams. Am J Phys 71(11):1196–1198

Chinn CA, Brewer WF (2001) Models of data: a theory of how people evaluate data. Cogn Instr 19 (3):323–393

Colin P, Viennot L (2000) Les difficultés d'étudiants post-bac pour une conceptualization cohérente de la diffraction et de l'image optique. Didaskalia 17:29–54

COMPADRE (2019). http://www.compadre.org/portal/items/detail.cfm?ID=8331

Corni F, Mascellani V, Mazzega E, Michelini M, Ottaviani G (1993) A simple on-line system employed in diffraction experiments. In: Pereira LC, Ferreira JA (eds) Light and information. Girep Book, Univ. do Minho, Braga, Portugal, pp 381–388

Erickson F (1998) Qualitative research methods for science education. In: Fraser BJ, Tobin KG (eds) International handbook of science education. Dordrecht, The Netherland, Kluwer A.P., pp 1155–1174

Gervasio M, Michelini M (2009) Lucegrafo. A simple USB data acquisition system. In: MPTL14 proceeding. http://www.fisica.uniud.it/URDF/mptl14/contents.htm

Grove TT (2003) A low-cost scanning Fabry–Perot cavity. Am J Phys 71:184

Hinrichsen PF (2001) Apparatus and demonstration notes. Am J Phys 69(8):917–919

Hirata K. (1998) How can we use microcomputers effectively in teaching and learning physics? In: Communicating physics. ICPE (IUPAP), p 132

Horn ME, Leisner A, Mikelskis HF (2002) Difficulties of model construction in optics probleme. In: German Physical Society. Spring Conference, Leipzig

Karmiloff-Smith A (1988) The child is a theoretician, not an inductivist. Mind Lang 3:183–195

Kryjevskaia M, Stetzer MR, Heron PR (2013) Student difficulties measuring distance in terms of wavelength. Phys Rev ST Phys Educ Res 9:010106

Mayes TW, Melton BF (1994) Fraunhofer diffraction of visible light by a narrow slit. Am J Phys 62:397–403

McDermott LC (1991) Millikan lecture 1990: what we teach and what is learned—closing the gap. Am J Phys 59:301

McDermott LC, Shaffer PS, Physics Education Group at the University of Washington (2012) Tutorials in introductory physics. Prentice Hall, Upper Saddle River

Michelini M (2006) The learning challenge: a bridge between everyday experience and scientific knowledge. In: Planinsic G, Mohoric A (eds) Informal learning and public understanding of physics. Girep Book, Univ. of Ljubljana, Ljubijana (SLO), pp 18–39

Michelini M, Stefanel A, Santi L (2004) Teacher training strategies on physical optics: experimenting the proposal on diffraction. In: Michelini M (ed) Quality development in the teacher education and training. Girep Book, Forum, Udine, pp 568–576

Michelini M, Santi L, Stefanel A (2014) Upper secondary students face optical diffraction using simple experiments and on-line measurements. In: Proceedings FFP14, Marseille 2014. http://pos.sissa.it/archive/conferences/224/240/FFP14_240.pdf

Romdhane I, Maurines L (2007) ò. Les étudiants et les interferences lumineuses: cohérence des sources et principe de superposition. Didaskalia 31:85–114

Santi L, Mazzega E, Michelini M (1993) Understand radiation interference by means of computer modelling. In: Pereira LC, Ferreira JA (eds) . GIREP Book, Univ. do Minho, Braga, Portugal, pp 372–380

Theodorakakos A, Psillos D (2010) PEC task explore. In: Constantinou C (ed) CBLIS 2010. Oelizk, Warsaw, pp 75–83

TIMMS (2015). http://timssandpirls.bc.edu/timss2015/encyclopedia/introduction/

Wittmann MC (2002) The object coordination class applied to wave pulses: analysing student reasoning in wave physics. Int J Sci Educ 24(1):97–118

Wosilait K, Heron PRL, Shaffer PS, McDermott LC (1999) Addressing student difficulties in applying a wave model to the interference. Am J Phys 67(7):S5–S15

Chapter 19
Research-Based Path Proposal on Optical Spectroscopy in Secondary School

Daniele Buongiorno and Marisa Michelini

Abstract Optical Spectroscopy (OS) represents a conceptual and historical bridge between classical and modern physics: it is one of the experimental evidences of the atomic structure as well as a key to understand light–matter interaction phenomena. Its importance is widely recognized on disciplinary, cultural and social plans, nevertheless very few efforts have been made in order to integrate those aspects in an equally considerable educational Scheme.

A research-based educational path on OS was designed for secondary school students in the theoretical framework of Model of Educational Reconstruction and step-by-step calibration by means of Design-Based Research. The Inquiry-Based Learning strategy involved students in experimental and interpretative tasks in order to recognize the link between the energy levels model of atoms and discrete luminous emission observed during laboratorial activities.

19.1 Introduction

All EU secondary school curricula take into account modern physics as an important content, but problems concerning modalities and specific contents to be addressed is still open as well as instruments and methods needed to teach these topics. In the praxis, the approach is very often limited to a narrative review of the conceptual problems and relative theoretical solution, focusing on the crucial aspects, that characterized the history of physics at the beginning of the XX century, rather than a conceptual, disciplinary, operative foundation of the theories building a culture linking new theories with instruments and methods of physics. (Michelini et al. 2014).

In the wider framework of modern physics, Optical Spectroscopy (OS) provides the experimental basis of the modern quantum theory and it represents a significant

D. Buongiorno (✉) · M. Michelini
Physics Education Research Unit, DMIF, University of Udine, Udine, Italy
e-mail: buongiorno.daniele@spes.uniud.it; marisa.michelini@uniud.it

B. G. Sidharth et al. (eds.), *Fundamental Physics and Physics Education Research*,
https://doi.org/10.1007/978-3-030-52923-9_19

methodological tool of how physics makes use of indirect measures based on energy in order to validate models.

OS is one of the main investigative tool in physics based on light–matter interaction and it is important on different plans: (a) it represents an epistemological contribution to physical culture (it is a bridge between classical and modern physics, it is a context in which the role of energy in physical investigation can be addressed, it represents a validation modality of interpretative micro and macro models through indirect measures); (b) it contributes on education on different plans (to build a coherent frames in the study of optics, to the conceptual basis of modern physics, to link instruments and methods linking experiment and theories); (c) it represents a socio-technological relevant contribution to various applications. The topic of atomic spectra in particular represents a fertile area to study some quantum concepts that students tend to form: the correct interpretation of atomic spectra requires a correct understanding of the quantized nature of light, of the quantized energy structure of matter as well as the way they interact.

We designed an educational path on optical spectroscopy for upper secondary school students, in which students are directly involved in experimental studies and interpretations founding the methodological basis of approaching microscopic phenomena related to light–matter interaction, in particular to gain a functional understanding of the conceptual link between discrete energy levels in atoms and discrete light emissions from atom themselves.

The educational path proposal is organized in a coherent teaching/learning flexible sequence studied to overcome the evidenced-in-literature conceptual knots, here described and commented.

History of physics has been used as an educational resource for developing the path: its role is not merely reduced to storytelling, but rather the history served as supporting the addressed concepts (Buongiorno and Michelini 2017).

19.2 The Research Perspective

The educational approach to teaching/learning OS, as well as to every topic that has to be addressed with a research approach in physics education, implies a reconstruction of the disciplinary contents from an educational point of view and an analysis of the most recurring conceptual knots (i.e. learning difficulties or spontaneous ideas) evidenced by students, as well as the main interpretative problems encountered by physicists themselves in the history of science development. The founding elements of the topic are selected in order to provide the designing of the educational proposal in the theoretical framework of the Model of Educational Reconstruction (MER) (Duit et al. 2005; Duit et al. 2012). The preliminary study of the literature, despite not very wide, evidenced some recurring learning difficulties related to spectra in secondary school and university students, mainly concerning the conceptual link between spectral lines and atomic levels and the experimental conditions causing the formation of atomic spectra. In particular, students read in a discrete spectrum directly the involved atomic energy levels, coupling in a 1:1 correspondence a single emission line with a single energy level, rather than associating a single emission with a quantum leap between a couple of levels (Rebello et al. 1998; Zollman et al. 2002). The problem of the conceptual link between spectral emissions and energy levels presents several sub-facets: the fundamental level is not considered as a level or it is involved in every transition (Korhasan and Wang 2016; Ivanjek 2012; Ivanjek et al. 2015a, b). In introductive astronomy courses, where spectroscopy plays a pivotal role, since absorption and emission of radiation represent the single tool to infer information concerning the physical properties of celestial objects, difficulties emerged in describing the process of luminous emission from atoms (Bardar et al. 2006). Among university students an idea emerges according to which the energy of the radiation is linked to the intensity rather than to the colour, represented by wavelength or frequency (Lee 2002). Students do not have a clear idea nor of the quantum model for atoms, or of the quantum model for light, so they struggle to predict the way they interact in emission and absorption processes (Savall-Alemany et al. 2016). An evidence regarding the existence of spontaneous models concerning the formation of discrete spectra and of their link with the quantized structure of atoms emerges. Those models have to be overcome in order to reach a scientific view of the topic (Gilbert et al. 1998a, b). Another plan in the conceptual knots is represented by the role of the experimental setup in forming and detecting a spectrum: among the most common conceptual difficulties, student believe that a prism produces always a continuous spectrum and a diffraction grating produces always a discrete spectrum independently by the role of the source (Ivanjek 2012; Ivanjek et al. 2015a, b).

The learning path embeds explorative activities through which students correlate macroscopic observations (spectral emissions) and microscopic interpretations (matter energy structure). This choice was guided by the observation that an educational approach based on the spontaneous angles of attack to specific contents starting from common sense ideas of students is a necessary condition in order to activate the learning process (Viennot 2003).

19.3 The Educational Path

The educational path is built iteratively designing conceptual micro-steps in which active learning strategies as inquiry-based learning and experimental explorations produce the overcoming of the identified conceptual knots and the appropriation of the founding disciplinary elements by means of Design-Based Research (DBR) methods (DBR Collective 2003; Collins et al. 2004; Van der Akker et al. 2006; Anderson and Shattuck 2012) according to which every conceptual micro-step is designed, evaluated and redesigned iteratively on the basis of students' responses. The path has different steps (phases) described in detailed in the following.

The first strategic focus takes into account the classification of optical phenomena in three big thematic areas: production, propagation and matter–light interaction

(Ph1). In this phase, optical phenomena are discussed in order to build a basic idea of light as a massless entity travelling in space, which possess energy which guides the description of processes in which light itself interacts with matter causing different effects, such as heating, penetration, ionization and fluorescence. The difference between Snell's law for refraction and the phenomenology of light transmission through a transparent medium is discussed with students in order to distinguish the plan of formalized phenomenological description in macroscopic terms and the plan according to which light–matter interaction processes are interpreted in terms of energy. A reflection with students on the difference between refraction and transmission of light stimulates an analysis of the two point of view under which those two apparently similar phenomena, are seen: refraction is a macroscopic phenomenological description of the light path across the interface between two mediums with different indexes of refraction, while, when speaking of transmission processes, the focus is put on the interaction between light and matter and implies a deeper analysis, in particular on the energetic plan.

The problem of what "to see" means is posed to students, with the aim of identify, from the discussion of spontaneous models, the role of the observer, of the observed object and of the light (Ph2).

A fundamental parameter enters in the description of the vision mechanism: the colour (Ph3). Radiation gains a significant role in this perspective and poses the question of what the colour physically represents, starting from the observation that what is seen depends on the type of light contextualizing in the case of different picture or painting appear when illuminated by different lights. Surface absorption and nano-structuring allows to understand the mechanism according to which different objects appear differently coloured, in particular the colour of an object result from the fraction of diffuse light, and this depends on the type of radiation that it is used to illuminate the object.

A review and classification of different natural and artificial sources helps in recognizing a light source as a system able to transform in radiant energy any other form of energy, as well as to recognize the need to interpret the emission mechanism (Ph4). This represents a valid context to reinforce the idea that the light emission process has to be read from an energetic point of view. Sources are analysed in structural and functional terms, and according to the emitted light. Incandescent, fluorescence, phosphorescence emissions as well as halogen, gas-discharge lamps, LED and LASERs are examined (Fig. 19.1) identifying the process of energy transformation and discussing the different characteristics. Taking into account different light sources represent a strategy to reinforce the idea that production of light is an energetic process, in which light is recognized to be an entity carrying energy, emitted as a consequence of an energetic transformation, rather than a wave or a particle, which limits the interpretations (Fig. 19.1). The presence of an energetic exchange between emitting system and emitted light is at the basis of a first interpretation of the emission processes.

Characteristics of light emitted by different sources are examined and discussed in terms of colour and intensity (Ph5). Concerning colour, the exploration of the modalities allowing the production of light of different colours leads to focus on

Fig. 19.1 Different kinds of lamps are presented in the path: incandescent, halogen, fluorescent, white and coloured LEDs, gas discharge. A reflection is also stimulated on the existence of natural mechanisms to emit light (thermonuclear reactions in stars, bioluminescence in animals) as well as artificial ones. A light source is thus seen as a system able to convert a form of energy in radiant energy, i.e. light

additive and subtractive mechanisms, exploring if the light could be generated inherently coloured, performing also flame essays of different chemical compounds/elements. The presence of an energy exchange within a light source emerges from the analysis of the emission of an incandescent bulb: as the electric power supplied to the system increases, emitted light appears of different intensities and colours. The Stefan-Boltzmann phenomenological laws, the identification of the emissive and absorbing power and of the important law of nature that their relationship is only a function of temperature, leads to the examination of the emission process with the increase of the power supplied to a source.

The description of the historical discovery of infrared and ultraviolet radiation consolidates the idea that the emission of radiation can occur beyond the visible, having different interactions with matter and different effects: heating, penetration and ionization. Such historical experiences are recalled with the role of identifying light in more general terms of broad-spectrum radiation, not only visible, and to identify differentiated effects regarding light-matter interaction. The discovery of infrared radiation lies in this context, in which the idea that the emission of radiation occurs at every temperature and that the radiation emitted by the bodies can be outside the visible is consolidated. The discovery of ultraviolet radiation confirms this idea experimentally and accustoms to see the colours of the visible spectrum as different radiations of different energy and with different effects (heating for the IR radiation and activation of chemical reactions for the UV radiation).

The question if white light can be considered a colour is answered by analysing Newton's double prism experiment, confirming that white light is composed of different colours. Newton's reasoning on white light dispersion by means of a prism has been exploited in terms of problematic issues in order to support the

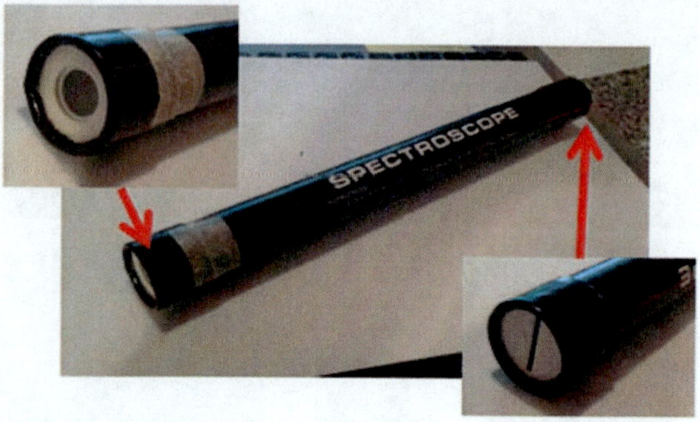

Fig. 19.2 A simple spectroscope: the slit allows light to enter the instruments, and the grating decomposes it in its colours

idea that a prism is able to separate the different chromatic components and does not operate a transformation. Students are elicited to propose an experiment able to decide between the two interpretations.

The evidence that light can be chromatically decomposed is used as a starting point to search for other dispersive mechanisms for light, as the diffraction, for which an experimental activity allows to identify its phenomenological laws and the dispersion role of a grating in the analysis of polychromatic lights.

The usage of simple spectroscopes implementing a slit, a small tube and a grating (Fig. 19.2) allows to characterize a source according to the specific spectrum. Students themselves classify different sources according to their spectra: continuous, discrete and band. Emission spectra of the different sources listed above are examined: the phenomenological exploration of the spectra allows their classification into three categories: continuous, discrete and band spectra (Ph6). The spectroscope is examined as an artefact (Ph7) to recognize its functions, structure and role of the individual components (slit, grating, tube) with the method of the artefacts, which brings an initial global description, the subsequent discussion of the functions of the described parts and constructive alternatives, so as to give meaning to the functional role of each component of the object. The role of each part of the spectroscope is analysed with the artefact method (Bartolini Bussi and Mariotti 1999) in order to give sense to the functional role of every component, in particular, the diffraction grating which seems to operate a sort of "decomposition", which paves the way to the study of diffraction. In this phase students observe, test and manipulate the object in order to grasp its way of functioning without having seen it before.

Diffraction phenomena are experimentally studied (Ph8) with data acquisition of luminous intensity as a function of the position. The starting point is the experimental study of monochromatic diffraction performed with digital acquisition of light intensity as a function of the position (Gervasio and Michelini 2009), at the beginning studying the case of single slit diffraction, and then ending with the analysis of the diffraction pattern produced by a grating, assigning to diffraction the role of dispersive mechanism able to highlight the chromatic structure of light, previously obtained by using a prism (Fig. 19.3, right). At this stage, the colour turns out to be a measurable quantity defining the type of light. The analysis of the diffraction pattern

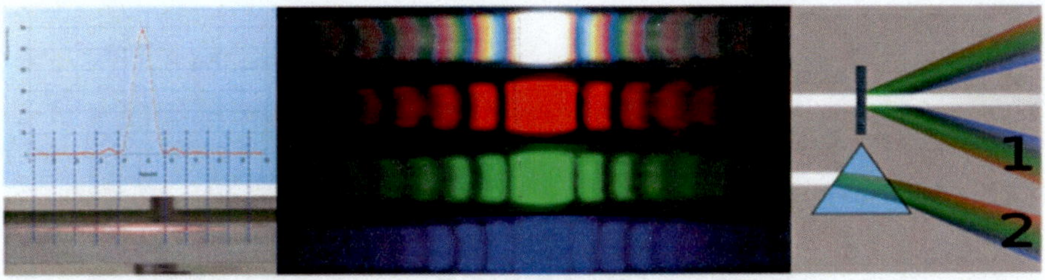

Fig. 19.3 Different dispersive phenomena: diffraction (left) and centre and dispersion (right) are used to highlight the chromatic structure of light. The colour turns out to be a parameter upon which such phenomena depend on

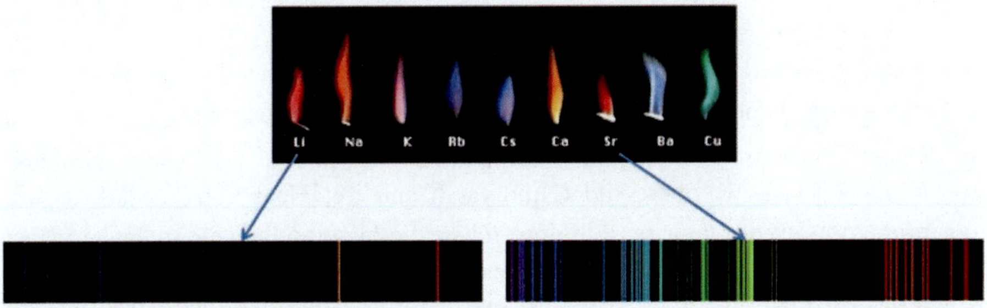

Fig. 19.4 Flame tests

took place with IBL strategy (Michelini and Stefanel 2015) in which students themselves (Ph8.1) design an experiment in order to obtain the laws describing the observed diffraction patterns (Fig. 19.3, left).

The possibility of identifying the presence of gaseous substances through flame tests highlights the important role of the spectrum in the recognition of the elements, having been presented as an experience to recognize the ways in which different light sources can be characterized by the colour of the light they emits and by the relative chromatic composition. An emblematic example is represented by the discovery of caesium, rubidium and helium thanks to spectroscopy. Flame tests are considered as an emblematic example of how the physical structure of the emitting source is intimately linked with the emitted light, and as a proof of the existence of atoms themselves and pose the problem of the interpretation: why are atomic spectra discrete? What is the specific coloured emission due to? (Fig. 19.4).

The quantum nature of light (Ph9) in terms of photons of energy corresponding to the colour and intensity corresponding to the number of photons relies on the analysis of the photoelectric effect. An analysis of the Balmer series of hydrogen (Fig. 19.5) is proposed as a context in which to look for regularities in the observed discrete spectra (Ph10). Once a spectrum is defined, as the series of colours that appears in a certain order of the diffraction pattern, an analysis of the Balmer series for hydrogen is proposed as a context to search for regularities in the observed spectrum. The historical reasoning is reconstructed by students that realize that the

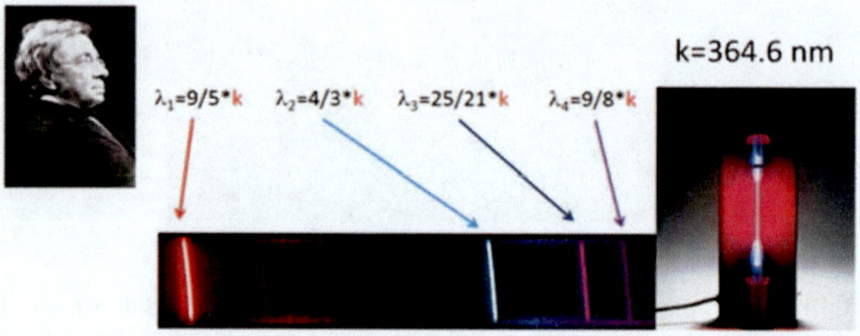

Fig. 19.5 Empirical analysis of the Balmer series in the hydrogen spectrum. Different wavelengths are multiple quantities of a constant quantity k, multiplied by ratios between integers

coefficients obtained by Balmer can be represented by the general empirical formula $\lambda_n = k \cdot \left(\frac{n^2}{n^2-4}\right)$ (Hindmarsh and Ter Haar 1967), that is subsequently re-elaborated for obtaining a more general and interpretable one, in terms of wave numbers: $\frac{1}{\lambda_n} = k' \cdot \left(\frac{1}{4} - \frac{1}{n^2}\right)$, as Rydberg did (Hindmarsh and Ter Haar 1967). Students are thus protagonists with their reasoning to relive the historical development of ideas: the coefficients obtained by Balmer allow the reading of the experimental results in which the wavelengths of the first four lines of the visible hydrogen spectrum are obtained with the empirical formula. According to Einstein's interpretation of photoelectric effect, the reading in energy (proportional to wavenumber) terms suggests how the energy of a specific light emission in a discrete spectrum is caused by an energetic variation at the microscopic level in the emitting system.

In the first versions of the path the empirical formula was proposed to the students in terms of wave number (Ph10.1) or directly in energy (Ph10.2) for the search for interpretations, in particular in light of the hypothesis of the photoelectric effect (Einstein 1917). The history of physics in support of concepts has here an essential role in making students relive the same experience as Balmer and Rydberg in identifying the rules with which one can describe the spectral lines and then look for an interpretation of the emission processes. Balmer's work inspires the operative proposal of the analysis of the regularity of the positions of the lines in the hydrogen visible spectrum and the preparation of the phenomenological law which describes the position of the first four lines of the series. The usage of wavenumbers, proportional to frequency, in turn proportional to energy, allows an energetic reading of Rydberg's formula inspiring the hypothesis of the description of the emitting systems in terms of permitted energetic states and of the emission as discrete energetic de-energization. The works of Balmer first, and then Rydberg, have been proposed as problem solving to retrace the meanings and to explain the process of emission in terms of energy through the link between levels and spectral lines (Ph11).

The semiclassical model of the Bohr atom allows to justify the negative value of the total energies that characterize a bound system. The emission process is explained by the link between energy levels and spectral lines (Ph12), whose nature

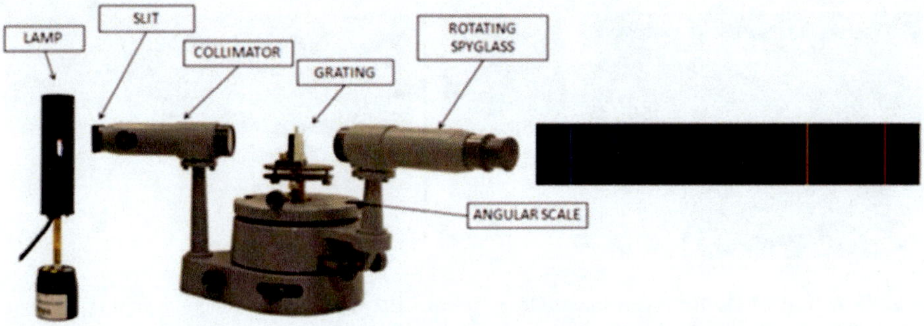

Fig. 19.6 The optical goniometer experiment: emitted light from the lamp is collimated and normally incides on the grating. Different chromatic components thus are angularly resolved and the rotating spyglass allows to measure the corresponding angles

in terms of line energy, corresponding to energy differences between atomic levels, requires consolidation exercises, in particular the design of energy levels and related emissions (Ph12.1) or vice versa with the reconstruction of a structure of energy levels starting from a discrete spectrum (Ph11.2). Knowingly, the Bohr model is avoided in order to avoid the idea that a single energy level correspond to a specific orbit (valid only for the hydrogen atom), and discussing, in more general terms, an atom as a system having access to only specific energetic values.

Beyond the lab exploration of diffraction phenomena described above, two experimental activities have been implemented in the path for enriching the educational proposals: thanks to them, students observe different kinds of spectra, the discrete one emitted by a gas-discharge lamp and a non-complete continuous one with an emission peak typical of LEDs. The two experiments described in the following, despite widely used in spectroscopy lessons, assumed an original role since they have been coupled with targeted questions to inquiry students' reasoning.

The optical goniometer experiment. The goal was to observe the light pattern produced by the interaction of the light emitted from a gas-discharge lamp (different lamps were available containing different elements (cadmium, helium, zinc, mercury) with a diffraction grating, in order to measure the wavelengths and energies corresponding to the various discrete emissions. The analysis was conducted at first at a qualitative level, observing the spectra corresponding to the various orders, then at a quantitative level by measuring the diffraction angle of every chromatic component and associating it with the corresponding wavelength and energy (Fig. 19.6). Starting from the measure of an angle, students evaluate the corresponding wavelength using the grating formula and then they convert it into energy using Einstein's formula for describing energy-frequency relation.

The LED-ruler experiment. It is assembled with low-cost materials and it allows to observe the spectrum of the light emitted from a LED and to evaluate the energy corresponding to the dominant colour: observing the LED through diffraction toy glasses it is possible to observe its spectrum projected along a ruler and with a simple trigonometric calculus the diffraction angle, and thus the corresponding colour

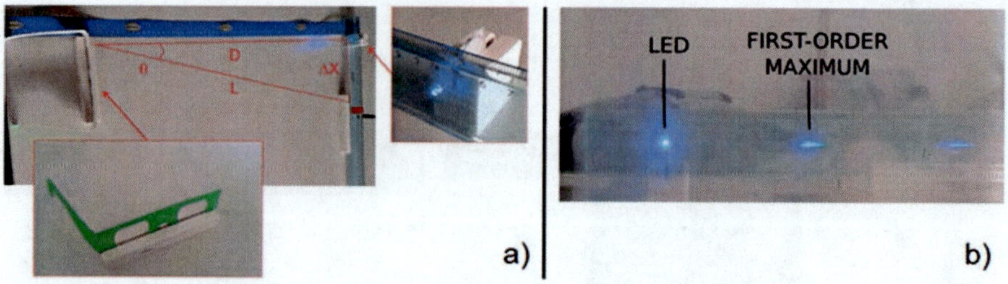

Fig. 19.7 The LED-ruler experiment: Observing a LED through a diffraction grating (**a**) it is possible to measure the diffraction angle corresponding to the peak emission. The spectrum appears projected along the ruler (**b**)

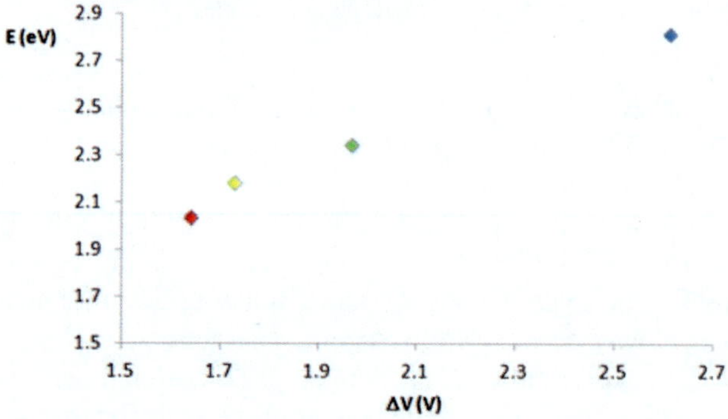

Fig. 19.8 Linear correlation obtained by students between the threshold voltage and the energy emitted as light by different LEDs

energy, is evaluated (Fig. 19.7). This quantity is put into relation with the triggering voltage of the LED (Fig. 19.8).

Simple diffraction toy glasses are used as dispersive element. Students can observe the general features of the emitted spectrum and they are guided in measuring the position of the light peak corresponding to the dominant colour. Simple geometrical measures allow to evaluate the diffraction angle of the peak; the grating formula associates it with a specific wavelength and thus energy. LEDs are supplied with a variable voltage ranging from 0 to 3 V, students have the possibility to measure the threshold voltage of each LED thanks to a potentiometer in the self-build supply board, allowing current-voltage measurements. Since the majority of secondary students involved in the experimentations did not have any confidence with the concept of electric voltage it has been spoken of energy for unit of charge, in order to give meaning to the energy supplied to the system. The linear correlation between threshold voltage and energy of the emitted colour, consequence of the inverse photoelectric effect which is the working mechanism thanks to which a LED produces light, highlight the energetic nature of colours.

19.4 Conclusions

The theoretical framework of the Model of Educational Reconstruction guided the design of the educational path from a research perspective on optical spectroscopy for upper secondary school students. Strategies as inquiry-based learning, experimental activities performed by students themselves and using the history of physics in supporting the concepts, helped in overcoming the main conceptual knots.

Different versions of the educational path have been experimented in different contexts according with a Design-Based Research approach to study how to overcome the main learning difficulties evidenced in the literature. The general structure here described may represent the basis for a coherent path overcoming the following difficulties: as the role of e spectroscope in generating and detecting a spectra, the non-distinction between a diffraction pattern (intensity distribution) and a spectrum (energy distribution), difficulties related to linking spectral lines and energy levels, that can be caused by erroneous models, or vice versa. Moreover, it emerges the need of clarifying strengths and weakness of different models (orbit, orbital, energy states, etc.) considering not only the specific case of the hydrogen atom, in which the correspondence between orbits and levels is direct, but misleading, but also more general cases.

Acknowledgements Authors are grateful to the PLS Italian Plan for the approval of the project and for the opportunities to implement the path in 24 secondary schools.

References

Anderson T, Shattuck J (2012) Design-based research: a decade of progress in education research? Educ Res 41(1):16–25

Bardar EM, Prather EE, Brecher K, Slater TS (2006) The need for a light and spectroscopy concept inventory for assessing innovations in introductory astronomy survey courses. Astron Educ Rev 4(2):20

Bartolini Bussi MG, Mariotti MA (1999) Semiotic mediation: from history to mathematics classroom. Learn Math 19(2):27–35

Buongiorno D, Michelini M (2017) The conceptual contribution of the history in learning physics: the case of optical spectroscopy. In: Proceedings of SISFA2017, Bari, Sept 26–29

Collins A, Joseph D, Bielaczyc K (2004) Design research: theoretical and methodological issues. J Learn Sci 13(1):15–42

DBR Collective (2003) Design-based research: an emerging paradigm for educational inquiry. Educ Res 32(1):5–8

Duit R, Gropengießer H, Kattmann U (2005) Towards science education research that is relevant for improving practice: the model of educational reconstruction. In: Fischer HE (ed) Developing standards in research on science education. London, Taylor & Francis, pp 1–9

Duit R, Gropengießer H, Kattmann U, Komorek M, Parchmann I (2012) The model of educational reconstruction – a framework for improving teaching and learning science. In: Jorde D, Dillon J (eds) Science education research and practice in Europe. Rotterdam, Sense Publishers, pp 13–37

Einstein A (1917) On the quantum theory of radiation. Phys Z 18:121

Gervasio M, Michelini M (2009) Lucegrafo. A USB data acquisition system. In: Lamboune B et al (eds) MPTL14 proc. http://www.fisica.uniud.it/URDF/mptl14/contents.htm

Gilbert JK, Boulter C, Rutherford M (1998a) Models in explanations: Part 1. Horses for courses? Int J Sci Educ 20(1):83–97

Gilbert JK, Boulter C, Rutherford M (1998b) Models in explanations: Part 2. Whose voice, whose ears? Int J Sci Educ 20(2):187–203

Hindmarsh WR, Ter Haar D (1967) Atomic spectra. Pergamon Press, Oxford

Ivanjek L (2012) An investigation of conceptual understanding of atomic spectra among university students. Ph.D. dissertation, University of Zagreb

Ivanjek L, Shaffer PS, McDermott LC, Planinic M, Veza D (2015a) Research as a guide for curriculum development: an example from introductory spectroscopy: I. Identifying student difficulties with atomic emission spectra. Am J Phys 83(1):85

Ivanjek L, Shaffer PS, McDermott LC, Planinic M, Veza D (2015b) Research as a guide for curriculum development: an example from introductory spectroscopy: II. Addressing student difficulties with atomic emission spectra. Am J Phys 83(2):171

Korhasan ND, Wang L (2016) Students' mental models of atomic spectra. Chem Educ Res Pract 17:743–755

Lee S (2002) Students' understanding of spectra. Ph.D. dissertation, Kansas State University

Michelini M, Stefanel A (2015) Upper secondary students face optical diffraction using simple experiments and on-line measurements. In: Triay R (ed) FFP14 proc

Michelini M, Santi L, Stefanel A (2014) Teaching modern physics in secondary school. In: Proceedings of science (FFP14) 231, 15–18 July 2014

Rebello NS, Cumaranatunge C, Escalada LT, Zollman DA, Donnelly D (1998) Simulating the spectra of light source. Comput Phys 12(1):28

Savall-Alemany F, Domènech-Blanco JL, Guisasola J, Martinez-Torregrosa J (2016) Identifying student and teacher difficulties in interpreting atomic spectra using a quantum model of emission and absorption of radiation. Phys Rev ST Phys EdUC Res 12(1):01013

Van der Akker J, Gravemeijer K, McKennedy S (2006) In: Nieveen N (ed) Educational design research. Routledge, London and New York

Viennot L (2003) Relating research in didactics and actual teaching practice: impact and virtues of critical details. In: Psillos D et al (eds) Science Education Research in the Knowledge-Based Society

Zollman DA, Rebello NS, Hoggs K (2002) Quantum mechanics for everyone: hands-on activities integrated with technology. Am J Phys 70(3):252

Chapter 20
A Combination of Historical Physics Documents and Other Teaching Tools for the Instruction of Prospective Teachers in Chaos and Complexity

Aristotelis Gkiolmas, Artemisia Stoumpa, Anthimos Chalkidis, and Constantine Skordoulis

Abstract This work presents and analyses a didactic methodology for teaching aspects of Chaos and Complex Systems in Physics to prospective Science educators. The methodology includes historical Physics texts and experimental instrumentation, as well as computer models and simulations. The objectives are mainly to help undergraduate teachers realise the way that Physics evolves through changes and standoffs, and the way in which scientists work, which is much related to teaching the Nature of Science (NoS). At the same time, through this teaching methodology, there is an attempt to instruct undergraduate students in basic elements of Chaos Theory and Complexity Theory, by avoiding heavy mathematical formalism incompatible with their age and their learning level and ability. This teaching sequence is intended to be applied in a pilot study involving undergraduate students of the Department of Primary Education, University of Athens, so as to generate initial qualitative and quantitative results.

20.1 Introduction

There is a growing tendency in contemporary Physics education to include aspects of (1) Chaos Theory, (2) Complexity and (3) Complex Systems' properties (Strogatz 2014) and comportment.

Several attempts have been made or are being made to bring Chaos and Complexity into school classrooms (Peitgen et al. 1991, 1992, 1999, 2004), to popularise these sciences (Briggs and Peat 1989; Ruelle 1993; Gleick 2008) and to didactically

---

A. Gkiolmas (✉) · A. Stoumpa · A. Chalkidis · C. Skordoulis
Department of Primary Education, National and Kapodistrian, University of Athens, Athens, Greece
e-mail: agkiolm@primedu.uoa.gr; artemisst@primedu.uoa.gr; achalkid@primedu.uoa.gr; kskordul@primedu.uoa.gr

B. G. Sidharth et al. (eds.), *Fundamental Physics and Physics Education Research*,
https://doi.org/10.1007/978-3-030-52923-9_20

transform certain concepts from them (Duit and Komorek 1997; Duit et al. 1997, 1998).

The reasons for this aforementioned tendency are multiple. Firstly, Chaos is a branch of modern Physics which concerns the average-size scales of Nature—the ones in which we live—and thus has phenomena that are relatively easy to notice and observe. This is in contrast to Relativity, which usually concerns very large scales, and to Quantum Mechanics, which mainly describes events in the microscopic world.

Secondly, Chaos, as a concept for instruction, could bring significant changes to the ideas about and the perceptions of everyday phenomena that the learning subjects may have. For instance, involvement with Chaos Theory abolishes the learner's belief that small causes have small effects and that, as the cause increases in size and significance, so does the effect (Lorenz 2005a; Smith 2007). Additionally, Chaos Theory deals a severe blow to the certainty that the same system, with the same or similar pre-existing conditions, will evolve totally similarly (identically) in time (Lorenz 1963, 1969; Prigogine 1997; Stewart 2002).

Thirdly, Chaos can easily be produced and studied in a school Physics laboratory. It requires only simple activities and equipment, such as a Chaos Pendulum (Skordoulis et al. 2014; PASCO 2017), that exists in many high schools.

Additionally, Complex Systems and Complexity, in a variety of their aspects, as well as the ideas that stem from these fields, such as cellular automata (Wolfram 2002), arise in more and more scientific fields and in an increasing number of events in daily life (Kaufmann 1995; Mitchell 2009; Holland 2014). The same is true of Fractals, the mathematical representation of chaotic systems (Mandelbrot 1982; Bountis 2004).

It is obvious that, if knowledge of Chaos, chaotic natural systems, Complexity and Complex Systems is to be diffused into school classrooms (mainly high school classrooms, but also primary), a necessary prerequisite is to teach future teachers, as well as in-service Science teachers, about these issues. This is the reason why the current research and the teaching methodology stemming from it focus mainly on undergraduate primary school teachers. The concepts related to Chaos and Complex Systems that are intended to be taught to prospective primary school teachers must, by necessity, be discharged from heavy mathematical formalism, and bring out mainly conceptual aspects of these fields of Physics. Such aspects are: the sensitive dependence on initial conditions, the limited predictability, the existence of rules in apparently chaotic natural systems (Kellert 1994), the emergence of complex patterns based on simple rules, the fact that "the whole is larger than the sum of its parts", the critical state (a small change in the cause can cause a great or—at least—an unpredictable change in the results), etc.

20.2 The Learning Subjects of This Teaching Method

The persons chosen here for instruction in Chaos are prospective primary school teachers (undergraduate students). This is an audience that possesses a mathematical and scientific background but is of great research interest, we believe, because these are the individuals who will transfer knowledge and stances about scientific fields to primary school students, thus placing the foundations for the latter's affection for and understanding of Science.

Additionally, it is considered a challenge to instruct in Chaos using extremely limited mathematical formulation and Physics terminology. A teaching sequence must therefore be created which relies on optical representations, simulations, activities and narrations.

20.3 Methodology and Tools of Instruction

Central to the methodology are texts and documents written by scientists themselves on facing Chaos for the first time. The two scientists studied for this purpose are Poincaré and Lorenz (Poincaré 1914, 2017a; Diacu 1996; Lorenz 2005a). Using these resources, students are guided to the contradictions, the gridlocks, the unexpected facts and the inexplicable results that led the scientists to the new scientific field (Chaos). A drawing similar to that of Poincaré which is used in our teaching

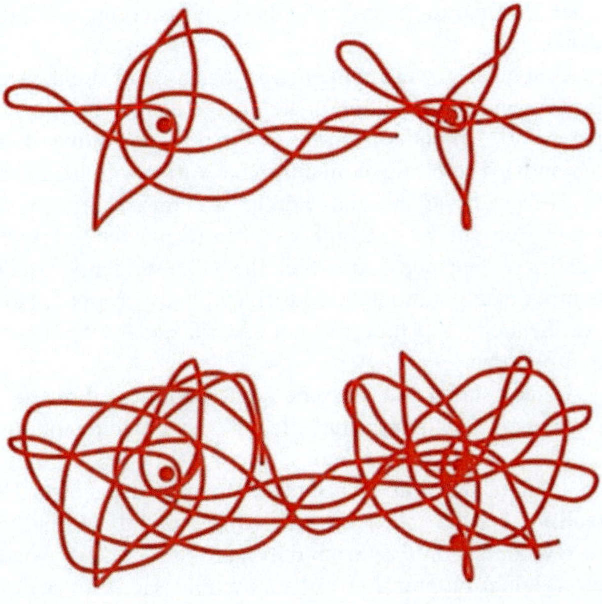

Fig. 20.1 A representation of the indicative orbits in the three-body problem of Poincaré (2017b). (Artist: Ben Satchel. www.benpics.com)

method is depicted in Fig. 20.1. Similar figures can be found in books such as the one by Stewart (2002).

The text of Poincaré used in the teaching sequence refers to a cone that one makes stand on its apex—after which it falls (Poincaré 1914). Poincaré's writing states that such a cone, with a vertical axis and only gravity acting upon it, would—if it had perfect symmetry—never fall. However, any degree of asymmetry would cause it to lean and, eventually, to fall. Similarly, even in a case of perfect symmetry, any outside force could be enough to disturb the balance.

Thus, this case shows that a small force can have a demonstrable effect—and the force may be so small that it passes unnoticed, thus the effect appears to be caused by nothing more than chance. Without full knowledge of the laws of nature and the state of the "universe" (the environment, the natural world that the object belongs to) at the moment of the trial, it would be impossible to predict exactly what the next second of the universe will contain. According to Poincaré, even if we possessed full knowledge of all the secrets of the "universe", it is likely that we would still only be able to gain an approximate understanding of its state at any given point (Poincaré 1948, 2001). This might allow us to create a similarly approximate prediction—but, as tiny differences can lead to noticeable differences at a later point, even a small mistake in the former could lead to a much greater one in the latter. This, Poincaré concludes, makes prediction impossible, and leaves us with what we call "chance" (Poincaré 1914, pp. 67–68).

After reading the excerpt from Poincaré, the students discuss the notions of predictability or non-predictability as presented in the text above, as well as the concept of sensitive dependence on initial conditions.

As a second example, students read in groups a famous extract from Lorenz (2005a), in which he describes how even small rounded-up approximations in very small decimal digits created Chaos in his computer's generation of meteorological values.

Graphically, what happened in the case of Lorentz is depicted in Fig. 20.2, which is a drawn representation of the image created by him. It shows a section of a time period of 15 months from the original point of time, divided into three segments of 5 months. The physical quantity that varies with time is a meteorological parameter.

The exact variable that Lorenz was measuring was the latitude of the strongest winds from the west; a high value corresponded to low latitude. The diagram shows a series of "episodes", in which the value suddenly rose, remained at a high level for approximately a month, and then suddenly dropped. However, these episodes were not identical, and they did not last for equal lengths of time; their behaviour was non-periodic.

Lorenz stated that, at some point, he repeated several computations to study this phenomenon in more detail. He notes that, while inputting numbers, he took a break of about an hour and returned to find that the computer had generated weather simulations that bore no resemblance to the usual ones. Upon studying the results, he realised that the values started to differ slightly at first, then more greatly, until they were doubling in size around every 4 days, leaving them drastically different from the original output after only 2 months. He soon realised what had happened; by

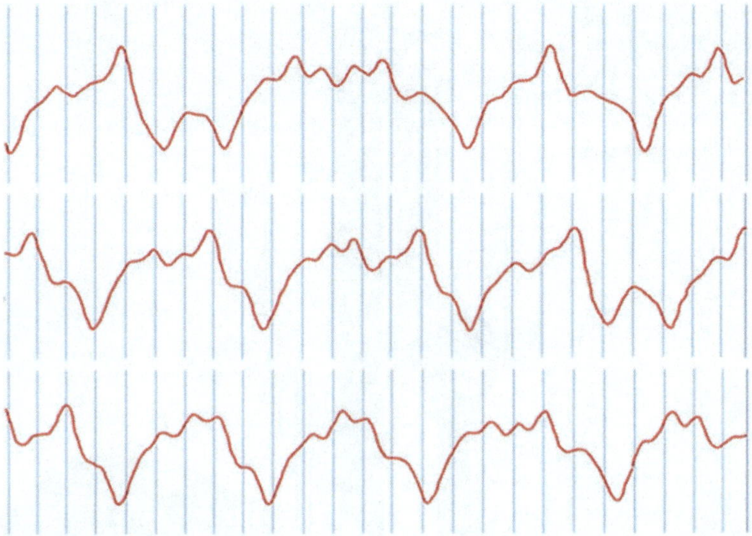

Fig. 20.2 A representative drawing of the 15-month sequence of the values of a meteorological parameter related to winds, similar to the one that appeared on the computer screen of E. N. Lorenz

inputting numbers that were rounded-off by the computer rather than the exact ones, he had introduced errors to the calculations and these had amplified—and created Chaos.

Lorenz considered the implications that the calculations in his model had on the real world—that not accurately measuring temperature, wind and so forth to three decimal places meant that long-range weather forecasting was impossible. He came to believe that it was this "amplification" of small differences that caused the lack of periodicity in his computations (Lorenz 2005a, pp. 134–135).

Then, in the worksheets delivered to the students, there is a discussion of the extract from Lorentz, and certain questions are posed to the students about this historical text.

As auxiliary tools, instruments such as a Chaos Pendulum (PASCO 2017) and forms of software such as NetLogo (Wilensky 1998, 1999) are utilised in order to represent chaotic (Head and Wilensky 2017) and/or Complex System's comportment in real circumstances, for didactical reasons. In Figs. 20.3 and 20.4, images of the aforementioned teaching tools are shown.

Finally, activities with pen, pencil and paper or with other simple materials (Peitgen et al. 1991) help the learning subjects to "create" chaotic conditions or time-evolutions and patterns by themselves. One such activity is the "chaos game" depicted in Fig. 20.5.

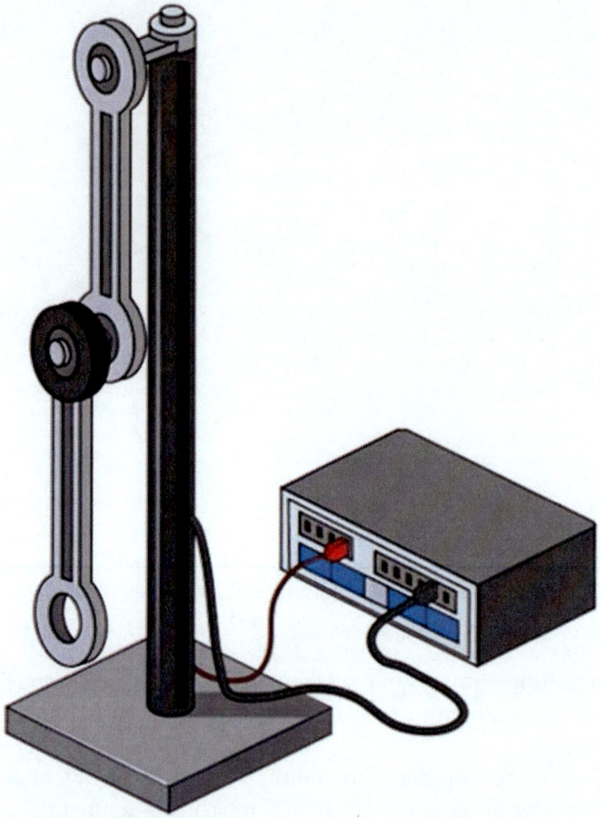

Fig. 20.3 A drawn image of a type of "Chaos Pendulum", similar to the one used for our didactic purposes. (Artist: Ben Satchel. www.benpics.com)

20.4 The Scheduled Teaching Intervention (Teaching Sequence) in Chaos and Complexity

There are *five steps* in the teaching sequence and in the worksheets of the present research:

- Initial interest is raised through questions and visual material.
- Students interact and read the documents (extracts) from the scientists.
- Students take measurements and make graphic representations with the Chaos Pendulum.
- NetLogo is used to realise and conceptualise chaotic behaviour, as well as certain aspects of Complex Systems, such as percolation (Wilensky 1997).
- Students draw conclusions, consolidate knowledge and make extensions to everyday life and their actual surroundings.

The overall teaching sequence is based on the inquiry-based learning model (Bybee et al. 2006).

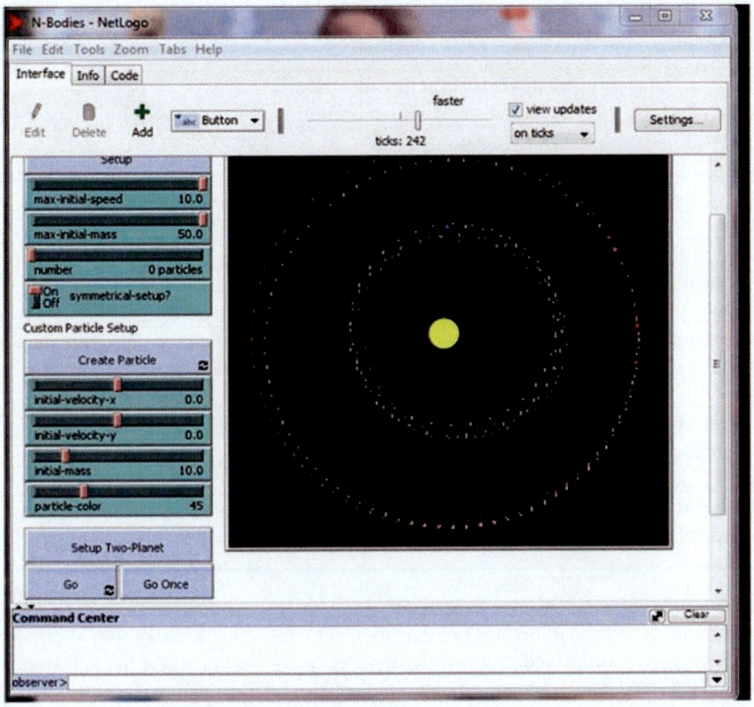

Fig. 20.4 A screenshot of an initial stage of the model of the N-bodies problem in NetLogo (Wilensky 1998)

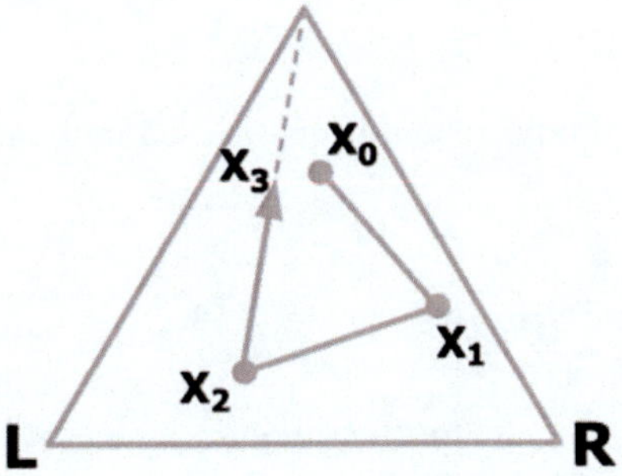

Fig. 20.5 A drawn representation of an early stage of the "Chaos Game"

20.5 Conclusion

The undergraduate students' pre-existing ideas about Chaos, Complex Systems and Complexity, the method of instruction used to rephrase them, and their new (targeted) ideas after the teaching sequence presented in this work, are shown in Table 20.1.

Table 20.1 Change in undergraduate students' perspectives through instruction in Chaos and Complex Systems

Students' initial idea and/or concept	Methods used in this research (teaching sequence) in order to change perspectives	Final idea and/or concept after instruction
Phenomena with similar starts will evolve similarly	Historical text, NetLogo, Chaos Pendulum	Phenomena with similar starts can evolve very differently
There are no rules in chaotic behaviour or chaotic orbits	Historical text, NetLogo, Chaos Pendulum	There is a certain "order" in Chaos
Science can make extremely accurate predictions	Historical text	Science can make predictions with limited accuracy, and this sometimes produces "Chaos"
Simple rules create simple aggregate outcomes	NetLogo	Simple rules can create very complex (emergent) outcomes
Small change in a cause results in a small change in the effect	Historical Text, Netlogo, Chaos Pendulum	Small change in a cause can result in tremendous or unpredictable changes in the effect

Acknowledgements The authors would like to thank: the graphic artist Ben Satchel for his beautiful drawings, the editor and proof-reader Bridget Leary-Georgiadi for her valuable contributions, as well as the Chaos' researcher and teacher Kostas Karamanos for the important and delightful discussions that we shared about chaotic systems!

20.6 Appendix

The exact excerpts from Poincaré and Lorentz, used in the teaching sequence.

20.6.1 Extract from Poincaré

...it seems that chance alone will decide. If the cone were perfectly symmetrical, if its axis were perfectly vertical, if it were subject to no other force but gravity, it would not fall at all. But the slightest defect of symmetry will make it lean slightly to one side or other, and as soon as it leans, be it ever so little, it will fall altogether to that side. Even if the symmetry is perfect, a very slight trepidation, or a breath of air, may make it incline a few seconds of arc, and that will be enough to determine its fall and even the direction of its fall, which will be that of the original inclination.

A very small cause which escapes our notice determines a considerable effect that we cannot fail to see, and then we say that that effect is due to chance. If we knew exactly the laws of nature and the situation of the universe at the initial moment, we could predict exactly the situation of that same universe at a succeeding moment.

But, even if it were the case that the natural laws had no longer any secret for us, we could still only know the initial situation approximately. If that enabled us to predict the succeeding situation with the same approximation, that is all we require, and we should say that the phenomenon had been predicted, that it is governed by laws. But it is not always so; it may happen that small differences in the initial conditions produce very great ones in the final phenomena. A small error in the former will produce an enormous error in the latter. Prediction becomes impossible, and we have the fortuitous phenomenon." (Poincaré 1914, pp. 67–68)

20.6.2 Extract from Lorentz

"*In Fig. 43* [note: corresponds to Fig. 20.2 of the chapter], *we see a copy of fifteen months of the somewhat faded original output, divided for display purposes into three five-month segments.*

The chosen variable is an approximate measure of the latitude of the strongest westerly winds; a high value indicates a low latitude. There is a succession of "episodes," in each of which the value rises abruptly, remains rather high for a month or so, and then drops equally abruptly, but the episodes are not identical and are not even equal in length, and the behavior is patently non-periodic. At one point I decided to repeat some of the computations in order to examine what was happening in greater detail. I stopped the computer, typed in a line of numbers that it had printed out a while earlier, and set it running again. I went down the hall for a cup of coffee and returned after about an hour, during which time the computer had simulated about two months of weather. The numbers being printed were nothing like the old ones. I immediately suspected a weak vacuum tube or some other computer trouble, which was not uncommon, but before calling for service I decided to see just where the mistake had occurred, knowing that this could speed up the servicing process. Instead of a sudden break, I found that the new values at first repeated the old ones, but soon afterward differed by one and then several units in the last decimal place, and then began to differ in the next to the last place and then in the place before that. In fact, the differences more or less steadily doubled in size every four days or so, until all resemblance with the original output disappeared somewhere in the second month. This was enough to tell me what had happened: the numbers that I had typed in were not the exact original numbers, but were the rounded-off values that had appeared in the original printout. The initial round-off errors were the culprits; they were steadily amplifying until they dominated the solution. In today's terminology, there was chaos. It soon struck me that, if the real atmosphere behaved like the simple model, long-range forecasting would be impossible. The temperatures, winds, and other quantities that enter our estimate of today's weather are certainly not measured accurately to three decimal places, and, even if they could be, the interpolations between observing sites would not have similar accuracy. I became rather excited, and lost little time in spreading the word to some of my colleagues. In due time, I convinced myself that the amplification of small differences was the cause of the lack of periodicity. Later, when I presented my results at the Tokyo meeting, I added a brief description of the unexpected response of the equations to the round-off errors." (Lorenz 2005b, pp. 134–135).

References

Bountis T (2004) The wonderful world of fractals: a tour in the new science of Chaos and complexity (Greek language). Leader Books, Athens, Greece

Briggs J, Peat FD (1989) Turbulent mirror: an illustrated guide to Chaos theory and the science of wholeness. Harper & Row, New York

Bybee RW, Taylor JA, Gardner A, Van Scotter P, Powell JC, Westbrook A, Landes N (2006) The BSCS 5E instructional model: origins, effectiveness, and applications. BSCS, Colorado Springs

Diacu F (1996) The solution of the n-body problem. Math Intell 18(3):66–70

Duit R, Komorek M (1997) Understanding the basic ideas of Chaos-theory in a study of limited predictability. Int J Sci Educ 19:247–264

Duit R, Komorek M, Wilbers J (1997) Studies on educational reconstruction of Chaos theory. Res Sci Educ 27:339–357

Duit R, Roth W-M, Komorek M, Wilbers J (1998) Conceptual change cum discourse analysis to understand cognition in a unit on chaotic systems: towards an integrative perspective on learning in science. Int J Sci Educ 20:1059–1073

Gleick J (2008) Chaos: making a new science. Penguin Books, New York

Head B, Wilensky U (2017) NetLogo Chaos in a box model. Center for Connected Learning and Computer-Based Modeling, Northwestern University, Evanston, IL. http://ccl.northwestern.edu/netlogo/models/ChaosinaBox

Holland J (2014) Complexity: a very short introduction. Oxford University Press, Oxford

Kaufmann S (1995) At home the universe: the search for the laws of self-organization and complexity. Oxford University Press, New York

Kellert SH (1994) In the wake of Chaos: unpredictable order in dynamical systems. The University of Chicago Press, Chicago

Lorenz EN (1963) Deterministic non-periodic flow. J Atmos Sci 20(2):130–141

Lorenz EN (1969) Three approaches to atmospheric predictability. Bull Am Meteorol Soc 50:345–349

Lorenz EN (2005a) The essence of chaos. University of Washington Press, Seattle

Mandelbrot B (1982) The fractal geometry of nature. W. H. Freeman and Company, Canada

Mitchell M (2009) Complexity: a guided tour. Oxford University Press, New York

PASCO (2017) Chaos experiment. https://www.pasco.com/prodCatalog/EX/EX-5523_chaos-experiment/index.cfm. Accessed 15 Oct 2017

Peitgen H-O, Jürgens H, Saupe D, Maletsky E, Perciante T, Yunker L (1991) Fractals for the classroom: strategic activities, vol 1. Springer-Verlag, New York

Peitgen H-O, Maletsky EM, Jürgens H, Perciante T, Saupe D, Yunker L (1992) Fractals for the classroom: strategic activities, vol 2. Springer, New York

Peitgen H-O, Maletsky EM, Jürgens H, Perciante T, Saupe D, Yunker L (1999) Fractals for the classroom: strategic activities, vol 3. Springer, New York

Peitgen H-O, Jürgens H, Saupe D (2004) Chaos and fractals: new frontiers of science, 2nd edn. Springer-Verlag, New York

Poincaré H (1948) Science and method. Dover, New York

Poincaré H (2001) The value of science: essential writings of Henri Poincaré. Modern Library Science, New York

Poincaré H (2017a) Science and hypothesis. CreateSpace Independent Publishing Platform, Scotts Valley, CA

Poincaré H (2017b) The three-body problem and the equations of dynamics: Poincaré's foundational work on dynamical systems theory. Translation: Bruce D. Popp. Springer International Company, Switzerland

Prigogine I (1997) The end of certainty: time, Chaos and the new laws of nature. The Free Press, New York

Ruelle D (1993) Chance and Chaos. Princeton University Press, Princeton, NJ

Skordoulis C, Tolias V, Stavrou D, Karamanos K, Gkiolmas A (2014) Teaching Chaos with a pendulum to Greek secondary school students. Adv Syst Sci Appl 14(2):158–169

Smith L (2007) Chaos: a very short introduction, 1st edn. Oxford University Press, Oxford

Stewart I (2002) Does god play dice? The new mathematics of chaos, 2nd edn. Blackwell Publishing, Oxford

Strogatz SH (2014) Nonlinear dynamics and chaos: with applications to physics, biology, chemistry and engineering (studies in nonlinearity), 2nd edn. Westview Press, Boulder, CO

Wilensky U (1997) NetLogo Fire model. Center for Connected Learning and Computer-Based Modeling, Northwestern University, Evanston, IL. http://ccl.northwestern.edu/netlogo/models/Fire

Wilensky U (1998) NetLogo N-Bodies model. Center for Connected Learning and Computer-Based Modeling, Northwestern University, Evanston, IL. http://ccl.northwestern.edu/netlogo/models/N-Bodies

Wilensky U (1999) NetLogo. Center for Connected Learning and Computer-Based Modeling, Northwestern University, Evanston, IL. http://ccl.northwestern.edu/netlogo/

Wolfram S (2002) A new kind of science. Wolfram Media, Champaign, IL

Historical Physics' Textbooks (Whose Extracts Have Been Used)

Lorenz E (2005b) The essence of Chaos. UCL Press, London. ISBN: 0-203-27116-5 (Adobe e-Reader Format). First published in 1993 by the University of Washington Press

Poincaré H (1914) Science and method [translated in English by Francis Maitlad]. Thomas Nelson and Sons, London, Edinburgh, Dublin & New York

Chapter 21
Design, Construction and Use of a Quantitative Spectroscope for Science Dissemination

Francisco Savall-Alemany, Jesús Carnicer Murillo, Rocío Espinosa Gutierrez, Esther Perales Romero, and Hernán A. Supino Graña

Abstract In this work we present a spectroscope that we have constructed for science dissemination in museums. The materials used are economical and easy to handle, allowing all attendees in the spectroscopy workshop to build, use and take home a small but useful spectroscope. We also present the steps followed to construct it and the STEM activities that are carried out in the museum to learn how to use it.

21.1 Introduction

Spectroscopy has been a technique of great scientific relevance, practically since its appearance. It contributed to the discovery of elements as caesium, rubidium or helium; it was used to determine the composition of the Sun and other celestial objects, to establish the structure of atoms, etc. Nowadays, its use is common in the detection of chemical substances or in the study of celestial objects, among many other applications. In addition, since the popularization of LEDs as a light source, several studies have been carried out which highlight their impact on the reconciliation of sleep, an effect that is related to the spectre of the light they emit (Gooley et al. 2011). Despite its scientific importance, our experience shows us that it is

F. Savall-Alemany (✉)
Universidad de Alicante, Alicante, Spain

J. Carnicer Murillo · H. A. Supino Graña
Museo Didáctico e Interactivo de Ciencias de la Vega Baja del Segura de la Comunitat Valenciana (MUDIC-VBS-CV), Alicante, Spain

R. Espinosa Gutierrez
Stowmarket High School, Stowmarket, UK

E. Perales Romero
Grupo de Visión y Color, Departamento de Óptica, Farmacología y Anatomía, I. U. Física Aplicada a las Ciencias y la Tecnología (IUFACyT), Universidad de Alicante, Alicante, Spain

B. G. Sidharth et al. (eds.), *Fundamental Physics and Physics Education Research*,
https://doi.org/10.1007/978-3-030-52923-9_21

completely unknown to the general public. On the other hand, in the school and academic world, spectroscopy has an important presence due to its relationship with the atomic models studied in Physics and Chemistry at high school. However, didactic research has found that even university science students have difficulties to explain spectrum formation (Ivanjek et al. 2015; Savall et al. 2016).

At the Museo Didáctico e Interactivo de Ciencias de la Vega Baja del Segura de la Comunitat Valenciana (MUDIC-VBS-CV), we have set the objective of bringing spectroscopy closer to the general public through hands-on and STEM activities. Visitors are involved in activities in which they have to build a simple spectroscope and use it to observe and interpret the spectra of common light sources such as energy-saving bulbs, LEDs or old tungsten filament bulbs.

21.2 Project Description

Building a spectroscope is a simple task that most science professors have carried out and that has been presented in numerous publications (De Luís and Martínez 2010; Remón et al. 2016). However, the main part of the proposals is based on qualitative spectroscopes, being the quantitative spectroscopes difficult to construct due to the materials used. Our work has focused on that objective: the construction of a spectroscope with economical and easy-to-find materials that allows us to measure the wavelengths of spectral lines.

The Museo Didactico e Interactivo de Ciencias de la Vega Baja del Segura de la Comunitat Valenciana (MUDIC-VBS-CV) is mainly visited by schoolchildren (and when the accompanying teachers request it they carry out a spectroscopy workshop) as well as general public that visits the museum during open days or when offering other activities. In this workshop, visitors construct the spectroscope with the help of a museum monitor and use it to analyse the most common light sources: incandescent lamps, fluorescent lamps, LED lamps, etc.

The main aim of the workshop is that many families from the provinces of Alicante and Murcia build the spectroscope and use it to analyse the different light sources. By that, we will contribute to achieve the objectives of the MUDIC-VBS-CV, which refer to the dissemination of scientific content outside the school. Visiting teachers will also be able to borrow spectroscopes for their classrooms. The spectroscope will also be provided to teachers in the training courses organized by the MUDIC-VBS-CV, so that it can be used to deal with the curricular contents related to spectroscopy.

21.3 Methodology and Materials

For the construction of the spectroscope, we use an A4-size cardboard, light coloured or preferably white. On the cardboard we print the template of the spectroscope, which is provided to the attendants, together with a small diffraction grating (Fig. 21.1).

We have designed the template from other spectroscope models (see, for example, Savall et al. 2013). In this case, we have reduced the spectroscope length and we have calibrated the scale to be able to read the values of the wavelengths of the spectral lines directly, a useful improvement that allows us to read the wavelength directly avoiding the use of mathematical functions.

Once mounted, we obtain the spectroscope shown in Fig. 21.2. It consists of a box of 18 cm × 8 cm × 4 cm. In one of the sides it has a rectangular hole, the ocular, which is covered with a diffraction grid of 465 lines/mm. On the opposite side, aligned with the grid, it has a narrow slit that is the objective of the spectroscope. Inside the box, to the right of the slit, it has a scale that allows the user to measure the

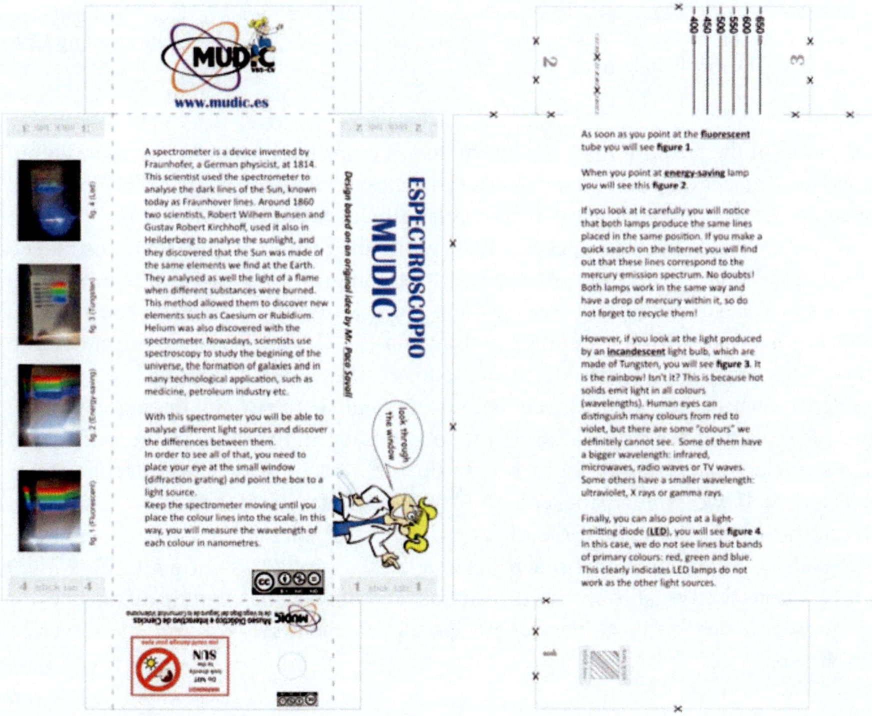

Fig. 21.1 On the left it is shown the outer part of the spectroscope which includes a brief explanation of spectroscopy history. The right image shows the inner side, which includes an explanation of the mounting and use of this spectroscope. It is offered to the attendees printed and die-cutting. Its real dimensions are 18 cm × 8 cm × 4 cm

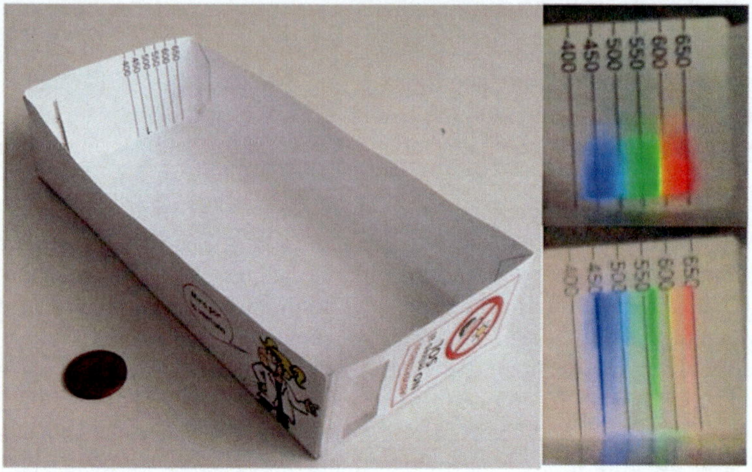

Fig. 21.2 On the left it is shown the spectroscope once mounted. At the bottom we can see the slit behind which the light source is located and to the right we appreciate the calibrated scale for measuring the wavelength. On the front, we see the ocular, covered by a diffraction grid. Next to the spectroscope there is a 2-cent euro coin which serves as a reference for appreciating its small size. On the right side of the image we show two spectra, corresponding to a white light emitting LED (top) and a low-energy light bulb (bottom)

wavelength of the spectral lines in nanometres. To observe the spectrum of any light source, the diffraction grid and the slit must be aligned with the source. The spectrum is seen on the scale, which gives the wavelengths of the spectral lines.

To calibrate the spectroscope, we analyse the diffraction phenomenon produced by the grid. When a wavefront reaches a diffraction grid, each of the slits in the grid becomes a new source of new wavefronts. On each point of a screen located at a distance L from the diffraction grid (as shown in Fig. 21.3), there will be an overlap of these secondary wavefronts. At those points on the screen where the waves arrive in phase, constructive interference will occur and a maximum intensity will be detected. In fact, if we point a laser (monochromatic light) through the diffraction grid we observe a series of light points on the screen where the interference is constructive. If we change the colour of the laser, the light points are placed in different positions, because of the different laser's wavelength.

Considering the waves that reach point X on the screen (as shown in Fig. 21.3) from two consecutive slits in the diffraction grid, the paths' length difference for both waves corresponds to the length of the segment Δ. We can then set the relationships:

$$\sin \alpha = \frac{\Delta}{d}$$

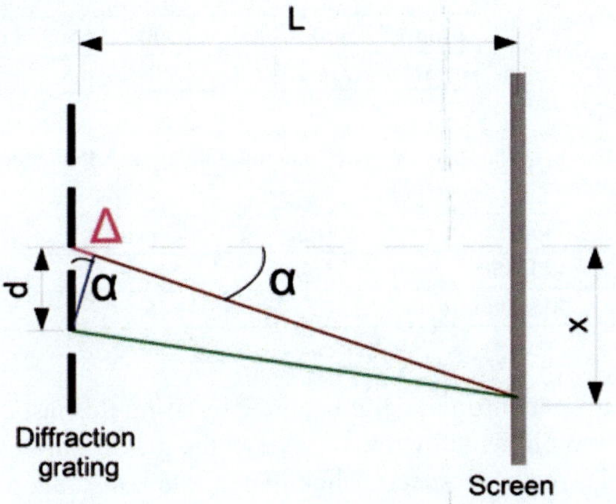

Fig. 21.3 Diffraction grid containing slits at a distance d. At a distance L from the grid, we set a screen on which the light spectrum is observed. We consider an X-point where constructive interference occurs. Red and green lines represent the paths followed by the waves from two consecutive slits to X-point and Δ represents the difference in both paths' length

and

$$\sin\alpha = \frac{x}{\sqrt{x^2 + L^2}}$$

By combining both expressions we find that

$$\frac{\Delta}{d} = \frac{x}{\sqrt{x^2 + L^2}}$$

that relates the paths' length difference to the X-point of the screen.

The interference will be constructive where the path difference is a multiple of the wavelength:

$$\Delta = n\lambda$$

In our case, we will work only with $n = 1$, therefore:

$$\frac{\lambda}{d} = \frac{x_{\max}}{\sqrt{x_{\max}^2 + L^2}}$$

So the positions of the maximums will be:

$$x_{\max} = \frac{\lambda L}{\sqrt{d^2 - \lambda^2}}$$

Table 21.1 Position where each spectral line is observed

λ (nm)	400	450	500	550	600	650	700
x (cm)	3,4	3,9	4,3	4,8	5,2	5,7	6,1

Table 21.2 Measured wavelengths for each spectral line of the low-energy bulb and values tabulated in the literature

Line	Violet	Blue	Green	Orange	Red
$\lambda_{measured}$ (nm)	400–450	500	550	600	650
$\lambda_{tabulated}$ (nm)	405 and 436	492	546	577	612

If we use a non-monochromatic light source, by using this last expression we can determine the position in which we will observe the maximum of intensity for each wavelength, that is, for each spectral line. Then, we can draw a scale to directly measure the wavelength of the radiation. In our spectroscope, we have drawn a line on the right of the slit in the positions shown in Table 21.1 for the seven values of the wavelength.

When we use the energy-saving bulb as the light source, we obtain the spectrum shown in Fig. 21.2 and the scale allows us to measure the wavelength of each spectral line. In Table 21.2, we show the wavelengths measured using the spectroscope for each spectral line of the energy-saving bulb. Some references (Sansonetti et al. 1996; Kraftmakher 2010, 2012) allow us to obtain the wavelengths emitted by these bulbs. We observe in Table 21.2 that the measured and the tabulated wavelength are similar, within the limits that can be expected for an inaccurate instrument like this.

We have also tested the spectroscope with three laser pointers (red, green and blue) with wavelengths of 635, 532 and 405 nm, respectively. By measuring the wavelength with the spectroscope, we confirm that the wavelength measured by using the spectroscope matches the value provided by the laser devices, within the precision expected.

21.4 Teaching and Learning Sequence Developed in the Workshop

The teaching and learning sequence (TLS) is carried out with 20 attendees in two different rooms, each of which has a maximum of 10 attendees. Each group is guided by a museum monitor. The workshop is developed through an inquiry-based strategy.

At the very beginning, the TLS sets a problem and involves the attendees in a research aiming at finding an answer using their own previous knowledge and with the assistance of the monitor. In our case, the problem is "*Which visible light sources do we currently have? What light do they emit?*" After setting the problem, we carry out some activities to familiarize the attendees with the problem: we show the attendees some light sources (an image of a star, a flame, some different bulbs or a TV screen, among others) and we ask them to think what they have in common, whether they emit light of the same colour and what colour of light would be most useful for their purpose.

Once we have familiarized the attendees with the light sources and the colour of the light emitted by each one, we introduce the possibility of "breaking down" the light into spectra to analyse its chromatic composition. To do so, we show them an image of a natural rainbow and ask if a similar analysis of artificial light sources could be done using artificial instruments. After showing them a diffraction grid and stick it into a spectroscope that they will later mount, we use it to observe the light spectra of the different artificial sources previously introduced. We asked the audience about the similarities and differences between the observed spectra.

The assistants observe that there are light sources that have the same spectrum (the fluorescent light tube, the energy-saving bulb and the mercury Crookes tube) and associate it with the similarity between the bulbs. The monitor explains that sources with the same spectrum contain the same gas, in this case mercury. Then, we try to explain why a substance always emits the same light spectrum. We reflect on the nature of light and introduce the idea that it is a beam of photons and that changes in the energy of the atom produce these photons. Therefore, if the changes are limited (which implies that the energy in the atom is quantized) the light emitted will always be the same. By reflecting on this explanation, the assistant can explain (in a non-formal way) how the other sources emit light.

After completing this first part, the attendees move on to the other room where they will build a spectroscope (Fig. 21.2) with the help of the monitor. The spectroscope is already printed and die-cast on a cardboard sheet, so not any additional instruments to assemble it is required. While assembling it, attendees need to reflect on its structure and determine where the slit, scale and diffraction grid should be located. As attendees construct the spectroscope, they discuss the usefulness of the spectroscope in everyday life, its scientific usefulness and industrial applications. Once the spectroscope has been built, we encourage the attendees to use it to prove it by observing the spectrum of light sources in the room and to explain them qualitatively using the concepts of electronic states and electronic transition introduced in the first part of the workshop.

21.5 Conclusions

By using simple instruments and economic materials, we have been able to design and build a quantitative spectroscope that can be used both in scientific dissemination activities and in the classroom. In fact, the work carried out in the classroom with the spectroscope shows an increment of the students' interest in this analysis method and a better learning of the academic contents. Likewise, its use in the museum has aroused the general public's interest in light-emitting processes, their technological application and their social and economic consequences.

References

De Luís M, Martínez E (2010) ¿Un espectroscopio casero? Alambique 65:106

Gooley JJ, Chamberlain K, Smith KA, Khalsa SBS, Rajaratnam SMW, Van Reen E, Zeitzer JM, Czeisler CA, Lockley SW (2011) Exposure to room light before bedtime suppresses melatonin onset and shortens melatonin duration in humans. J Clin Endocrinol Metabol 96(3):E463

Ivanjek L, Shaffer PS, McDermott LC, Planinic M, Veza D (2015) Research as a guide for curriculum development: an example from introductory spectroscopy. I. Identifying student difficulties with atomic emission spectra. Am J Phys 83(1):85

Kraftmakher Y (2010) Experiments with fluorescents lamps. Phys Teach 48:461

Kraftmakher Y (2012) Demonstrations of optical spectra with a video camera. Phys Educ 47:184

Remón L, Ferrando V, Furlan WD, Monsoriu JA (2016) Diseño físico-matemático de un espectrómetro óptico. Model Sci Educ Learn 9(2):5

Sansonetti CJ, Salt MJ, Reader J (1996) Wavelengths of spectral lamps in mercury pencil lamps. Appl Opt 35:74

Savall F, Domènech JL, Martínez Torregrosa J (2013) El espectroscopio cuantitativo como instrumento para la construcción de modelos en física cuántica. Rev Bras Ensino Fís 36(4):4302

Savall F, Domènech JL, Guisasola J, Martínez Torregrosa J (2016) Identifying teachers and students difficulties in interpreting atomic spectra. Phys Rev Phys Educ Res 12(1):010132

Chapter 22
SPETTROGRAFO: A Digital Spectrometer for Educational Lab Activities

Daniele Buongiorno, Mario Gervasio, and Marisa Michelini

Abstract Optical spectroscopy is a significant methodological example of how physics makes use of indirect measure in order to obtain information on physical systems and validate models. Physics Education Research Unit from Udine University (IT) designed an educational path on optical spectroscopy, in which students are directly involved in experimental activities focused on interpretative plan allowing to highlight the link between energy levels in atoms and discrete light emissions.

After analyzing several commercial devices and mobile APPs allowing spectroscopic measurements, a digital spectrometer using a simple webcam implementing the various functionalities offered by the existing proposal to be connected via USB to PCs has been designed and realized. The hardware allows to use different diffraction grating, optical filters, and an optical goniometer. The software is designed to allow calibration and qualitative and quantitative measures of wavelengths and energies. Here we describe in detail the system and some experimental activities to be carried out with secondary school students and in introductory physics courses.

22.1 Introduction

Teaching modern physics in secondary school requires innovative and effective approaches since its teaching has been integrated in European school curricula since few years. This approach needs to found a scientific culture linking new theories with instruments and methods typical of physics. Optical spectroscopy representing a conceptual, methodological, and historical bridge between classical and modern physics provides a fertile experimental basis of the modern atomic

D. Buongiorno (✉) · M. Michelini
Physics Education Research Unit, DMIF, University of Udine, Udine, Italy
e-mail: buongiorno.daniele@spes.uniud.it; marisa.michelini@uniud.it

M. Gervasio
DMIF-URDF, Università degli Studi di Udine, Udine, Italy
e-mail: mario.gervasio@uniud.it

B. G. Sidharth et al. (eds.), *Fundamental Physics and Physics Education Research*,
https://doi.org/10.1007/978-3-030-52923-9_22

theory. Its disciplinary contribution to the teaching of modern physics concerns the phenomena of quantized emission and absorption of radiation which are basilar concepts representing the main investigative tool based on light–matter interaction. Optical spectroscopy, from an epistemological point of view, is a methodological and experimental context in which the role of the energy is pivotal, a validation instrument of interpretative models through indirect measures, a way through which interpret a code, hidden in the emitted light, in order to obtain information concerning states and changes of a microscopic system, as the atom, allowing to highlight the link between light emissions and atomic energy levels. From an educational point of view, competence concerning specific inquiry modalities employed in physics can be gained during physics lectures. Existing educational proposals (Luo and Gerritsen 1993; Oupseph 2007; Scheeline 2010; Amrani 2014; Onorato et al. 2015) include simple experiments allowing qualitative and quantitative measures, but those proposals have been designed and implemented in limited contexts, since students obtain the bare measurement without focusing on the emission process or the functioning of the experimental setup. Obtaining optical spectra from luminous sources is quite easy: a CD or a cheap diffraction grating are easily available objects and the produced spectra can be collected with a digital camera or a smartphone and analyzed (quantitatively or qualitatively) with specific APPs. On the other side, commercial experimental devices are often implemented in expensive and excessively structured setups that limit students' understanding of their principles of functioning, limiting one again to the bare measure itself. The general problem is thus that laboratorial educational proposals on optical spectroscopy in secondary school are offered in the form of sterile commercial devices, leaving teacher having the task of integrating them in a coherent educational path embedding the physics of the emission process and the importance of controlling the measuring process.

The aim of Physics Education Research Unit from Udine University (IT) is to build an educational path on optical spectroscopy allowing students to be directly involved in experimental and interpretative tasks. The pivotal laboratorial activity to effectively teach and learn physics could be supported by the opportunities offered by ICT (Information and Communication Technologies). An example has been already discussed in Gervasio and Michelini (2009) and Michelini and Stefanel (2015) in the case of an educational path on single-slit optical diffraction implementing inquiries activities based on measurements performed by students themselves searching for models explaining the observed phenomena. With this in mind, our research group analyzed the most popular commercial devices and APPs performing spectroscopic measurements, in order to design a prototype for a digital spectrometer implementing some proposals based on ICT making use of a specifically designed software allowing qualitative and quantitative analysis of a digitalized spectrum. The developing of the complete low-cost technical solution will be illustrated in the following with examples of significant measures.

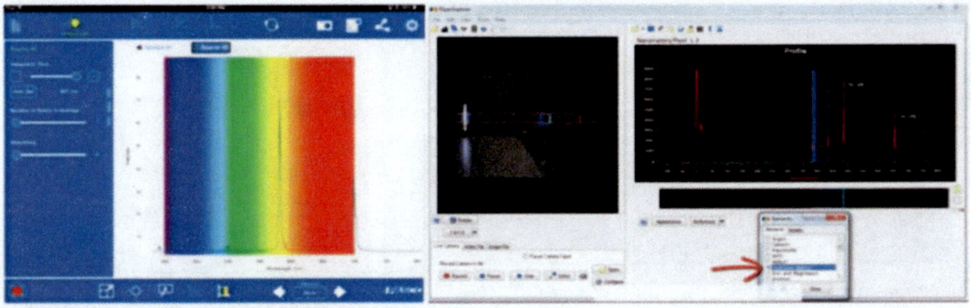

Fig. 22.1 *PASCO PS-2600* (left) and *RSPEC EXPLORER SYSTEM* (right) user interfaces

22.2 Some Existing Proposals

The *PASCO PS-2600* wireless spectrometer[1] contains a dispersive element (a diffraction grating) and a CCD sensor (sensibility between 380 and 980 nm). Light is guided from the source to be analyzed to the dispersive element with the aid of an optical fiber. The dispersive element creates the spectrum which is projected to the CCD sensor. Thanks to a USB or Bluetooth connection, the software automatically shows the emission spectrum as a function of the wavelength (Fig. 22.1, left). Obtained spectra, with a resolution of about 3 nm, can be compared with discrete reference spectra. The device allows also the analysis of absorption and fluorescence spectra: a test cuvette with the liquid sample can be inserted inside the device, in order to obtain absorption spectra: spectra are obtained by illuminating the sample with a reference white LED light and the fluorescence spectra are obtained by illuminating the sample with light of 405 or 500 nm. The absorbance curve is automatically shown on the user interface of the software. User has only to perform automatic actions without having access to the physics characterizing the measure or data analysis, in particular no calibration phase is required.

Another spectrometer, whose functioning has been studied, implementing a focusing element (a lens), rather than an optical fiber, is the *RSPEC EXPLORER SYSTEM* digital spectrometer.[2] A diffraction grating is placed in front of the webcam lens connected via USB to the PC. When the webcam is pointed towards the light source, its spectrum, in the range between 380 and 980 nm, is registered on the CCD sensor. Users limit to select the portion of the frame where the spectrum is visible. The software digitalizes it, providing a graph in which intensity (in arbitrary units) is shown as a function of the wavelength. The quality of the measures, which resolution is about 3 nm, allows to compare the recorded spectra with reference ones. The main limit of this device is represented by the fact that no calibration phase is required and the grating could not be removed, making impossible to change the dispersive power of the instrument (i.e., the pitch of the grating) and to appreciate the

[1]https://www.pasco.com/prodCatalog/PS/PS-2600_wireless-spectrometer/index.cfm

[2]https://www.fieldtestedsystems.com/

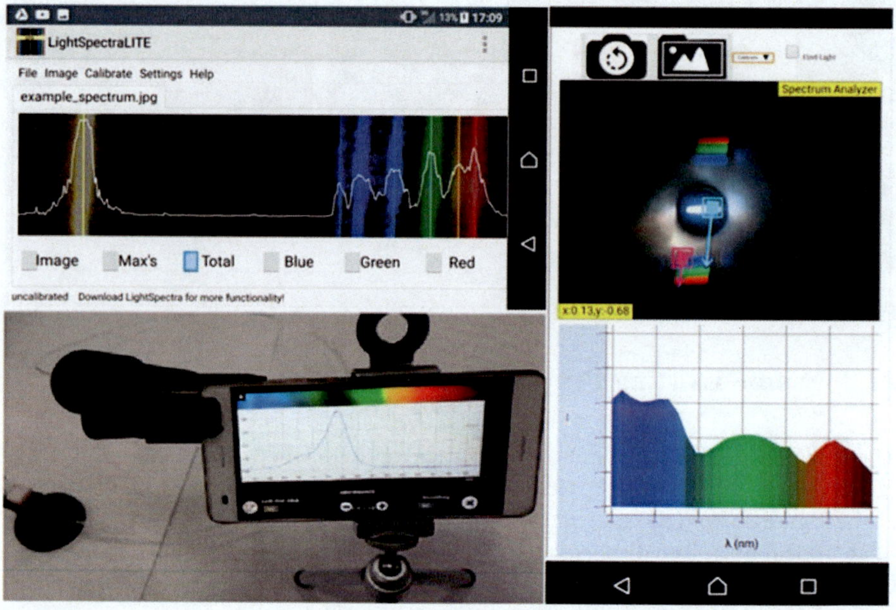

Fig. 22.2 User interfaces and logos of three APPs performing spectroscopic measurements. *LIGHT SPECTRA LITE* (top left), *SPECTRA UPB* (bottom left), and *SPECTRUM ANALYZER* (right)

effect of placing a grating in front of a light source. The user limits himself to line up the zeroth order of the diffraction pattern with a reference to automatically and uniquely associate a position along the pixel array with a specific wavelength (Fig. 22.1, right).

Both the quickly described devices are quite similar to "closed black boxes," in the sense that, despite their high quality performance, they do not allow user to approach the physics of the measure or to change the experimental conditions. Absorption measures are automatically generated by the software as well as the calibration, that is an important phase of the measuring process.

Other possibilities are available to record digitalized spectra, in a cheaper way with a Smartphone using mobile APPs, as *LEARNLIGHT SPECTROSCOPY*, *SPECTRA UPB*, *LIGHT SPECTRA LITE*, ASPECTRA *MINI*, and *SPECTRUM ANALYZER*. To perform those measures, mainly qualitative, which represent the main limitation of this possibilities, it is necessary to have a (self-build) spectroscope implementing a diffraction grating placed in front of the camera lens. In this way the spectrum is created and projected onto the camera sensor and it can be analyzed using the APPs. A precise calibration is possible only in few cases. Those APPs allow to obtain graphs of intensity as a function of the position along the spectrum, i.e., in different position of the digital image (Fig. 22.2).

22.3 The *SPETTROGRAFO* System

Our innovative device *SPETTROGRAFO* is a system implementing the simplicity of use of the existing commercial proposals and the feasibility of the analysis algorithms with new functionalities, missing in the analyzed devices, as the one to use grating with different pitches or to use colored filters to study selective absorption of light, or to have the possibility to dim the intensity of excessively luminous sources. Those little improvements allow the user to gain a better control of the measuring process, maintaining a good accuracy in the measured quantities using a low-cost instrument. The preliminary design of the prototype has been described in Buongiorno et al. (2018); here we report the final version of the system and relative specific technical characteristics with examples of employment in physics education.

22.3.1 The Hardware Components

In the *SPETTROGRAFO* system, a webcam is placed inside an aluminum case mounted on an adjustable tripod (Fig. 22.4). Technical data are shown in Table 22.1.

The dispersive element, a transmission diffraction grating could be placed in front of the lens of the webcam pointing towards the source whose light has to be analyzed. Light sources need to have small dimensions in order to produce spectra with enough separation between adjacent colors (we remind that a spectrum is the reproduction of the shape of the source at different angular positions with different colors). The device can be used also with extended light sources, making use of an external couple of black panels, working as a diaphragm with adjustable width.

Gratings with pitches of 1000 and 500 lines/mm can be used, both allowing spectral analysis in a range between 380 and 700 nm. The 1000 lines/mm grating allows to see only the first diffraction order with a resolution of 1.3 nm/pix, while with the 500 lines/mm the second order is visible, but the resolution decreases to 2.6 nm/pix. A couple of Polaroid filters can be inserted in front of the webcam to dim the light intensity of the source, avoiding the saturation of the sensor, in case it was too high. Colored filters can be also inserted in front of the webcam to study selective chromatic absorption (Fig. 22.3). The light passes through the grating and the diffraction pattern is registered on the CCD. A USB connection allows to send the digital image to the PC. A specifically designed ad-hoc software computes and analyzes the digital information.

Table 22.1 Technical data of *SPETTROGRAFO* system

Hardware part	Data
Webcam	Focus: 15 cm—infinity 60 frame per second Field of view is 60°
CCD	640×480 pix 1.2 Mpix
Case	$6\text{ cm} \times 6\text{ cm} \times 6\text{ cm}$

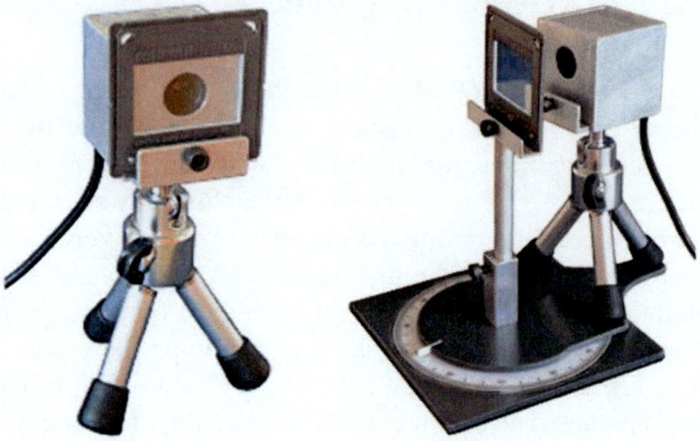

Fig. 22.3 The *SPETTROGRAFO* system (left). The horizontal support allows to place the accessories (diffraction gratings, colored and Polaroid filters) in front of the camera lens. The system could be placed on a rotating basis (right), allowing measures in "optical goniometer mode" (see below)

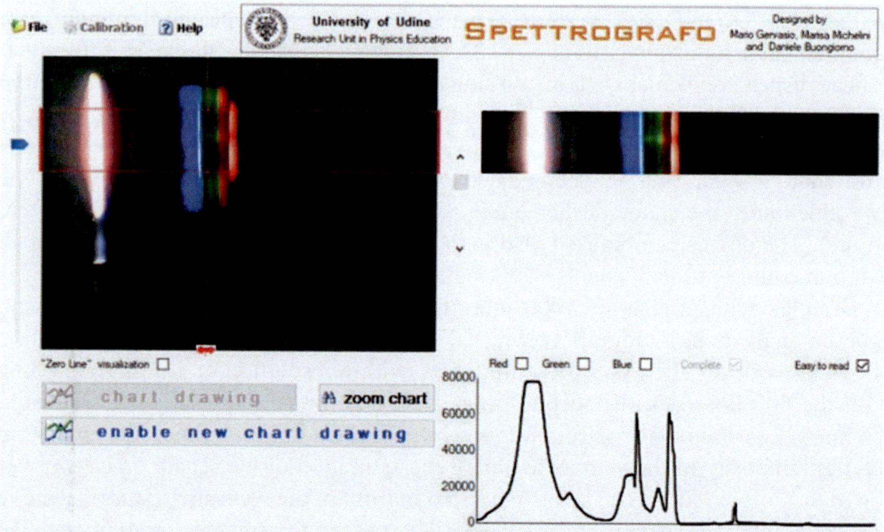

Fig. 22.4 Main user interface. Recorded image by the webcam (left) from which it is possible to select the area to be analyzed that is automatically digitalized in a graph (bottom right). In this example, the source is a hydrogen lamp observed through a 1000 lines/mm grating

22.3.2 *The Software Characteristics and Peculiarities*

The software was developed in a Microsoft "*framework.NET*" environment in C# language. As shown in Fig. 22.4, it allows to visualize the image as registered by the

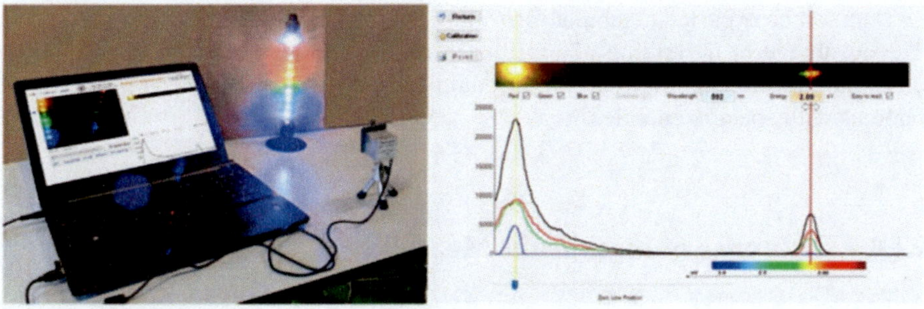

Fig. 22.5 *SPETTROGRAFO* system connected to PC, pointing the source (left) and calibrated yellow LED spectra (right). The yellow marker on the left targets the zeroth order, while the red marker, moving along the spectrum, allows to measure the energy or wavelength of the corresponding position

webcam, containing both the source and its spectrum (left area of the user interface). User can select the rectangular area to be analyzed (reproduced in the upper right area of the user interface) containing the zeroth order and the spectrum of the source extending from left to right, increasing the angle. Digitalization of the image occurs when the software operates a sum of the digital information of each pixel (proportional to the incident intensity) along every column of the selected area. A graph appears in the lower right area of the user interface where the intensity in arbitrary units and proportional to the mean intensity incident on the pixels of a same column is shown as a function of the position along the spectrum identified by the column number (from 1 to 640).

Of course, at this stage, the obtained graph does not contain any physical information (i.e., wavelength) on the spectrum: it is necessary to calibrate the graph in order to obtain calibrated spectra. The software allows to calibrate the measure: it is enough to select the type of used diffraction grating (the dimension of the rating pitch fixes the pixel–wavelength relationship) or, alternatively a calibration source can also be used: fixing the position of a known wavelength allows to calibrate the measure making the hypothesis of a linear relation between position along the sensor and relative wavelength. After those operations, a calibrated graph appears in which the horizontal axis is in wavelength (nm) or in energy (eV), since the code is equipped with the energy-wavelength inverse proportionality relation: a reference spectrum appears under the graph showing an energy scale. Two movable markers allow to sign the position of the image (zeroth order) and a generic position along the spectrum of the first order, resulting in a univocal measure of wavelength (expressed in nm) or energy (expressed in eV) (Fig. 22.5, right). In this way, student appreciates that every linear position along the CCD sensor corresponds to an angular position α that can be univocally coupled to a wavelength λ with the grating formula (where d is the pitch and m the order of the spectra):

$$d \cdot \sin \alpha = m \cdot \lambda$$

Data can be exported in tabular form allowing further analysis with a spreadsheet. The advantage of describing a spectrum in terms of energy or wavelength allows different educational proposals: light could be seen as a wave or as a stream of photons with specific energies.

22.4 Examples of Significant Measures

In principle, every light source could be coupled with *SPETTROGRAFO* system: discrete spectra from gas-discharge or fluorescent lamps, continuous spectra from incandescent lamps, or band spectra from LEDs and absorption spectra could be detected and analyzed. To perform a measure, it is enough to point the webcam lens towards the source (Fig. 22.5, left). One advantage is that no optical bench is needed: operatively it is enough to assure the alignment between source, grating, and sensor in a way that the spectrum is horizontal with respect to the array of pixels. Preliminary tests showed that the distance between the grating and the source does not affect the precision of the measures.

22.4.1 Analysis of Discrete Emissions from Gas-Discharge Lamps in "Static-Mode"

Using a thin gas-discharge tube, or a thin slit placed in front of an extended lamp in order to make the shape of the source as thin as possible, the spectrum is created reproducing the shape or the source in different positions on the CCD sensor in different colors. Once digitalized, it is possible to observe and measure the position (in wavelength or energy), width, and relative intensity of the various discrete emissions (Fig. 22.6). Lines peculiar features such as intensity and width could be discussed as consequences of quantum mechanical principles. Moreover, changing the grating's pitch, it is made clear how the resolving power of the measuring

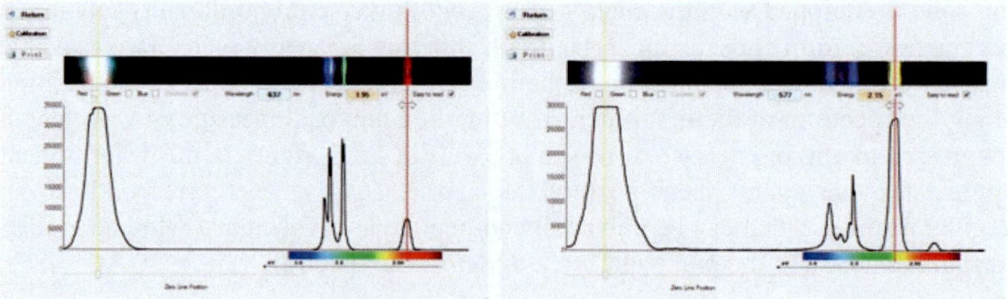

Fig. 22.6 Spectra of a cadmium (left) and helium (right) gas-discharge lamp taken with the *SPETTROGRAFO* system

Table 22.2 Measure of the main emission lines of Cd and He performed with *SPETTROGRAFO* system compared with values taken from official database (https://www.nist.gov/pml/atomic-spectra-database)

Cadmium			Helium		
λ_{meas} (nm)	λ_{std} (nm)	$\Delta\%$	λ_{meas} (nm)	λ_{std} (nm)	$\Delta\%$
639	643.85	0.75	656	667.82	1.77
506	508.58	0.51	577	587.56	1.80
480	480.00	0.00	496	501.57	1.11
469	467.81	−0.25	471	471.31	0.07
–	–	–	447	447.15	0.03

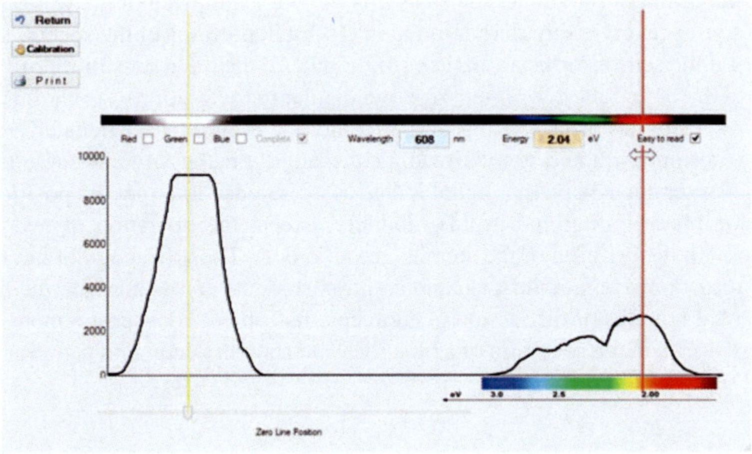

Fig. 22.7 Spectra of an incandescent lamp taken with the *SPETTROGRAFO* system

instruments changes: in particular doubling the pitch, the resolving power decreases of a factor two, and some lines can merge together and be no more angularly separated. As an example, some visible emissions in cadmium and helium spectra have been measured and they are reported in Table 22.2, compared with standard values obtained from official databases.

As an example, in Fig. 22.7 the spectra of a common incandescent bulb is shown. The black-body shape of the spectra is quite recognizable, allowing the measure of the wavelength peak and relating it with the temperature of the incandescent body via the Stefan–Boltzmann law.

Table 22.3 Measure of the wavelength associated to the peak emission of a blue LED

$m = 1$	$m = 2$		$m = -1$			$m = -2$	
A	λ (nm)	A	λ (nm)	α	λ (nm)	A	λ (nm)
13°	449.9	27°	454.0	13°	449.9	27°	454.0

22.4.2 *Analysis of Discrete Emissions from Gas-Discharge Lamps in "Optical Goniometer Mode"*

The LUCEGRAFO system could be placed on a rotating base (Fig. 22.3, right) making it more similar to an optical goniometer in which the dispersive element is fixed at the center of rotation of the basis (not more fixed in front of the webcam lens) and the sensor revolves circularly around it. Different portions of the spectra are thus observed if the sensor forms an angle α (the angular scale has a sensibility of 1°) with respect to the direction of symmetry (perpendicular to the grating). At various orders m, the wavelength corresponding at a specific angle could be evaluated with the grating's formula, quoted above, reading the angle on the angular scale (a fixed vertical marker appears on the digital image as a reference to target the position). No calibration phase is required in this modality, except the operation of making the 0° angle with the position of the source (zeroth order). The advantage of having this second measurement modality available is that students can appreciate the angular symmetrical features of diffraction phenomena. In Table 22.3, measurements of the luminosity peak in the spectrum of a blue LED are shown, taken with a grating with a 500 lines/mm.

22.4.3 *Selective Absorption of Colors and Evaluation of Transmissivity Curve*

With *SPETTROGRAFO* system it is possible to appreciate in real time, how do colored filters modify the spectrum of a reference source containing all visible colors (for example a white LED, Fig. 22.8, left) evidencing the phenomenon of selective absorption (Fig. 22.8, right). In order to do this, it is enough to record the reference spectra and then place colored filters in front of the reference source; the resulting spectrum would be deprived of some colors, since some of them are absorbed by the filter, which is transparent to others. The percentage of absorption (or transmission) at different colors could be visualized and evaluated via the software itself, which extracts data in a tabular form (intensity vs wavelengths), that can be further analyzed with the aid spreadsheet in order to quantitatively evaluate the absorbance as a function of the color. In particular, named $I_0(\lambda)$ the intensity of the reference spectrum as a function of the wavelength and $I(\lambda)$ the intensity of the absorption spectrum, the quantity representing the transmittance of the filter $T(\lambda) = I(\lambda)/I_0(\lambda)$ can be evaluated and displayed graphically (Fig. 22.9).

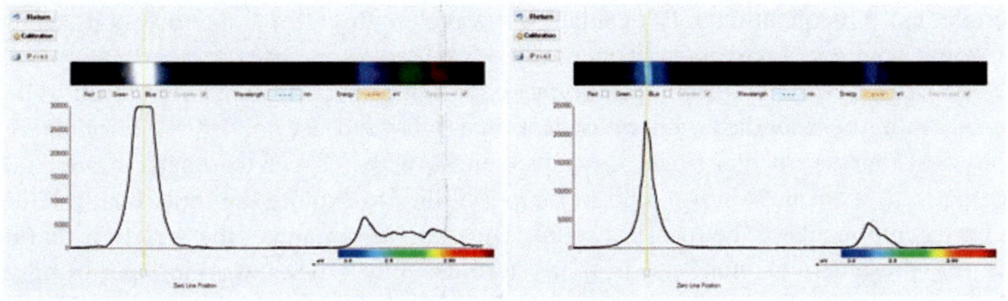

Fig. 22.8 Reference spectra (left) and absorption spectrum having placed a blue filter in front of the reference source (right)

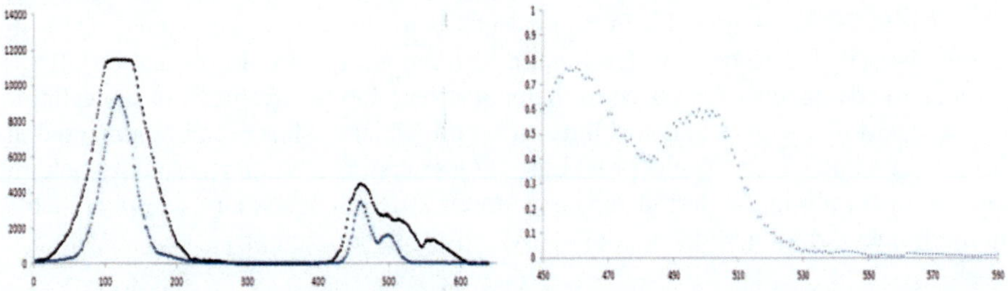

Fig. 22.9 Elaboration with a spreadsheet: light spectrum from a white LED compared with the spectrum of the same light passing through a blue filter. Abscissa values refer to the column of pixel; on the left the peak due to the source is visible (left) and transmittance of the blue filter (right)

22.5 Conclusions

The main required features for an effective educational lab system for spectroscopic measurements, i.e., the importance of the calibration process, the insight into the physical processes, the possibility to change the accuracy, and the data acquisition modality emerged after having analyzed potentialities and limits of some available digital spectrometers and mobile APPs. A cheap but affordable digital device, *SPETTROGRAFO*, has been thus designed and realized in order to implement in a simple setup all the emerged needs. The original device allows analysis of optical spectra of different light sources and it consists of a USB webcam with the possibility to place in front of it different diffraction gratings, and colored filters to study selective absorption of colors. Virtual images of spectra are recorded on the CCD sensor of the webcam and observed with the aid of a specifically designed software. Calibration occurs via software itself by selecting the used diffraction grating or with the aid of a calibration source, and the tests showed that precision of the measures have uncertainties less than 5%. Recorded spectra are digitalized in a graph representing luminous intensity as a function of the wavelength, or energy. The device can be also mounted on a rotating base allowing to measure the diffraction angle and thus quantitatively evaluate the wavelength with hand-and-pencil calculations, as in the classical experiment of the optical goniometer.

SPETTROGRAFO system, prototype for a digital spectrometer, offers itself to be used both in secondary school educational labs and in university introductive physics courses, thanks to its inexpensiveness with respect to other commercial devices, to it easiness in use, and to the possibility to explore the functional role of every components of the measure setup. The main advantage of the system is that it is not presented to students as a mysterious "black box" working performing measurements without any awareness of the inner processes by students, rather, it allows students to develop a functional understanding of the measuring process, which is one of the main goal of an educational lab.

Up to now the device has been presented to a group of ten secondary school teachers and used by them in the context of "National School for Teachers on Modern Physics" within IDIFO6 project[3] held in Udine University (IT) in September 2017, a teacher professional development activity. In the same IDIFO6 project, the device has been used in the educational lab for freshmen in biotechnology as a part of a wider didactical innovation project and it has been implemented in Masterclass and CLOE (Conceptual Lab of Operative Explorations) activity held in Udine University in the period January–March 2018 for secondary school students in the framework of a wider schools-university collaboration project.

References

Amrani D (2014) Hydrogen Balmer series measurements and determination of Rydberg's constant using two different spectrometers. Eur J Phys 35(4):045001

Buongiorno D, Gervasio M, Michelini M (2018) The spectrograph: a prototype for a digital spectrometer. In: Borg Farrugia C (ed) Junior College multi-disciplinary conference: research, practice and collaboration: breaking barriers: conference proceedings. University of Malta, Junior College, Malta. ISBN: 9789995714369

Gervasio M, Michelini M (2009) Lucegrafo. A USB data acquisition system. In: Lamboune B et al (eds) MPTL14 proc. http://www.fisica.uniud.it/URDF/mptl14/contents.htm

Luo WW, Gerritsen HJ (1993) Seeing the Fraunhofer lines with only a diffraction grating and a slit. Am J Phys 61(7):632

Michelini M, Stefanel A (2015) Upper secondary students face optical diffraction using simple experiments and on-line measurements. In: Triay R (ed) FFP14

Onorato P, Malgieri M, De Ambrosis A (2015) Measuring the hydrogen Balmer series and Rydberg's constant with a homemade spectrophotometer. Am J Phys 61(7):632

Oupseph PJ (2007) CD rainbows. Phys Teach 45:11

Scheeline A (2010) Teaching, learning, and using spectroscopy with commercial, off-the-shelf technology. Appl Spectrosc 64(9):256A

[3]http://www.fisica.uniud.it/URDF/laurea/idifo6.htm

Chapter 23
Time as Topic Transversal Perspective in Teacher Professional Development Carry Out in the Master in Science Education

Marisa Michelini and Emanuela Vidic

Abstract Teachers professional development in the scientific field is a complex problem. With the aim to improve scientific learning, eight Italian Universities worked together to offer qualified and innovative formative interventions in science for kindergarten, primary and low secondary school teachers through a biennial Master in Science Education. An important characteristic of the model implemented in Udine University is the interdisciplinary approach as a result of preliminary disciplinary exploration of topics and problems with cultural and social relevance. This work presents the educational project on the theme of time, designed and experimented in primary school, which develops transversal perspective on the subject more and which is proposed as reproducible and innovative for learning.

23.1 Introduction

Teachers professional development in the scientific field is a complex problem where different aspects are important: limited disciplinary competences, traditional ways of addressing scientific issues, lack of expertise in grasping the opportunities and situations that allow to engage scientific learning on phenomenological contexts (Berger et al. 2008; Borko 2004; Buckberger et al. 2000). Professional development of in-service teachers, therefore, represents the possibility of improving student learning, renewing the curriculum, and introducing didactic and methodological innovation research based results (Ball and Cohen 1999; Calderhead 1996).

With the aim to improve scientific learning, the eight Italian Universities of Genova, Milano, Torino, Modena, Reggio Emilia, Napoli, Palermo and Udine

M. Michelini (✉)
Physics Education Research Unit, DMIF, University of Udine, Udine, Italy
e-mail: marisa.michelini@uniud.it

E. Vidic
Primary School of Attimis, Istituto Comprensivo di Faedis, Udine Tutor of the Faculty of Primary Education, University of Udine, Udine, Italy
e-mail: emanuela.vidic@uniud.it

B. G. Sidharth et al. (eds.), *Fundamental Physics and Physics Education Research*,
https://doi.org/10.1007/978-3-030-52923-9_23

worked together to offer qualified and innovative formative interventions in science for kindergarten, primary and low secondary school teachers through a biennial Master in Science Education.

23.2 The Master: Key Aspects Chosen for Professional Development

An important characteristic of the model implemented in Udine University is the interdisciplinary approach as a result of preliminary disciplinary exploration of topics and problems with cultural and social relevance. School activity is prepared by means of interdisciplinary workshops where teachers from three different scientific areas discussed with the participants of the Master of different culture, competence and professionalism.

The grafts on that interdisciplinary laboratories, of a planning activity for research based, were the most significant and qualifying elements for professional development of participants and activate the master thesis processes. Relevant was the choice in the Master organization to organized around the same topic some (3–4) thesis of teachers of different school level to build a vertical path and to compare parallel plan with different perspective. The multi-perspective of the resulting planning produces a territory of educational path with different aspect for learning trajectories of students.

Project work and a thesis were planned in interdisciplinary terms. This is really important in particular for the primary school that is the privileged place where we have the foundation of the transversal knowledge. Another relevant aspect is the situated learning of teachers: each participant teacher implements micro-teaching and intervention modules planned and monitored during the Master in research perspective. During the whole Master activities were paid attention to two dimensions of teacher professional development: the analysis of learning processes on micro-module teaching and curricular design in a vertical and transversal perspective. The matrix structure adopted for the design of the didactic work has allowed a vertical profile design on the issues addressed; the contents have been deepened in different fields for each subject, maintaining a common core (Fig. 23.1).

23.3 Theoretical Framework

The theoretical framework in which the training is framed is the Model of Educational Reconstruction MER (Duit 2006; Niedderer 2001) which integrates knowledge and content analysis (CK), pedagogical knowledge with the investigation of students' ideas (PK) and the ability to design learning environments for a teaching professional (PCK) (Shulman 1986) on basic scientific education in which physics has its own specific role in a transversal perspective.

A ASPETTI METODOLOGICI GENERALI	8	
A1 Strumenti e metodi di indagine dei processi di apprendimento		2
A2. Il gioco e l'educazione informale: metodologie e tecniche del gioco		2
A3 Aspetti storici ed epistemologici della matematica e delle scienze		2
A4 Aspetti metodologici nelle scienze sperimentali		2
B. APPROFONDIMENTI DISCIPLINARI E ASPETTI DIDATTICI	20	
B1 ***Matematica***		8
B2 ***Fisica, chimica e astronomia***		5
B3 ***Scienze biologiche, naturali ed ambientali***		7
C ATTIVITA' LABORATORIALI	22	
C1 ***Laboratorio di matematica***		4
C2 ***Laboratorio di astronomia e chimica***		2
C3 ***Laboratorio di fisica***		2
C4 ***Laboratorio di scienze biologiche e naturali***		2
C5 ***Laboratorio di scienze della terra***		2
C6 ***Laboratorio di educazione ambientale***		2
C7 ***Laboratorio di progettazione, sperimentazione, verifica e analisi di carattere pluridisciplinare***		8
D APPROFONDIMENTI METODOLOGICI E DIDATTICI	4	4
D1 ***Percorsi individualizzati di formazione***		
E FORMAZIONE SITUATA E PROVA FINALE	6	6
TOTALE CUF	60	60

Fig. 23.1 General educational plan—in addition to disciplinary workshops, interdisciplinary workshops were conducted

From our research (Michelini and Stefanel 2014), a successful condition for professional development is the integration of the three models of teacher education: Metacultural, Experiential and Situated (PCK in action), including the personal involvement of teachers on the conceptual nodes (PCK test), the multi-perspective individual and group reflections, the design and experimentation of didactic paths, and the documentation and the discussion/analysis of learning progression. These are the indispensable conditions of success for the professional development that consolidates partial competences (Davis and Smithey 2009; Michelini and Stefanel 2014). In particular, the cultural model provided for the critical discussion between researchers and teachers on a disciplinary, didactic and pedagogical level; in the experiential model, the trainee year teachers personally tested the didactic proposals, through tutorials, to reflect on the conceptual nodes, recognize the characteristics of coherent paths that promote the reasoning and the involvement of children in the learning process, build activities based on problematization (inquiry-based learning).

23.4 Interdisciplinary Workshops to Promote Transversal Knowledge

School activity is prepared by means of interdisciplinary workshops where professors from three different scientific areas discussed with the participants of the Master. Learning environment was enriched by the different competences of the participants having different cultural and professional experiences, coming from the kindergarten, primary and secondary school with following degrees: natural science, math, physics, pedagogy, psychology, human science, languages.

The design looks a vertical curriculum, with attention to the interdisciplinary nature of the contexts and the transverse dimension of the conceptual elements. Strategies, instruments and methods adopted are multiple. The experiential methodology involves the active role of the children and the exploratory nature of the activities, pointed to the construction of formal thinking in terms of models and interpretative ideas, as well as representations and simultaneous use of different languages: graphics and iconographic language. Inquiry-Based Learning (IBL) produce the personal involvment children with active role hands-on and minds-on in a sequence of activities in which they have held exploratory, interpretive and creative role. Discovery learning has been promoted through guided discussions through questions—stimulus and meta-reflection has been activated through the worksheet. Common elements of the planning phase are: wide look at the subject; interpretation in different perspectives, integration and complementarity between scientific and humanistic disciplines, phenomenological-operational approach, education to the plural dimension of knowledge and its unity, analysis of the relapses according to the mode of action-research.

23.5 The Choice of the Topic Time: Projects

Time is a comprehensive, naturally transversal topic that offers the opportunity to face different kinds of experimental and exploratory activities. Time is part of many common experience and is a transversal topic that plays a fundamental role of conceptual referent that can develop in multidisciplinary way.

The theme of time was one of this transversal topic engaging different primary participant teachers. For its particular nature, time was a common matrix for three thesis articulated according to different perspectives: one proposes the integration between the literary, scientific and artistic aspects, thinking about the measure of time; another one wants to integrate philosophy, physics and astronomy, thinking of the concept of time and its measure, both thesis through human artefacts and by observations of geo-astronomical phenomena; the third develops in particular time concept education.

Time was chosen because it is a multidisciplinary and complex subject as well as a fundamental concept in anybody experience, allowing to construct a bridge between common sense and scientific knowledge, which is one of the main objectives indicated by the research literature on scientific learning processes. It is important to develop in this context for children the habit to look in terms multi-perspective knowledge, but there is the need to develop this competence for teachers by means of examples and good praxis. In this setting, we designed and tested a research-based experiment with this goal, starting with a reflection on the concept of time in the philosophical, scientific, technological, literary and artistic perspective. The theoretical framework of the teacher planning is the Model of Educational Reconstruction (MER) (Duit 2009) and the dynamical research referent is the designed based research with the approach of inquiry-based learning. Strategies, instruments and methods used are IBL, with personal involvement of children with active role hands-on and minds-on in a sequence of activities in which they have held exploratory, interpretive and creative role. The design looks a vertical curriculum, with attention to the interdisciplinary nature of the contexts and the transverse dimension of the conceptual elements. The active role of the children and the exploratory nature of the activities pointed to the construction of formal thinking in terms of models and interpretative ideas, as well as representations and simultaneous use of different languages: graphics and iconographic language. Data on learning processes are analysed in the perspective of argumentation and the reasoning in the interpretative phases by means of monitoring materials.

This work presents the educational project on the theme of time, designed and experimented in primary school, which develops transversal perspective on the subject more and which is proposed as reproducible and innovative for learning.

23.6 An Educational Project on the Theme of Time: Understand and Measure Time

The theme of time was introduced to five classes of the primary school (from the first class to the fifth class of primary school) in a vertical curriculum.

The learning path on time involved educational approach to time in different contexts including philosophy, poetry, art and history and looking at different aspects: irreversible phenomena, cyclic phenomena, sequence of actions in everyday life, words related to time. Built, explored and experimented various instruments for time measurement: gnomons, hourglasses, oscillating fluids, pendulum (Fig. 23.2).

For the path, research tools have been designed: grid for the collection of spontaneous ideas; photographic journey to recall prior knowledge; programme of the discussion on the synthesis of time ideas with work sheets and small narratives (history of time, measurement, sundial, consumption watches, water clocks, watches sand); analysis of artefacts (height ring, solar ring, notturlabio, diptych, Galileo's clock, clock counter with gears); construction and calibration of instruments;

Activity
1. Investigation of children's spontaneous ideas on the concept of time. Analysis of the meanings of time and construction of a map.
2. Test-in.
3. TIME AND PHILOSOPHERS. Analysis of the thought of six philosophers on the concept of time.
4. THE TIME IN POETRY. Analysis of six poems containing various meanings of time.
5. CREATIVE ACTIVITIES. Drawings.
6. ASTRONOMY. The sun's motion and the study of the shadows.
7. EDUCATION TO IMAGE. Graphic representation of the sun. The sun of the artists - Reading of images that represent the sun.
8. CALIBRATION OF INSTRUMENTS FOR MEASURE TIME. Candle, incense stick, the Indian hourglass water
9. PERIODIC MOTIONS FOR MEASURING TIME. The pendulum and spring
10. CONSTRUCTION AND CALIBRATION OF INSTRUMENTS FOR MEASURING THE TIME. Hourglass with the sand, the Egyptian hourglass water, the Merket.
11. STUDY OF PHENOMENA IN TIME: time measurement with the pendulum. In trip-viscosity liquids (oil, syrup, honey and detergent). Melting ice on different media. Heating a mass of water.
12. WHEELS AND GEARS AND Pascalina
13. HISTORY OF THE MEASURE OF TIME THROUGH ANCIENT DEVICES
14. Test-out

Fig. 23.2 Activities

measures, reflection and formalization; homework, out-in test. Work sheets: stimulus according to the strategy PEC (Previsione, esperimento, confronto); investigation of the ideas of the philosophers; selection of poems for listening, comprehension, poetry analysis (form, structure, content); image education; siding with math: Pascaline and gears.

23.7 Learning Outcomes

Significant results in terms of learning they are highlighted with the analysis of the test-in and test-out which is divided into the following nine questions:

1. *Write three sentences that contain the word time.*
2. *It has two events that illustrate the flow of time.*
3. *What time begins recreation? What time does it end? How long does it last?*
4. *Shows three ways to measure a time interval.*

5. *Write two ways to measure time.*
6. *It indicates three phenomena that are periodic.*
7. *Explain how you can recognize that a phenomenon is periodic.*
8. *Draw a picture illustrating your vision of irreversibility.*
9. *Name two contemporary phenomena.*

Test-in results have shown that many pupils answer to each question giving different meanings to the time concept, i.e.: whether time, action time interval. In test out the continuity in time meaning was present in the majority of pupils, as well as the concepts of instant, time interval and that periodic event can be used for clock building. In addition this concepts are well distinguished by those of whether or perceived time interval of actions. A conceptual organization of the concepts emerge from the pupils sentences.

Children's learning is evident from the results compared to the main research questions in the fields: methodological, cognitive and curricular.

On methodological framework, we note that observing the evolution of the phenomena helps us to understand the passage of time and the artefact influences the conceptualization of time.

In cognitive domain, we can see that spontaneous language of children develops into scientific language in relation to the acquired concepts and that time for children is both objective and subjective experience.

In the area of curriculum, it is highlighted that the time is not a unique concept but consists of many meanings, most of which arise from the experience linked to family background or education. Furthermore, the idea of periodic time interval is not spontaneously restricted to pupils, but you can build it successfully.

On the setting, we can state that interdisciplinary approach was the most valuable aspect of this work.

Integration between disciplines returns a unified image of knowledge that, although in its specialist articulations, contributes to the organ of the cognitive experience. Transversality of ideas, contents and concepts has allowed to reconcile the humanist to the scientific field and to preserve the specificity of the various disciplines by putting it at the service of understanding.

Verticality of the course in the primary school has allowed to understand the complexity of the concept of time and the meanings to which they refer and distinguish the time's measurement from the concept of time.

23.8 Concluding Remarks

With respect to the entire experience, the elements presented by the participants in the final thesis report have highlighted the main aspects of professional and personal growth in: mature knowledge and skills in the scientific field; acquire skills for examining learning processes; design by looking at the vertical curriculum; pay attention to the interdisciplinary nature of contexts and the transversal dimension of conceptual elements.

References

Ball DL, Cohen DK (1999) Developing practice, developing practitioners: toward a practice-based theory. In: Darling-Hammond L, Sykes G (eds) Teaching as the learning profession. Jossey-Bass, San Francisco, pp 3–31

Berger H, Eylon B-S, Bagno E (2008) Professional development of physics teachers. J Sci Educ Technol 17:399–409

Borko H (2004) Professional development and teacher learning. Educ Res 33(8):3–15

Bosatta G, Michelini M, Stefanel A (2001) Games, experiments, ideas from low-cost material to the computer on-line. In: Bandiera M, Caravita S, Vicentini M et al (eds) 120 simple experiments to do and not only tosee, in Research in Sciene Education in Europe: the picture expands, Roma, p 481

Buckberger F, Campos BP, Kallos D, Stephenson J (2000) Greenpaper on teacher education in Europe. TNTEE-European Commission (DG XXII)

Calderhead J (1996) Teachers: beliefs and knowledge. In: Berliner DC, Calfee RC (eds) Handbook of educational psychology. Macmillan, New York, pp 709–725

Colin P, Chauvet F, Viennot L (2002) Reading images in optics: student's difficulties and teacher's views. Int J Sci Educ 24(3):313–332

Davis A, Smithey J (2009) Beginning teachers moving toward effective elementary science teaching. Sci Educ 93(4):745–770

Duit R (2006) Science education research. ESERA Summer School, Braga, July 2006. http://www.naturfagsenteret.no/esera/summerschool2006.html

Duit R (2009) Bibliography—STCSE, students' and teachers' conceptions and science education. http://www.if.ufrj.br/~marta/aprendizagememfisica/fv09.pdf

Duit R, Gropengießer H, Kattmann U (2005) Towards science education research that is relevant for improving practice: the model of educational reconstruction. In: Fischer HE (ed) Developing standards in research on science education. Taylor & Francis, London, pp 1–9

Elbaz F (1983) Teacher thinking: a study of practical knowledge. Nichols, New York, p 11

Michelini M (1994) Games, experiments, ideas—from low cost materials to computer on-line. Forum, Udine

Michelini M (2004a) L'educazione scientifica nel raccordo territorio/università a Udine. Forum, Udine

Michelini M (ed) (2004b) Quality development in the teacher education and training. Girep book of selected papers. Forum, Udine

Michelini M (2004c) Physics in context for elementary teacher training. In: Quality development in the teacher education and training. GIREP book of selected papers. Forum, Udine, pp 389–394

Michelini M, Stefanel A (2014) Prospective primary teachers and physics pedagogical content knowledge, in teaching and learning physics today: Challenges? Benefits? In: Kaminski W, Michelini M (eds) Selected paper books of the International Conference GIREP-ICPE-MPTL 2010, Reims 22–27 August 2010. Udine, Lithostampa, pp 149–157. ISBN: 978-88-97311-32-4

Michelini M, Santi L, Stefanel A (2013) La formación docente: un reto para la investigación. Revista Eureka sobre Enseñanza de las Ciencias 10:846–870

Niedderer H (2001) Physics learning as cognitive development. In: Evans RH, Andersen AM, Sørensen H (eds) Bridging research methodology and research aims. The Danish University of Education, Student and faculty contributions from the 5th ESERA summer school in Gilleleje, Danmark, pp 397–414. ISBN: 87-7701-875-3

Shulman LS (1986) Those who understand: knowledge growth in teaching. Educ Res 15(2):4–14